The Spinal Cord during the Middle Second Trimester through the 4th Postnatal Month 130- to 440-mm Crown-Rump Lengths

This last of 15 short atlases reimagines the classic 5 volume *Atlas of Human Central Nervous System Development.* This volume presents serial sections of the spinal cord from specimens between 130 mm and 440 mm with detailed annotations. The presentation of these specimens emphasizes the sequence of myelination in various fiber tracts.

The Glossary (available separately) gives definitions for all the terms used in this volume and all the others in the *Atlas.*

Key Features

- Classic anatomical atlases
- Detailed labeling of structures in the developing spinal cord offers updated terminology and the identification of unique developmental features, such as myelination gliosis and gradients of myelination in the spinal cord white matter
- Appeals to neuroanatomists, developmental biologists, and clinical practitioners
- A valuable reference work on brain development that will be relevant for decades

ATLAS OF
HUMAN CENTRAL NERVOUS SYSTEM DEVELOPMENT
Series

Volume 1: The Human Brain during the First Trimester 3.5- to 4.5-mm Crown-Rump Lengths

Volume 2: The Human Brain during the First Trimester 6.3- to 10.5-mm Crown-Rump Lengths

Volume 3: The Human Brain during the First Trimester 15- to 18-mm Crown-Rump Lengths

Volume 4: The Human Brain during the First Trimester 21- to 23-mm Crown-Rump Lengths

Volume 5: The Human Brain during the First Trimester 31- to 33-mm Crown-Rump Lengths

Volume 6: The Human Brain during the First Trimester 40- to 42-mm Crown-Rump Lengths

Volume 7: The Human Brain during the First Trimester 57- to 60-mm Crown-Rump Lengths

Volume 8: The Human Brain during the Second Trimester 96- to 150-mm Crown-Rump Lengths

Volume 9: The Human Brain during the Second Trimester 160- to 170-mm Crown-Rump Lengths

Volume 10: The Human Brain during the Second Trimester 190- to 210-mm Crown-Rump Lengths

Volume 11: The Human Brain during the Third Trimester 225- to 235-mm Crown-Rump Lengths

Volume 12: The Human Brain during the Third Trimester 260- to 270-mm Crown-Rump Lengths

Volume 13: The Human Brain during the Third Trimester 310- to 350-mm Crown-Rump Lengths

Volume 14: The Spinal Cord during the First and Early Second Trimesters 4- to 108-mm Crown-Rump Lengths

Volume 15: The Spinal Cord during the Middle Second Trimester through the 4th Postnatal Month 130- to 440-mm Crown-Rump Lengths

The Spinal Cord during the Middle Second Trimester through the 4th Postnatal Month 130- to 440-mm Crown-Rump Lengths

Atlas of Human Central Nervous System Development, Volume 15

Shirley A. Bayer
Joseph Altman

CRC Press
Taylor & Francis Group
Boca Raton London New York

CRC Press is an imprint of the
Taylor & Francis Group, an **informa** business

Designed cover: Shirley A. Bayer and Joseph Altman

First edition published 2024
by CRC Press
6000 Broken Sound Parkway NW, Suite 300, Boca Raton, FL 33487-2742

and by CRC Press
4 Park Square, Milton Park, Abingdon, Oxon, OX14 4RN

CRC Press is an imprint of Taylor & Francis Group, LLC

LCCN no. 2022008216

ISBN: 978-1-032-22913-3 (hbk)
ISBN: 978-1-032-22910-2 (pbk)
ISBN: 978-1-003-27475-9 (ebk)

DOI: 10.1201/9781003274759

Typeset in Times Roman
by KnowledgeWorks Global Ltd.

Publisher's note: This book has been prepared from camera-ready copy provided by the authors.
Access the Support Material: www.routledge.com/9781032229133

CONTENTS

ACKNOWLEDGEMENTS

We thank the late Dr. William DeMyer, pediatric neurologist at Indiana University Medical Center, for access to his personal library on human CNS development. We also thank the staff of the National Museum of Health and Medicine that was at the Armed Forces Institute of Pathology, Walter Reed Hospital, Washington, D.C. when we collected data in 1995 and 1996: Dr. Adrianne Noe, Director; Archibald J. Fobbs, Curator of the Yakovlev Collection; Elizabeth C. Lockett; and William Discher. We are most grateful to the late Dr. James M. Petras at the Walter Reed Institute of Research who made his darkroom facilities available so that we could develop all the photomicrographs on location rather than in our laboratory in Indiana. Finally, we thank Chuck Crumly, Neha Bhatt, Kara Roberts, Michele Dimont, and Rebecca Condit for expert help during production of the manuscript.

AUTHORS

Shirley A. Bayer received her PhD from Purdue University in 1974 and spent most of her scientific career working with Joseph Altman. She was a professor of biology at Indiana-Purdue University in Indianapolis for several years, where she taught courses in human anatomy and developmental neurobiology while continuing to do research in brain development. Her lengthy publication record of dozens of peer-reviewed, scientific journal articles extends back to the mid 1970s. She has co-authored several books and many articles with her late spouse, Joseph Altman. It was her research (published in *Science* in 1982) that proved that new neurons are added to granule cells in the dentate gyrus during adult life, a unique neuronal population that grows. That paper stimulated interest in the dormant field of adult neurogenesis.

Joseph Altman, now deceased, was born in Hungary and migrated with his family via Germany and Australia to the US. In New York, he became a graduate student in psychology in the laboratory of Hans-Lukas Teuber, earning a PhD in 1959 from New York University. He was a postdoctoral fellow at Columbia University, and later joined the faculty at the Massachusetts Institute of Technology. In 1968, he accepted a position as a professor of biology at Purdue University. During his career, he collaborated closely with Shirley A. Bayer. From the early 1960s-2016, he published many articles in peer-reviewed journals, books, monographs, and free online books that emphasized developmental processes in brain anatomy and function. His most important discovery was adult neurogenesis, the creation of new neurons in the adult brain. This discovery was made in the early 1960s while he was based at MIT, but was largely ignored in favor of the prevailing dogma that neurogenesis is limited to prenatal development. After Dr. Bayer's paper proved new neurons are added to granule cells in the hippocampus, Dr. Altman's monumental discovery became more accepted. During the 1990s, new researchers "rediscovered" and confirmed his original finding. Adult neurogenesis has recently been proven to occur in the dentate gyrus, olfactory bulb, and striatum through the measurement of Carbon-14—the levels of which changed during nuclear bomb testing throughout the 20th century—in postmortem human brains. Today, many laboratories around the world are continuing to study the importance of adult neurogenesis in brain function. In 2011, Dr. Altman was awarded the Prince of Asturias Award, an annual prize given in Spain by the Prince of Asturias Foundation to individuals, entities, or organizations globally who make notable achievements in the sciences, humanities, and public affairs. In 2012, he received the International Prize for Biology - an annual award from the Japan Society for the Promotion of Science (JSPS) for "outstanding contribution to the advancement of research in fundamental biology." This Prize is one of the most prestigious honors a scientist can receive. When Dr. Altman died in 2016, Dr. Bayer continued the work they started over 50 years ago. In her late husband's honor, she created the Altman Prize, awarded each year by JSPS to an outstanding young researcher in developmental neuroscience.

INTRODUCTION

ORGANIZATION OF THE ATLAS

This is the 15th volume in a series of Atlases on the development of the human central nervous system. This volume deals with the development of the spinal cord from the middle second trimester through the 4th postnatal month. These specimens were presented in Volume 1 of the original Atlas Series (Bayer and Altman, 2001). Parts II to VIII feature photographs of transversely cut spinal cords from normal specimens ranging in age from gestational week (GW) 19 through the 4th postnatal month with crown-rump (CR) lengths from 130- to 440-mm. All specimens were selected from the Yakovlev Collection housed in the National Museum of Health and Medicine in Silver Springs, MD.[1]

An *overview plate* that shows thumbnail photographs of all sections of each specimen at the same scale to show the size differences between levels. The overview plate is followed by *companion plates* designated as **A** and **B** on facing pages. The **A** part of each plate on the left page shows the full-contrast, high-magnification photograph[2] of the sevtion without any labels; the **B** part of each plate on the right page shows a low-contrast copy of the same photograph with superimposed outlines and unabbreviated labels. **Parts II–VIII** feature multiple levels of the spinal cord in six specimens, many showing matched cell-body-stained sections paired with myelin-stained sections.

1. The *Yakovlev Collection* (designated by a **Y** prefix in the specimen number) is the work of Dr. Paul Ivan Yakovlev (1894–1983), a neurologist affiliated with Harvard University and the Armed Forces Institute of Pathology (AFIP). Throughout his career, Yakovlev collected many diseased and normal human brains. He invented a giant microtome that was capable of sectioning entire human brains. Later, he became interested in the developing brain and collected many human brains during the second and third trimesters. The normal brains in the developmental group were cataloged by Haleem (1990) and were examined by us during 1996 and 1997 when we spent time at the AFIP.

2. All sections were photographed using a Wild photomakroskop with Kodak technical pan black and white negative film #TP442. The film was developed for 6–7 minutes in dilution F of Kodak HC-110 developer, followed by stop bath for 30 seconds, Kodak fixer for 5 minutes, Kodak hypo clearing agent for 1 minute, running water rinse for 10 minutes, and a brief rinse in Kodak photo-flo before drying. Each specimen was photographed at the magnification that filled the microscopic field with the largest cross section of the spinal cord. Negatives were scanned at 2700 dots per inch (dpi) using a Nikon coolscan-1000 35-mm film scanner interfaced to a Macintosh PowerMac G3 computer. Normal-contrast Adobe Photoshop files are in **Part A** of each plate; low-contrast copies are in part **B** with labels put in using Adobe Illustrator.

DEVELOPMENTAL HIGHLIGHTS

The important developmental events that occur during the second and third trimesters include cumulative areal growth in the spinal cord white matter and gray matter (**Figure 1**). Only the cervical enlargement is quantified, but these changes are played out at all levels. Note that the overview plate introducing each specimen lists areas at each level. The spinal cord always has its larger cross-sectional areas in the cervical and lumbar enlargements, respectively. **Figure 2** deals with the changes in relative volumes throughout the entire span of prenatal and early postnatal development. By the middle of the second trimester and throughout further growth, the spinal canal, neuroepithelium/ependyma, roof plate, and floor plate are too small to be actually visible on the pie graphs. **Figure 3** deals with the relative volumes from cervical to sacral levels at selected times from the first trimester through the 4th postnatal month.

The second developmental event that we analyze in this volume is the sequence of myelination in various fiber tracts. **Figure 4** summarizes these changes at the cervical level. Note that comments in various plates point out gradients of myelination in several fiber tracts, including the dorsal columns and the corticospinal tract.

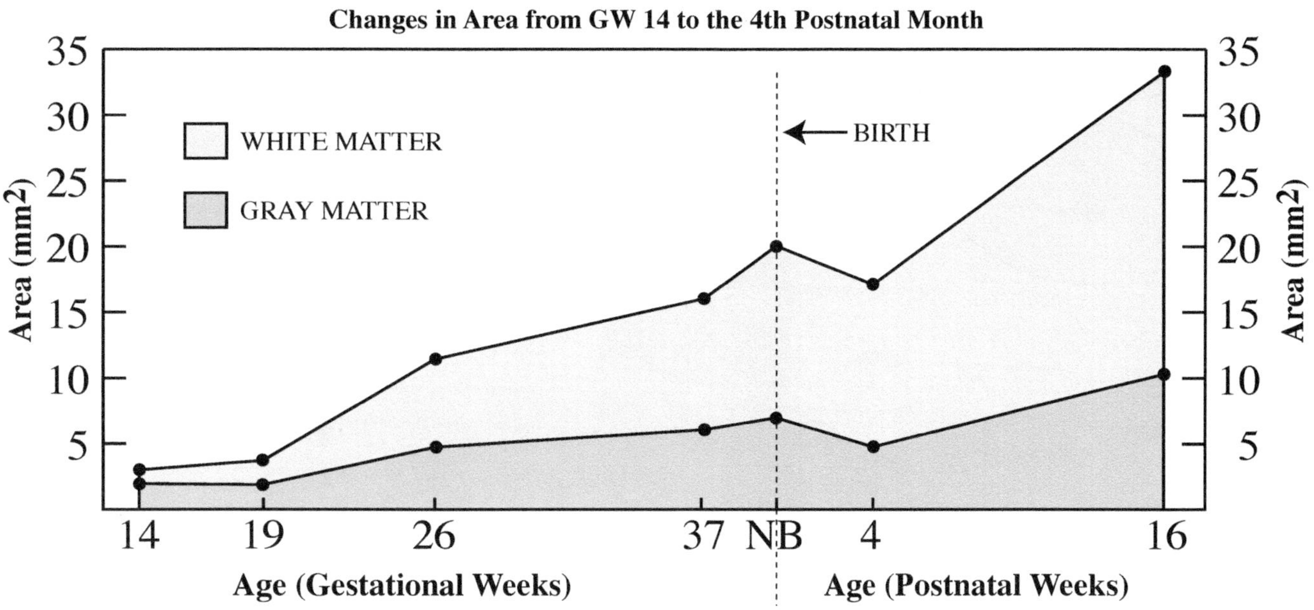

Figure 1. The area of the gray matter (*dark gray*) and white matter (*light gray*) in the cervical enlargement from the beginning of the second trimester (GW 14), birth (NB-newborn), to 16 weeks after birth. From GW 26 on, data from the myelin-stained sections area are used. The area graphs are stacked so that the scale on the Y axis indicates the total area of the spinal cord. These areas are measured in fixed tissue, so the actual values in the living spinal cord will be higher by an unknown amount. The areas of the ependyma and spinal canal are so small that they do not register above the baseline. Since neurogenesis finishes in the first trimester, the gradual increase in the area of the gray matter throughout this period is due to the growth and differentiation of the neurons themselves and the proliferation and maturation of supporting glial populations. The white matter increases more rapidly, especially between GW 19 and GW 26. There is gradual growth from GW 26 to the perinatal period (between GW 37 and postnatal week 4), then more rapid growth between postnatal weeks 4 and 16. The first growth spurt is probably due to the accumulation of axons in all of the major fiber tracts, including both parts of the corticospinal tract. The second growth spurt is probably due to the completion of myelination throughout the cervical enlargement at the 4th postnatal month. Only the outermost crescent of the corticospinal tract is still myelinating at this time (*see* **Plate 69**). From GW 26 on, the area of the white matter exceeds the area of the gray matter.

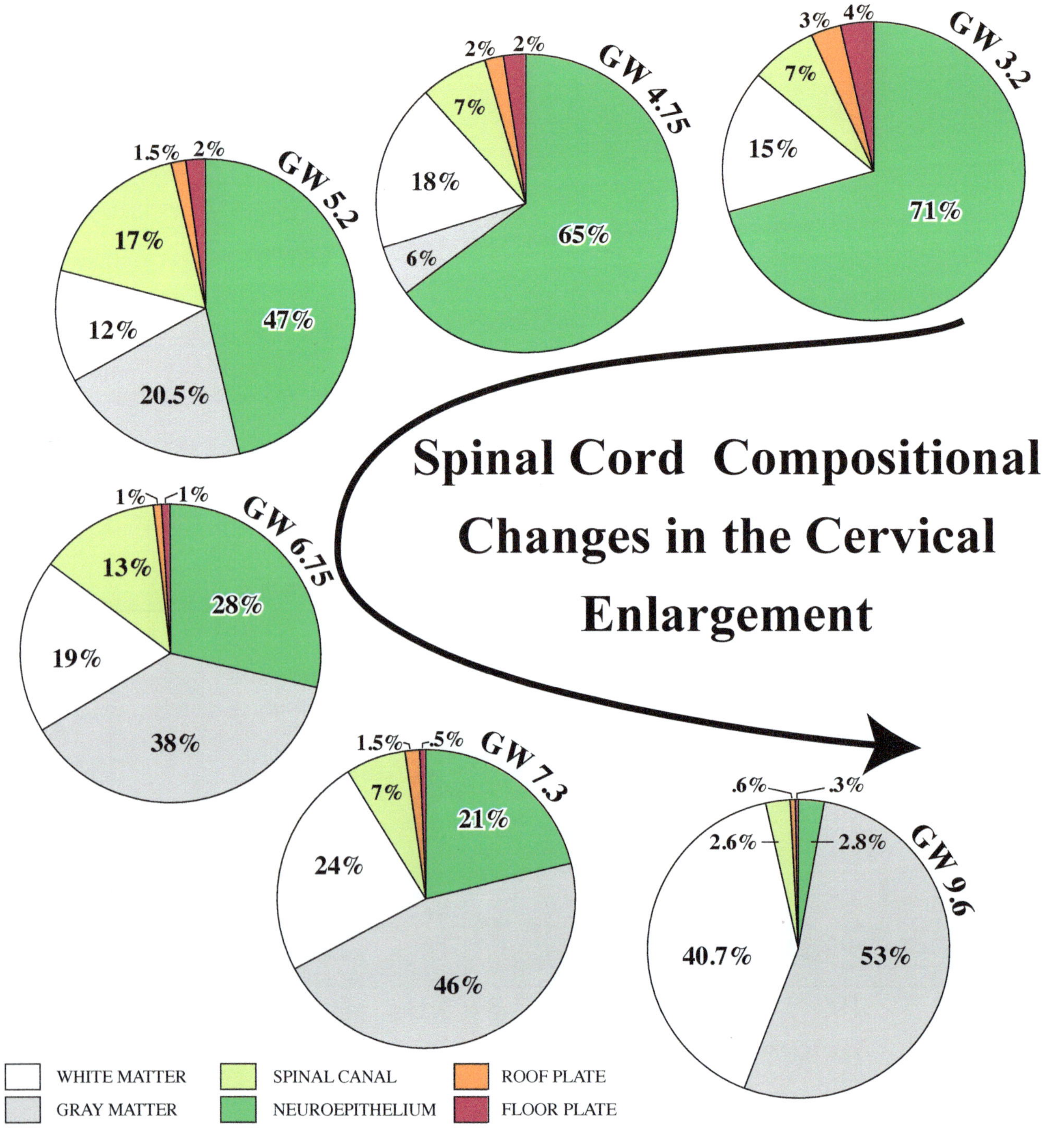

Figure 2. A series of pie graphs showing the proportional areas of the white matter (*white*), gray matter (*gray*), spinal canal (*light green*), neuroepithelium (**dark green**), roof plate (**orange**), and floor plate (**red**) at the *cervical enlargements* of specimens from gestational week (GW) 3.2 (*upper right graph*) to GW 9.6 (*lower right graph*). The section measured from the GW 3.2 specimen is illustrated in Volume 14, Plate 2, GW 4.75 in Volume 14, Plate 3, GW 5.2 in Volume 14, Plate 4, GW 6.75 in Volume 14, Plate 7, GW 7.3 in Volume 14, Plate 8, and GW 9.6 in Volume 14, Plate 12. The roof and floor plates have their largest proportional areas early (GW 3.2–5.2) supporting experimental evidence in animals for a prominent role in the overall dorsal-to-ventral organization of the gray matter and growth of axons parallel to the dorsal midline and across the ventral midline (*see* Chapter 4 in Altman and Bayer, 2001). The neuroepithelium dominates from GW 3.2 to GW 5.2 when stem cells are rapidly increasing in preparation for successive waves of neurogenesis. Even though the neuroepithelium declines in its proportion throughout the entire period, it is actually increasing its absolute area up to GW 7.8, CR 17.5 (*see* Figure 1, Volume 14). In contrast, the peak proportional areas of the spinal canal (GW 5.2–6.75) occur at the same time as the largest absolute neuroepithelial area. When neurons are produced and migrate out of the neuroepithelium, the area of the gray matter proportionally increases. The first proportional increase between GW 5.2–6.75 is almost exclusively due to growth of the ventral horn. Later proportional increases are due to continual growth of the ventral horn, the emerging intermediate gray, and the rapid expansion of the dorsal horn. The white matter measured at GW 3.2–5.2 is primordial and contains only a few axons; that area is the same as the mantle layer described in early developmental studies. From GW 6.75 onward, the proportional area increases are due to accumulating axons, mainly in the dorsal funiculus, ventral funiculus, and ventral commissure. By GW 9.6, the proportional areas of the gray matter and especially the white matter sharply increases at the expense of the rapidly shrinking neuroepithelium.

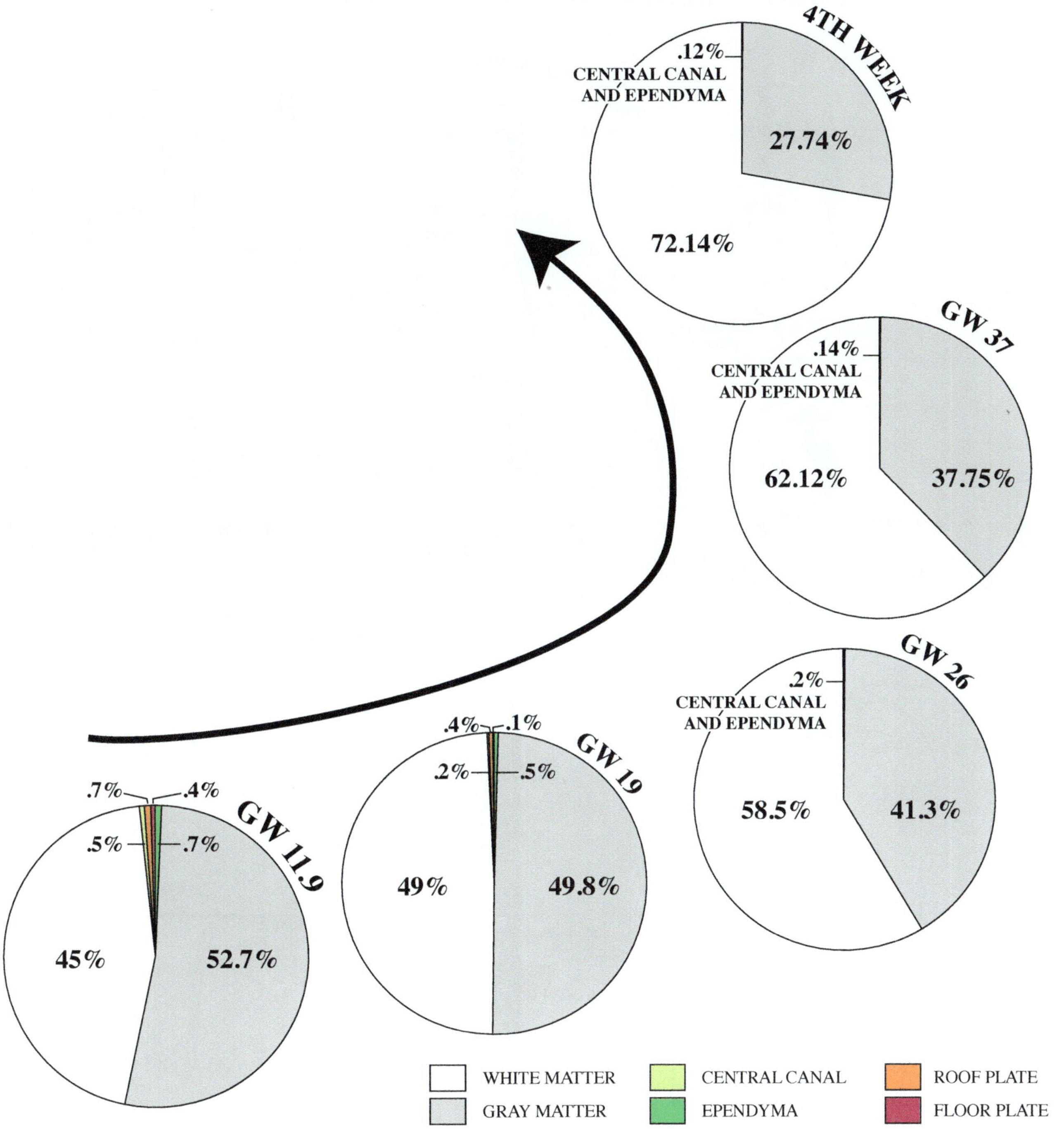

Figure 2, continued. Measurements in specimens at GW 11.9 (Volume 14, Plate 23), GW 19 (Volume 15, **Plate 3**), GW 26 (Volume 15, **Plate 10**), GW 37 (Volume 15, **Plate 28**), and the 4th postnatal week (Volume 15, **Plate 58**). During this period, the gray matter and white matter make up most of the spinal cord. The roof and floor plates disappear by GW 26, and the proportional areas of the central canal and ependyma are so small that they appear as a slightly thicker black line separating the gray matter and white matter at the 12:00 position of each graph. The gray matter continually loses proportional area throughout this period but the absolute area does increase gradually as neurons grow dendrites and some local circuit neurons grow elaborate terminal axon arbors (especially in the dorsal horn). The proportion of the white matter continually increases during this time, and it is the largest proportional area from GW 26 onward.

Compositional Changes in all Levels of the Spinal Cord

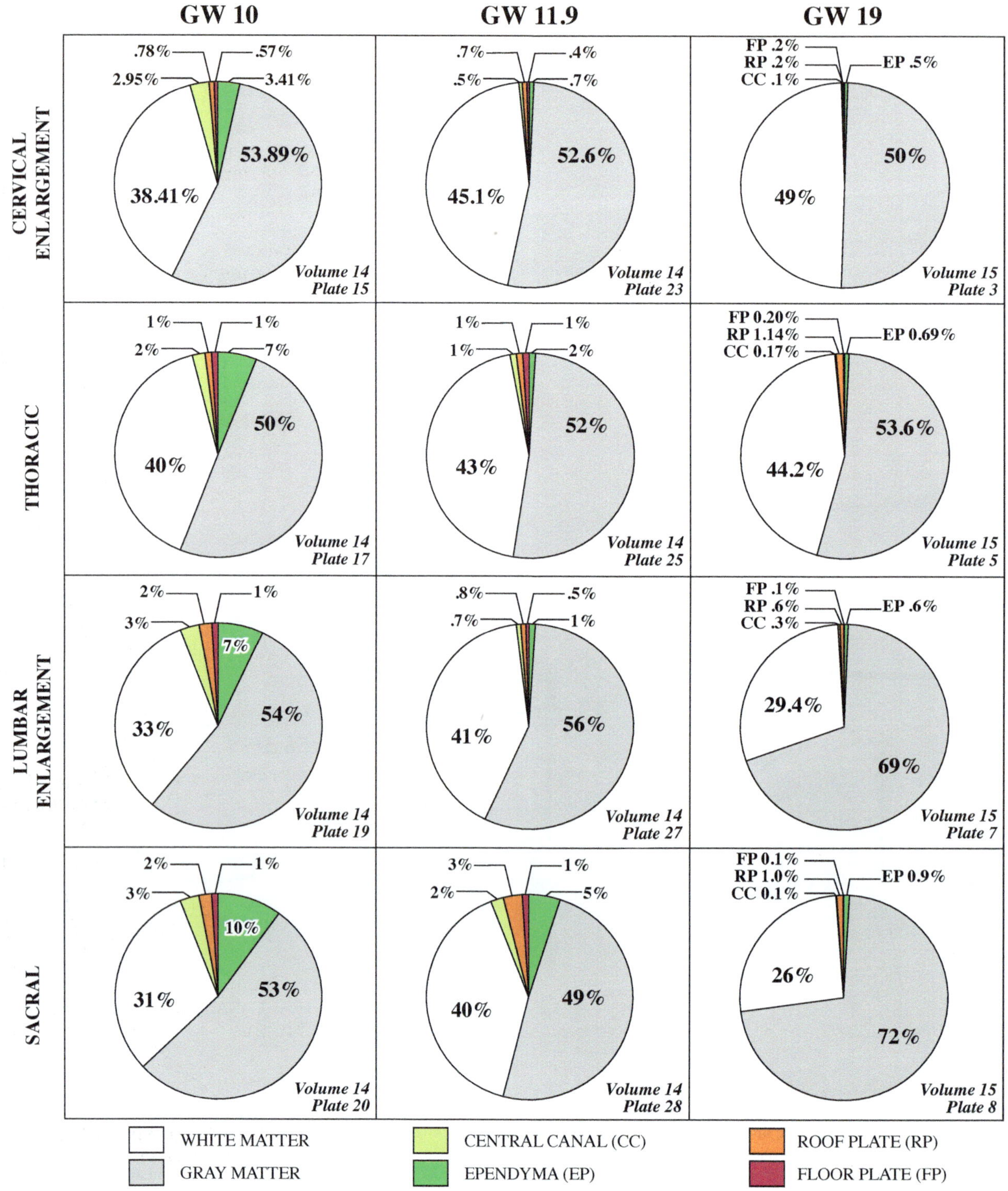

Figure 3. Pie graphs showing the proportional areas of the white matter (*white*), gray matter (*gray*), central canal (*light green*), ependyma (*dark green*), roof plate (*orange*), and floor plate (*red*) at the cervical enlargement level (*row 1*), thoracic level (*row 2*), lumbar enlargement level (*row 3*), and the sacral level (*row 4*). Each column contains measured sections from the same specimen; *column 1* is M2050, GW 10, *column 2* is Y380-62, GW 11.9, and *column 3* is Y52-61, GW 19. The *plate numbers* in each square refer to the illustration of the measured section in Volumes 14 and 15. By following the shifting values of the proportional areas within a column, a **gradient of maturation** is evident from the cervical enlargement to sacral levels. For example, the size of the ependyma increases posteriorly in the first two columns, indicating that the greatest degree of maturation is at the cervical enlargement level followed by declining levels of maturation at thoracic, lumbar enlargement, and sacral levels. By GW 19, the central canal, ependyma, roof plate, and floor plate have very small proportional areas indicating the greater maturation of the spinal cord throughout its length at this time.

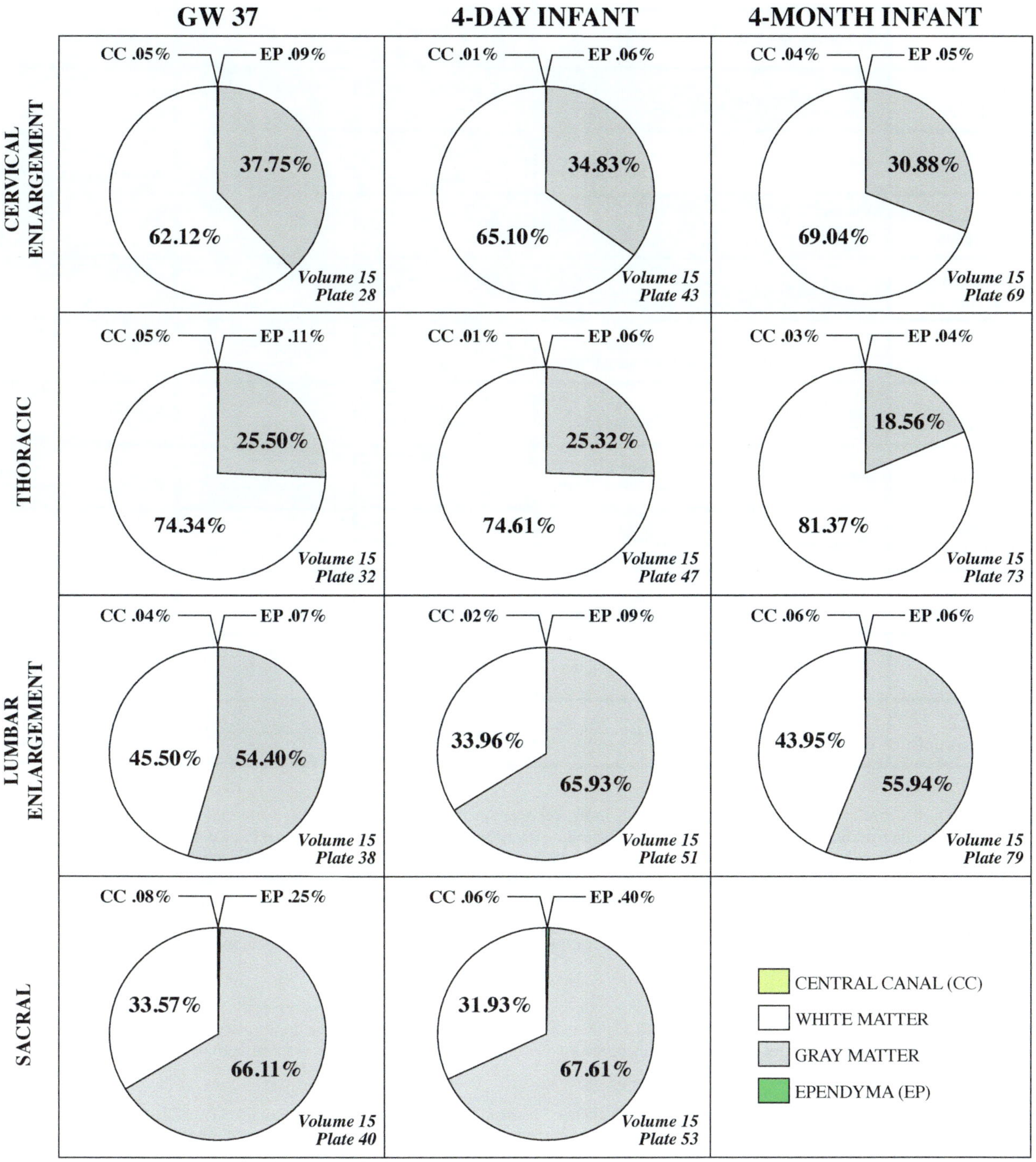

Figure 3 (continued). Measirements in specimens at GW 37 (*column 1*), 4th postnatal day (*column 2*), and 4th month (*column 3*). A section at the sacral level is not available in the 4-month infant. The roof and floor plates are absent at these ages, and the proportional areas of the central canal and ependyma appear on the graphs as slightly thicker black lines separating the gray matter and white matter at the 12:00 position in each graph. The smaller proportional areas of the white matter at the sacral level perinatally (*bottom row, first two columns*) indicate the paucity of axons that descend to this level. In contrast, the cervical enlargement and thoracic levels have high (*row 1*) to very high (*row 2*) proportions of white matter to gray matter because many axons are just "passing through" to lower levels. The very high proportion of white matter at the thoracic level is because the ventral horn is quite small at these levels because only axial musculature is innervated here, just as it is in the adult spinal cord. At the ages shown, the lower proportional area of white matter at the lumbar enlargement is due to the greater amount of gray matter dedicated to innervating large skeletal muscles in the lower limbs.

Myelination Sequences in Major Fiber Tracts (Cervical Enlargement)

TRACT	GW 26	GW 37	Birth	4th Week	4th Month
Sup. Fas. Gracilis	sparse reactive glia	some myelinating fibers	myelinated	myelinated	myelinated
Deep Fas. Gracilis	dense reactive glia	myelinated	myelinated	myelinated	myelinated
Sup. Fas. Cuneatus	sparse reactive glia	some myelinating fibers	myelinated	myelinated	myelinated
Deep Fas. Cuneatus	dense reactive glia	myelinated	myelinated	myelinated	myelinated
D.R. Col. Zone	dense reactive glia	myelinated	myelinated	myelinated	myelinated
D.R. Bif. Zone	some myelinating fibers	some myelinating fibers	myelinated	myelinated	myelinated
Spino-cerebellar	dense reactive glia	some myelinating fibers	many myelinating fibers	myelinated	myelinated
Intraspinal	dense reactive glia	some myelinating fibers	myelinated	myelinated	myelinated
Spino-thalamic	sparse reactive glia	dense reactive glia	sparse reactive glia	some myelinating fibers	myelinated
Ventral Commissure	some myelinating fibers	many myelinating fibers	many myelinating fibers	some myelinating fibers	myelinated
Ventral Corticospinal	no reactive glia	no reactive glia	sparse reactive glia	dense reactive glia	myelinated
Lateral Corticospinal	no reactive glia	no reactive glia	sparse reactive glia	dense reactive glia	many myelinating fibers

Figure 4. Steps in the myelination of some major fiber tracts in the cervical enlargement of the spinal cord from the second trimester to the fourth postnatal month. The unmyelinated Lissauer's tract and some descending tracts (tectospinal, medial longitudinal fasciculus, and vestibulospinal) are not shown. *Pale violet* indicates the **first step** when glial cells react with the myelin stain prior to the production of true myelin sheaths; these areas in the myelin stained sections contain a light dusting of punctate stain. *Dark violet* indicates the **second step** when the punctate staining is more plentiful, but few, if any, fibers are myelinating. *Pale purple* indicates the **third step** when the density of the stain increases and glia produce myelin sheaths around some fibers. *Medium purple* indicates the **fourth step** when the density of the stain increases yet again and more glia produce myelin sheaths around many fibers. *Dark purple* indicates the **fifth step** when glia produce myelin sheaths around nearly all the fibers in the tract; the truly myelinated areas appear solid black in the stained sections.

Both the **fasciculus gracilis** and the **fasciculus cuneatus** show gradients in the myelination steps from deep (early) to superficial (late) parts. If the sequence of myelination is linked to the time when axons first enter these fasciculi, the conclusion can be drawn that the first axons in the tract occupy the deepest areas, and fibers that enter the tract later occupy progressively more superficial parts. This order of ingrowing axons sets the stage for a topographic projection to the brain.

The **spinothalamic tract** does not have an even sequence of myelination because the dense reactive glial staining at GW 37 is bracketed by periods of sparse reactive glial staining. That may be due to more mature axons in the cervical level being invaded by waves of immature axons from parts of the spinal cord that are less mature, such as sacral and lumbar levels; the overall density of reactive glial staining dilutes as these immature axons fill in the tract.

The **ventral commissure** also has an uneven sequence of myelination. From the very first, some fibers are myelinated. However, more fibers are myelinated at GW 37 and postnatal day 4 than at the 4th postnatal week. That apparent regression is possibly due to the growth of unmyelinated axons from the ventral corticospinal tract into the commissure, thus diluting the number of myelinated fibers.

The **lateral corticospinal tract** is the last fiber tract to myelinate. Indeed, some parts of it are still myelinating during the 4th postnatal month, especially the outer crescent closest to the spinocerebellar tracts. That part of the tract contains fibers that will terminate at sacral levels of the cord. The lateral corticospinal tract in thoracic, lumbar, and sacral levels is progressively less mature at the 4th postnatal month. That gradient of maturation suggests that cortical axons myelinate first proximal to the cell body then sequentially myelinate the axons distal to the cell body.

This analysis departs from previously published accounts of myelination sequences which describe various fiber tracts as either "not myelinated" or "myelinated." It is a novel idea that myelination is a progressive event rather than an "all or none" event within a fiber tract. For a complete review of the literature on myelination in the spinal cord and for a more thorough analysis of the progressive steps in myelination, see Chapter 8 and Chapter 9, Section 9.4 in Altman and Bayer (2001).

REFERENCES

Altman J, Bayer SA (2001) *Development of the Human Spinal Cord: An Interpretation Based on Experimental Studies in Animals*. New York: Oxford University Press.

Bayer SA, Altman J (2002) *Atlas of Human Central Nervous System Development*, Volume 1: *The Spinal Cord from Gestational Week 4 to the 4th Postnatal Month*. Boca Raton, FL: CRC Press/Taylor & Francis Group.

Bayer SA, Altman J (2024) *The Human Spinal Cord diring the First and Early Second Trimesters. 4- to 108-mm Crown-Rump Lengths*, Volume 14: *Atlas of Human Central Nervous System Development*. Boca Raton, FL: CRC Press/Taylor & Francis Group.

Haleem M. (1990) *Diagnostic Categories of the Yakovlev Collection of Normal and Pathological Anatomy and Development of the Brain.* Washington, DC: Armed Forces Institute of Pathology.

Ranson SW, Clark SL. (1959) *The Anatomy of the Nervous System, its Development and Function.* (10th Ed.) Philadelphis, PA: Lea & Febinger.

PART II: Y52-61 CR 130 mm (GW 19)

Plate 1 is a survey of sections from Y52-61, a specimen in the Yakovlev Collection with a crown-rump length of 130 mm. All sections are shown at the same scale. The boxes enclosing each section list the approximate level and the total area (post-fixation) in square millimeters (mm^2). Full-page normal-contrast photographs of each section are in **Plates 2A–8A.** Low contrast photographs with superimposed labels and outlines of structural details are in **Plates 2B–8B.** In this specimen, the section numbers are not given because they are placed on large glass plates without any numbers. Twenty sections were photographed ranging from upper cervical to sacral levels. Low sacral and coccygeal levels were not preserved. To determine the approximate level, the 20 photographs were intuitively arranged using features such as the appearance and progressively smaller rostral-to-caudal size of the ventral and lateral corticospinal tracts.

The most notable characteristic of this specimen is the dense accumulation of glial cells in specific regions of the white matter. These are assumed to be glial cells proliferating prior to myelination (myelination gliosis). The myelination of fiber tracts in the spinal cord follows a progressive sequence (*see* Chapter 6, Section 6.3 in Altman and Bayer, 2001). Differences in the concentration of glia allow several major fiber tracts to be tentatively identified by this age (**Table 1**). The lateral and ventral corticospinal tracts (the last to myelinate) stand out as very sparse regions. Other regions in the ventral and lateral funiculi have different densities of proliferative glia in bands and clumps. A densely populated outer band at the cervical level (only in the lateral funiculus) is postulated to be the spinocerebellar tracts. A sparsely populated band just inside that (the outermost band in the ventral funiculus) is postulated to be the vestibulospinal tract and the spinocephalic tracts. A very dense to dense inner band adjacent to the ventral gray matter is postulated to contain the medial longitudinal fasciculus, the tectospinal tract (only at the cervical levels), and the intraspinal (propriospinal) tracts. Both of these fiber tracts contain density gradients of proliferating glia (see notes beneath **Table 1**). In the dorsal funiculus, the dorsal root collateralization and bifurcation zones contain very dense to dense proliferative glia all the way down to the lumbar enlargement. At cervical and middle thoracic levels, the cuneate fasciculus can be distinguished from the gracile fasciculus by the greater concentration of proliferating glia. However, the gracile fasciculus, especially its deep part, contains proliferating glia that is most dense in the lumbar enlargement and gradually declines through thoracic levels. It is least dense at cervical levels, except for the deep wedge. Generally, the lower concentration of proliferating glia in the white matter at lumbosacral levels (*Column 3*, **Table 1**) reflects the gradient of maturation from upper cervical to coccygeal levels.

Within the gray matter, columns of motoneurons continue to show progressive segregation in the ventral horn. The accumulation of lateral horn (autonomic) motoneurons is obvious at thoracic levels, and possibly at the sacral level. Large neurons in Clarke's column are evident at the low thoracic level, and may also be present at the middle thoracic level.

Table 1: Density of proliferating glia in the white matter at GW 19

Name	Cervical	Thoracic	Lumbosacral
DORSAL ROOT	sparse	---	sparse
VENTRAL ROOT	very dense	---	very dense
DORSAL FUNICULUS:			
dorsal root bif. zone	dense	dense	sparse
dorsal root col. zone	very dense	very dense	*dense
deep fas. gracilis	dense	very dense	---
superficial fas. gracilis	very sparse	sparse	sparse
deep fas. cuneatus	very dense	dense	---
superficial fas. cuneatus	sparse	very sparse	---
Lissauer's tract	very sparse	very sparse	very sparse
LATERAL and VENTRAL FUNICULI:			
lat. corticospinal tract	very sparse	very sparse	very sparse
ven. corticospinal tract	very sparse	very sparse	---
rubrospinal tract	dense	---	---
spinocerebellar tracts	dense	sparse	---
ven. commissure	very dense	dense	sparse
**intraspinal tracts	†gradient	†gradient	--
***spinocephalic tracts	††gradient	††gradient	--
med. long. fasciculus	very dense	---	---
tectospinal tract	very dense	---	---
vestibulospinal tract	sparse	sparse	---

* Sparse at the most lumbosacral level (**Plate 8**).

** Overlaps with the medial longitudinal fasciculus, the tectospinal tract, and the lateral reticulospinal tract.

† Very dense medial to the ventral horn, dense around the remaining ventral horn, sparse around the intermediate gray and lateral dorsal horn.

*** Contains anterior and lateral parts; overlaps with the vestibulospinal tract, the spinotectal tract, and the spino-olivary tract.

†† Deep parts (adjacent to intraspinal tracts) are more dense than superficial parts (adjacent to the ventrolateral pial membrane).

PLATE 1

CR 130 mm, GW 19, Y52-61

Plates 2A, 2B
Upper Cervical
Total area:
4.1202 mm^2

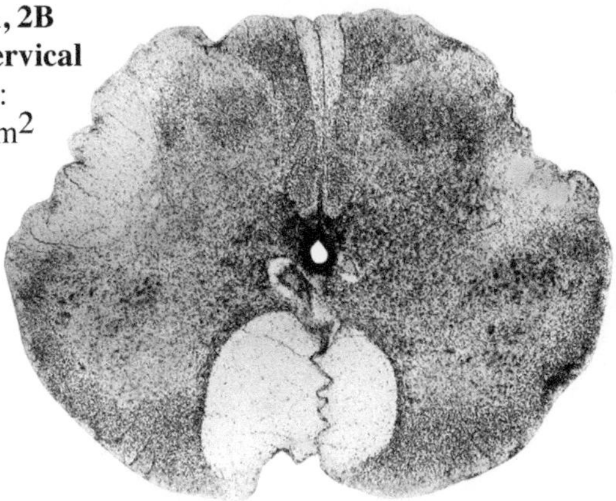

Plates 3A, 3B
Cervical Enlargement
Total area:
3.7511 mm^2

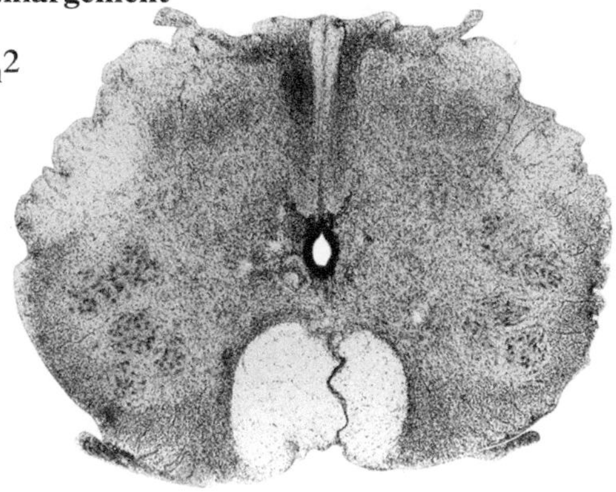

Plates 4A, 4B
Middle Thoracic
Total area:
1.3996 mm^2

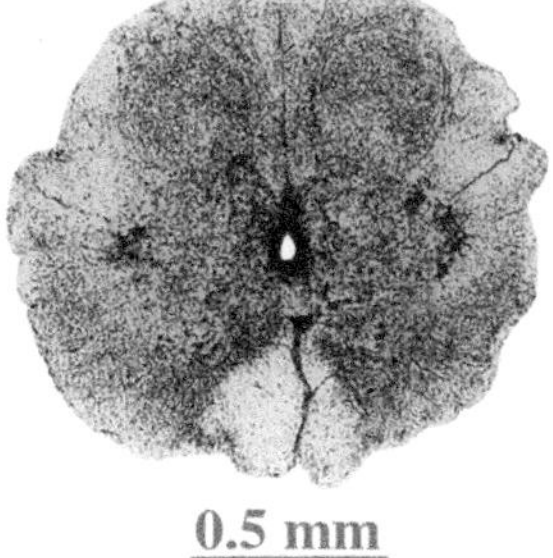

0.5 mm

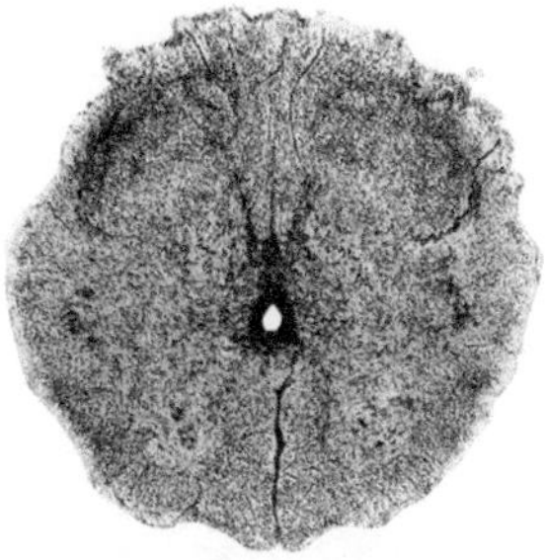

Plates 5A, 5B
Lower Thoracic
Total area:
1.4929 mm^2

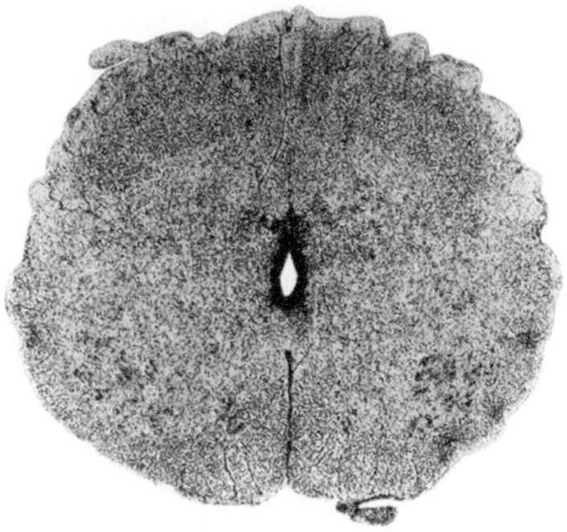

Plates 6A, 6B
Upper Lumbar
Total area:
2.2128 mm^2

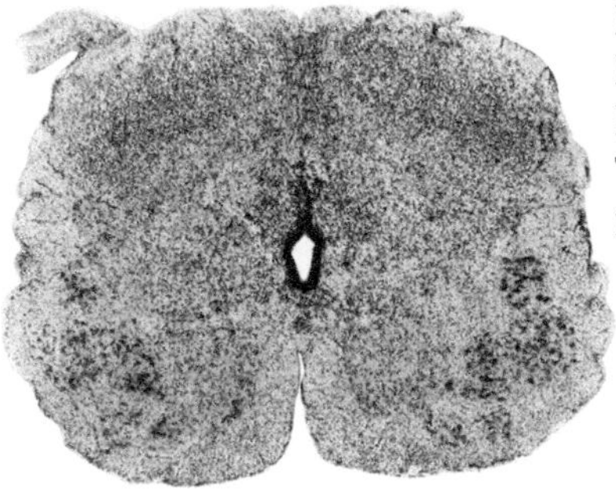

Plates 7A, 7B
Lumbar Enlargement
Total area:
2.6480 mm^2

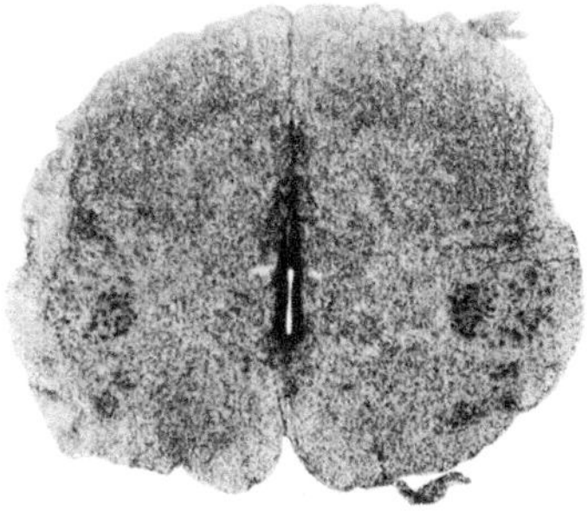

Plates 8A, 8B
Lumbosacral
Total area:
2.0164 mm^2

0.5 mm

PLATE 2A

CR 130 mm
GW 19
Y52-61
Upper Cervical
Cell body stain

Areas (mm^2)

Central canal	.0050
Neuroepithelium	.0170
Roof plate	.0152
Floor plate	.0054
Gray matter	1.8505
White matter	2.2271

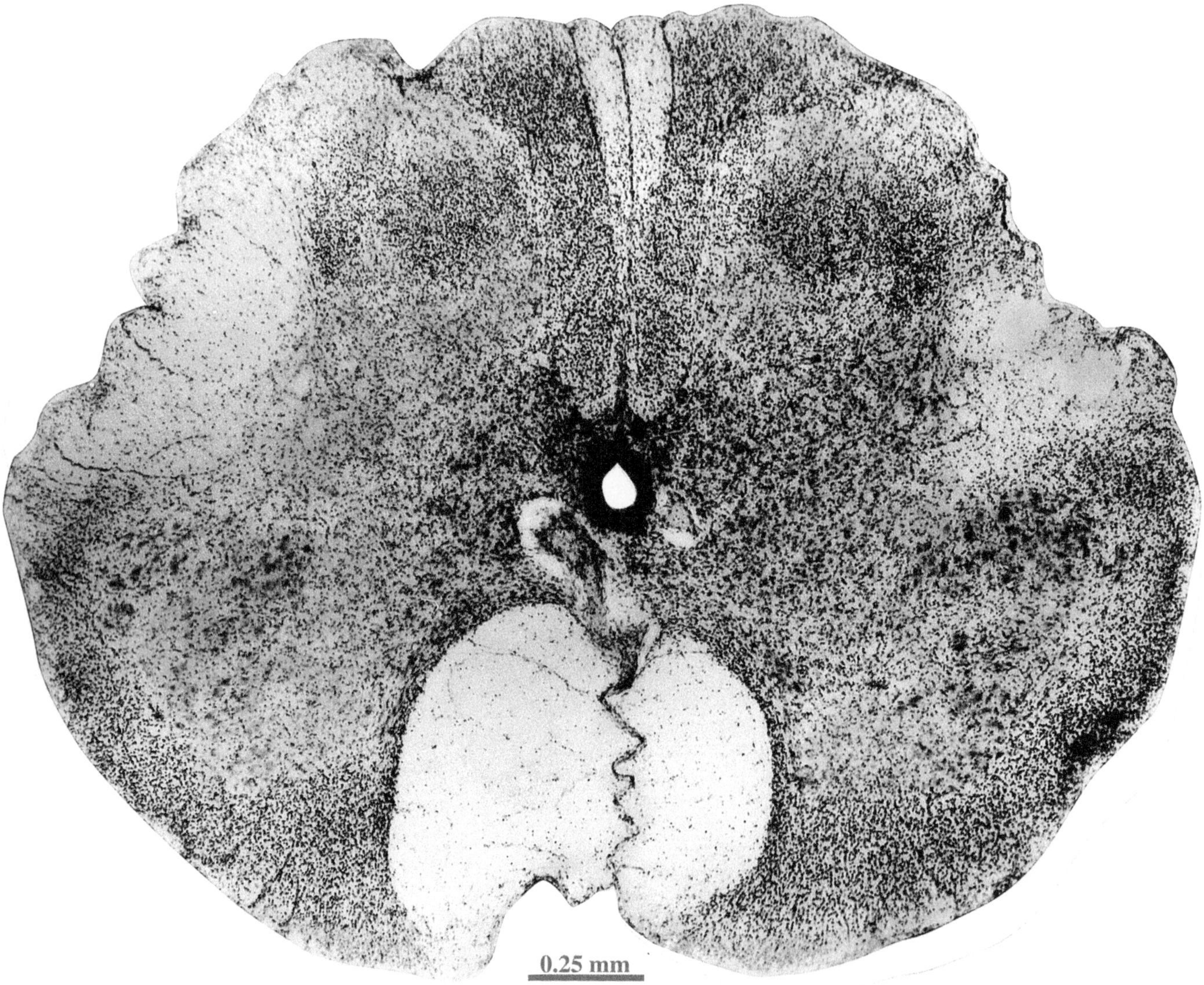

PLATE 2B

SEE TABLE 1 FOR THE LIST OF GLIAL DENSITIES IN EACH FIBER TRACT. EXAMPLES OF GLIAL DENSITIES IN EACH CATEGORY ARE LABELED IN THE LOWER-LEFT CORNER OF THIS PLATE.

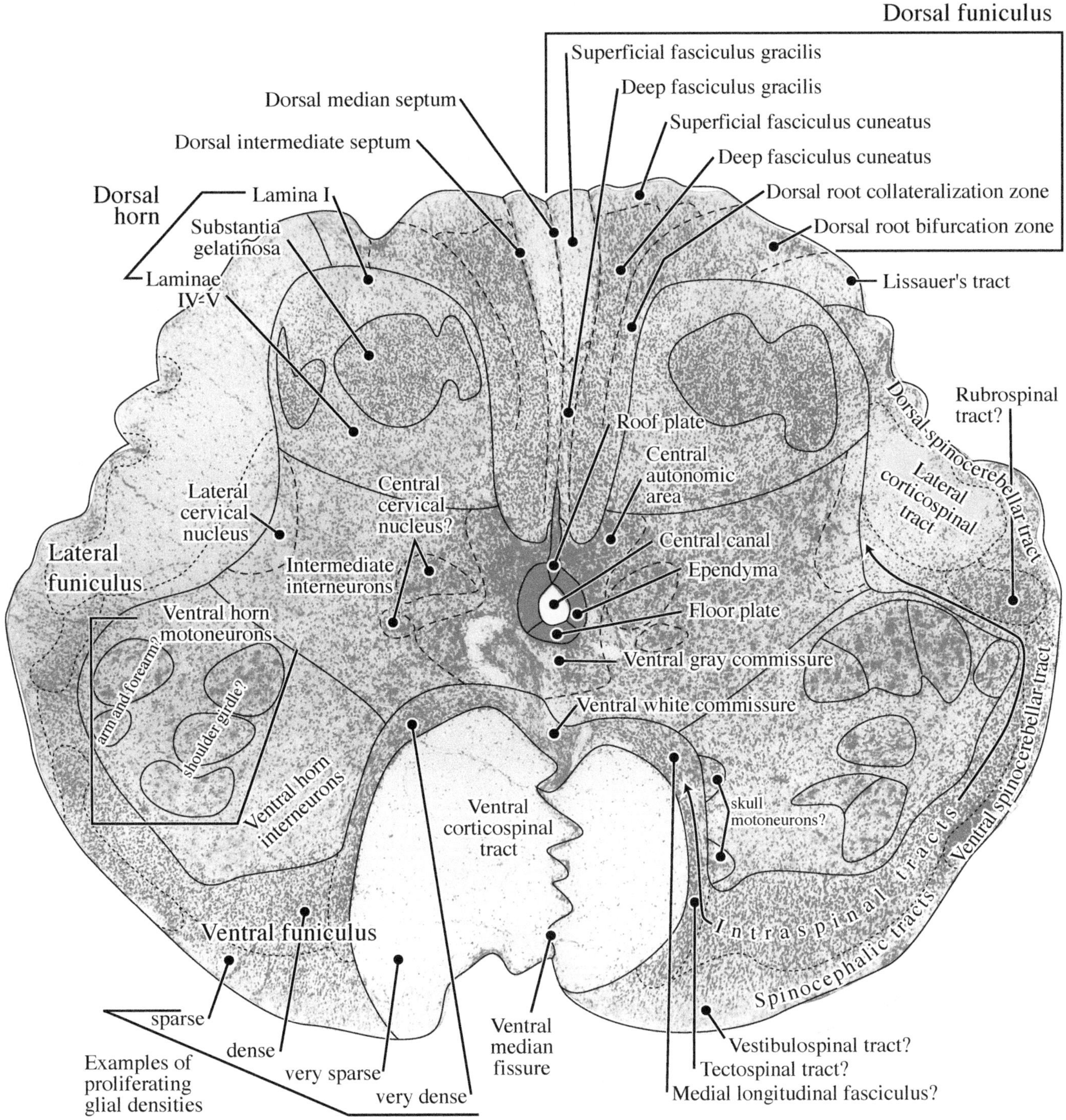

The lines in the lateral and ventral funiculi segregate regions of differing densities of proliferating glia, not the borders of fiber tracts. These lines may correspond to the borders of some fiber tracts, such as the spinocerebellar and the lateral corticospinal tracts. Other lines run across fiber tracts, such as the very dense region adjacent to the ventral corticospinal tract that includes the medial longitudinal fasciculus, the tectospinal tract, and the medial part of the intraspinal tracts. Many of the fiber tracts in the ventral and lateral funiculi overlap and do not have distinct borders.

PLATE 3A

CR 130 mm
GW 19
Y52-61
Cervical Enlargement
Cell body stain

Areas (mm^2)

Central canal	.0069
Neuroepithelium	.0172
Roof plate	.0154
Floor plate	.0037
Gray matter	1.8661
White matter	1.8419

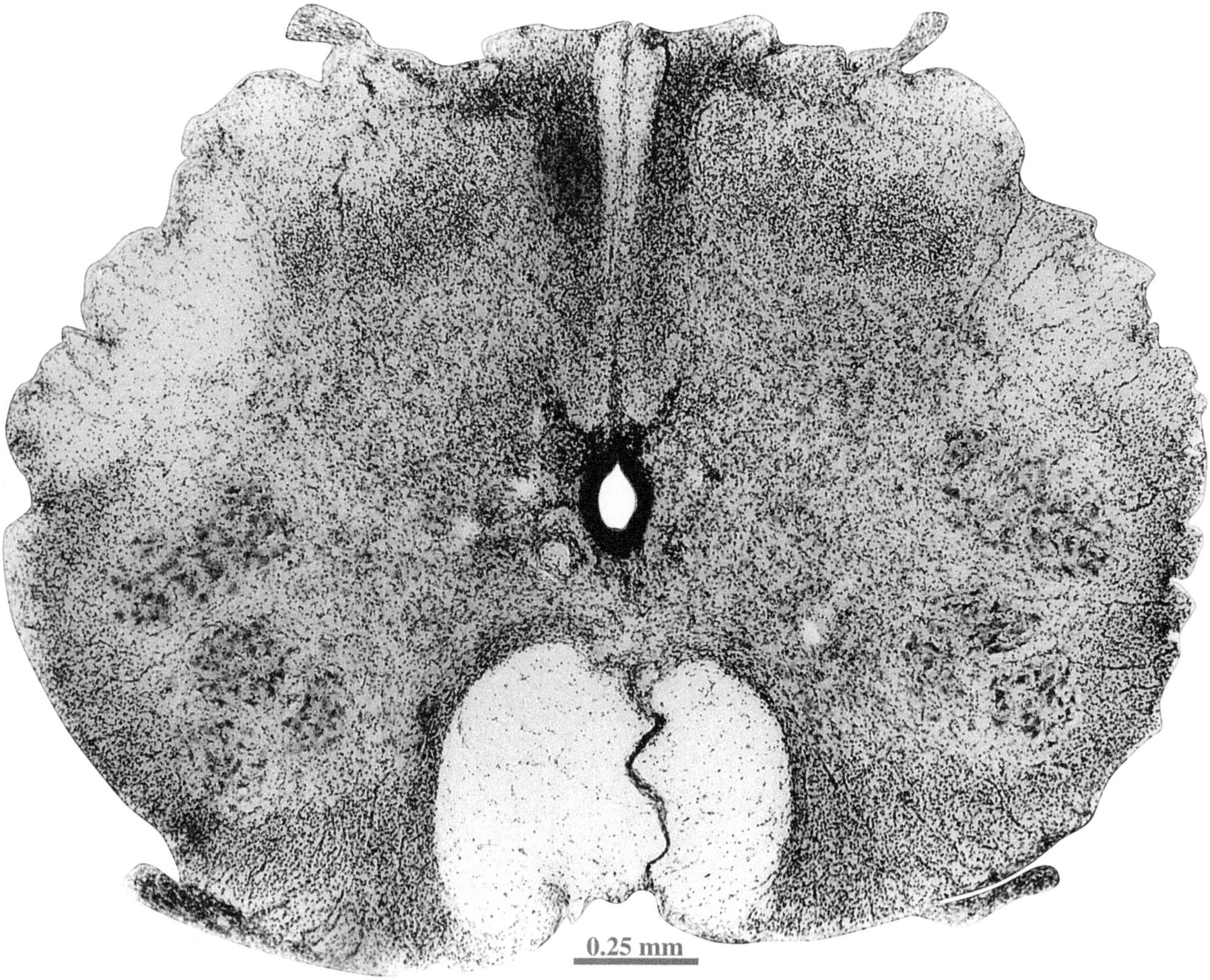

PLATE 3B

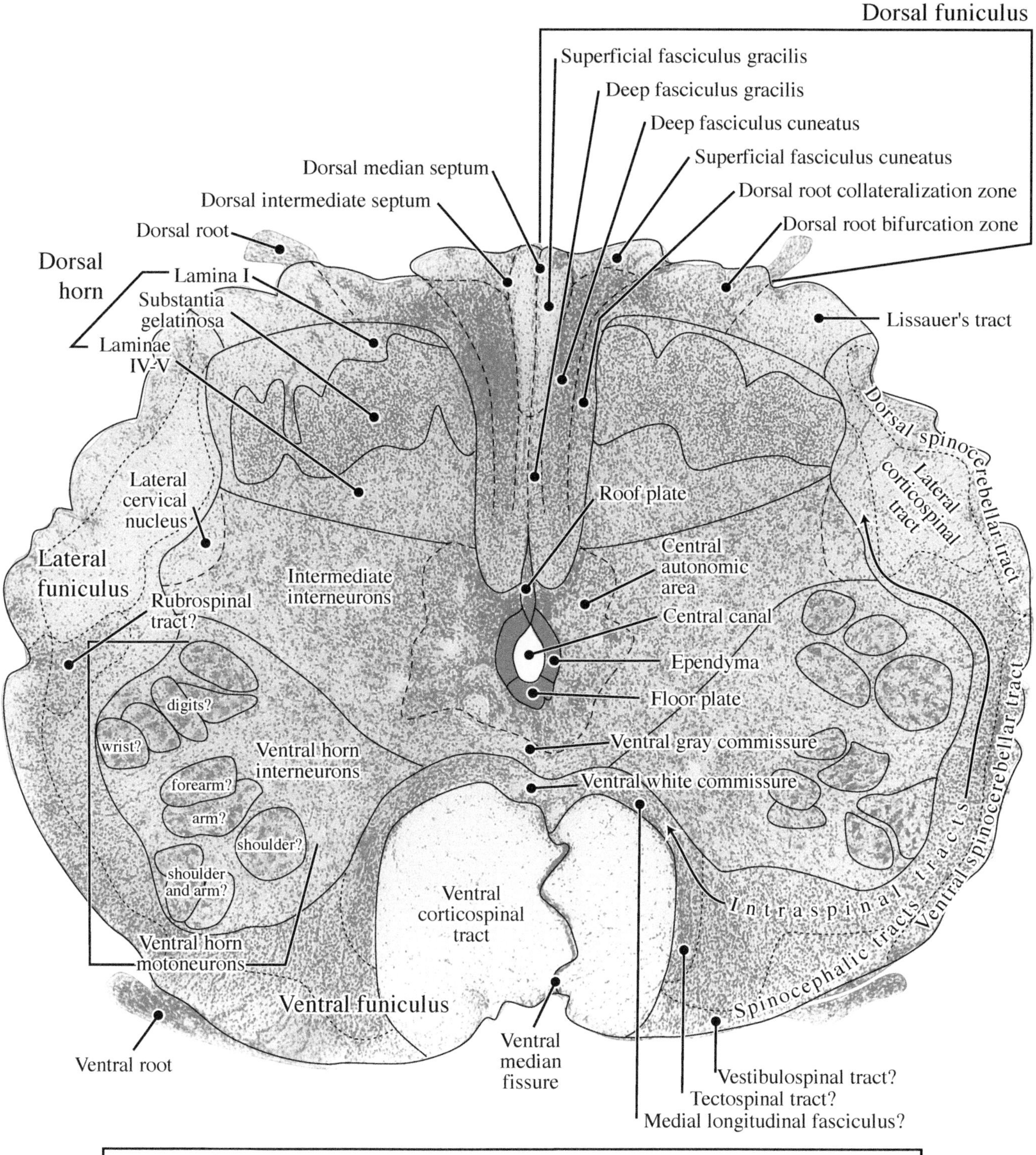

Note the larger lateral and ventral corticospinal tracts on one side of the spinal cord and the smaller ones on the other (see also **Plates 2 and 4**). Axons in the lateral corticospinal tract cross the midline, while axons in the ventral corticospinal tract remain ipsilateral. The *larger lateral* corticospinal tract (crossed component) is linked to the *smaller ventral* corticospinal tract (uncrossed component) and vice versa. It is unknown which is the true right or left side. This specimen is the only one in the Atlas to show such a pronounced asymmetry.

PLATE 4A

CR 130 mm
GW 19
Y52-61
Middle Thoracic
Cell body stain

Areas (mm^2)

Central canal	.0019
Neuroepithelium	.0092
Roof plate	.0156
Floor plate	.0029
Gray matter	.6000
White matter	.7701

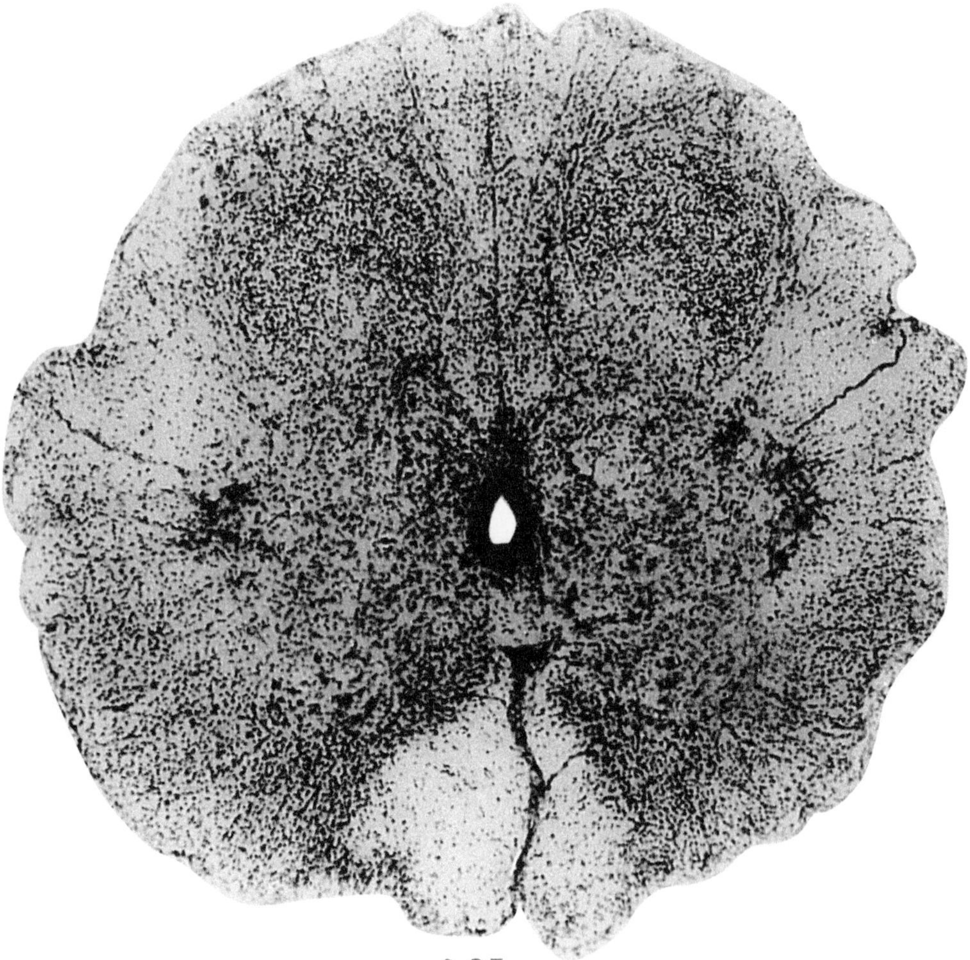

0.25 mm

PLATE 4B

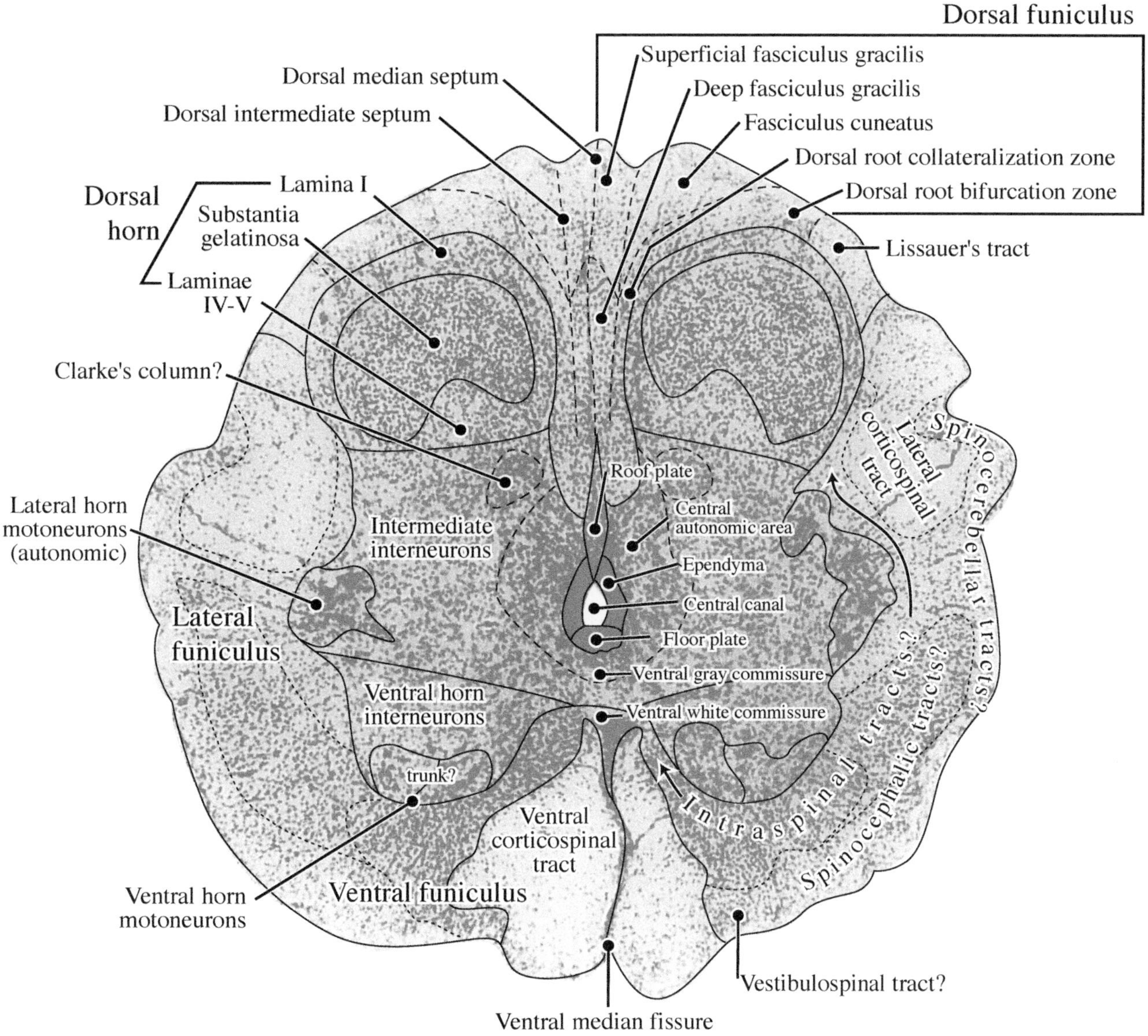
Dorsal funiculus
Superficial fasciculus gracilis
Deep fasciculus gracilis
Fasciculus cuneatus
Dorsal root collateralization zone
Dorsal root bifurcation zone
Lissauer's tract
Dorsal median septum
Dorsal intermediate septum
Dorsal horn
Lamina I
Substantia gelatinosa
Laminae IV-V
Clarke's column?
Lateral horn motoneurons (autonomic)
Lateral funiculus
Intermediate interneurons
Roof plate
Central autonomic area
Ependyma
Central canal
Floor plate
Ventral gray commissure
Ventral white commissure
Ventral horn interneurons
trunk?
Ventral horn motoneurons
Ventral corticospinal tract
Ventral funiculus
Ventral median fissure
Vestibulospinal tract?
Lateral corticospinal tract
Spinocerebellar tracts?
Spinocephalic tracts?
Intraspinal tracts?

PLATE 5A

CR 130 mm
GW 19
Y52-61
Lower Thoracic
Cell body stain

Areas (mm^2)

Central canal	.0025
Neuroepithelium	.0103
Roof plate	.0170
Floor plate	.0031
Gray matter	.7999
White matter	.6600

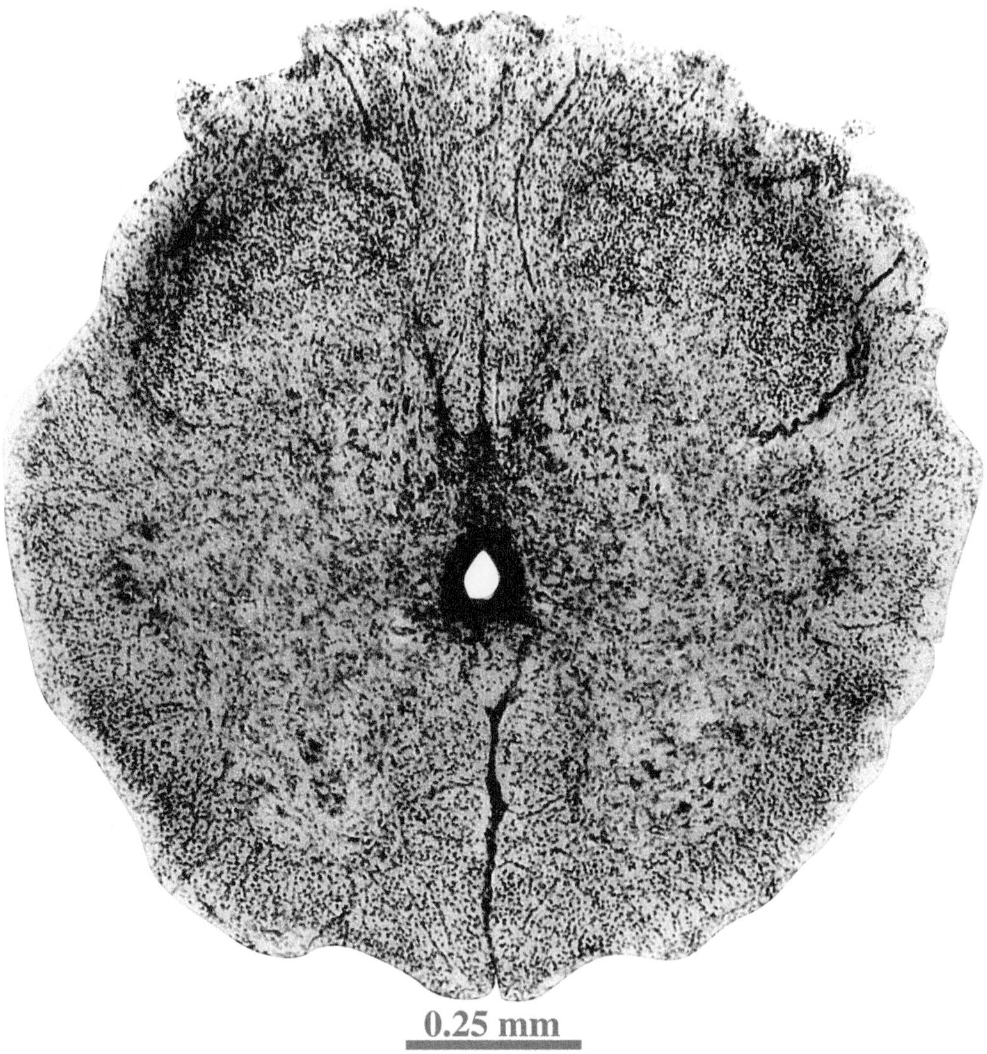

PLATE 5B

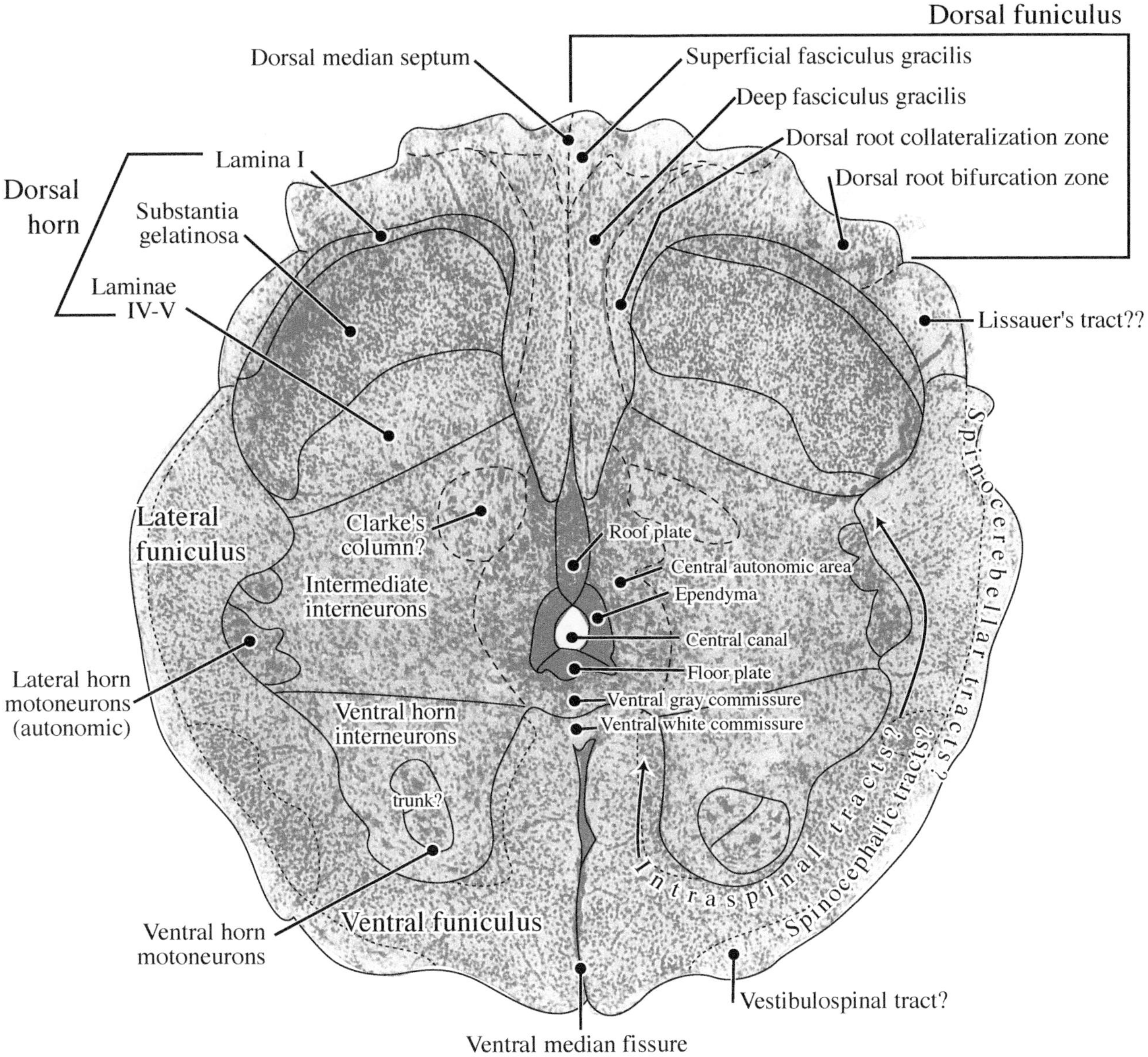
Dorsal funiculus
Dorsal median septum
Superficial fasciculus gracilis
Deep fasciculus gracilis
Dorsal root collateralization zone
Dorsal root bifurcation zone
Dorsal horn
Lamina I
Substantia gelatinosa
Laminae IV-V
Lissauer's tract??
Spino-cerebellar tracts?
Lateral funiculus
Clarke's column?
Intermediate interneurons
Roof plate
Central autonomic area
Ependyma
Central canal
Floor plate
Ventral gray commissure
Ventral white commissure
Lateral horn motoneurons (autonomic)
Ventral horn interneurons
trunk?
Intraspinal tracts?
Spinocephalic tracts?
Ventral funiculus
Ventral horn motoneurons
Vestibulospinal tract?
Ventral median fissure

PLATE 6A

CR 130 mm
GW 19
Y52-61
Upper Lumbar
Cell body stain

Areas (mm^2)

Central canal	.0049
Neuroepithelium	.0152
Roof plate	.0175
Floor plate	.0031
Gray matter	1.3115
White matter	.8607

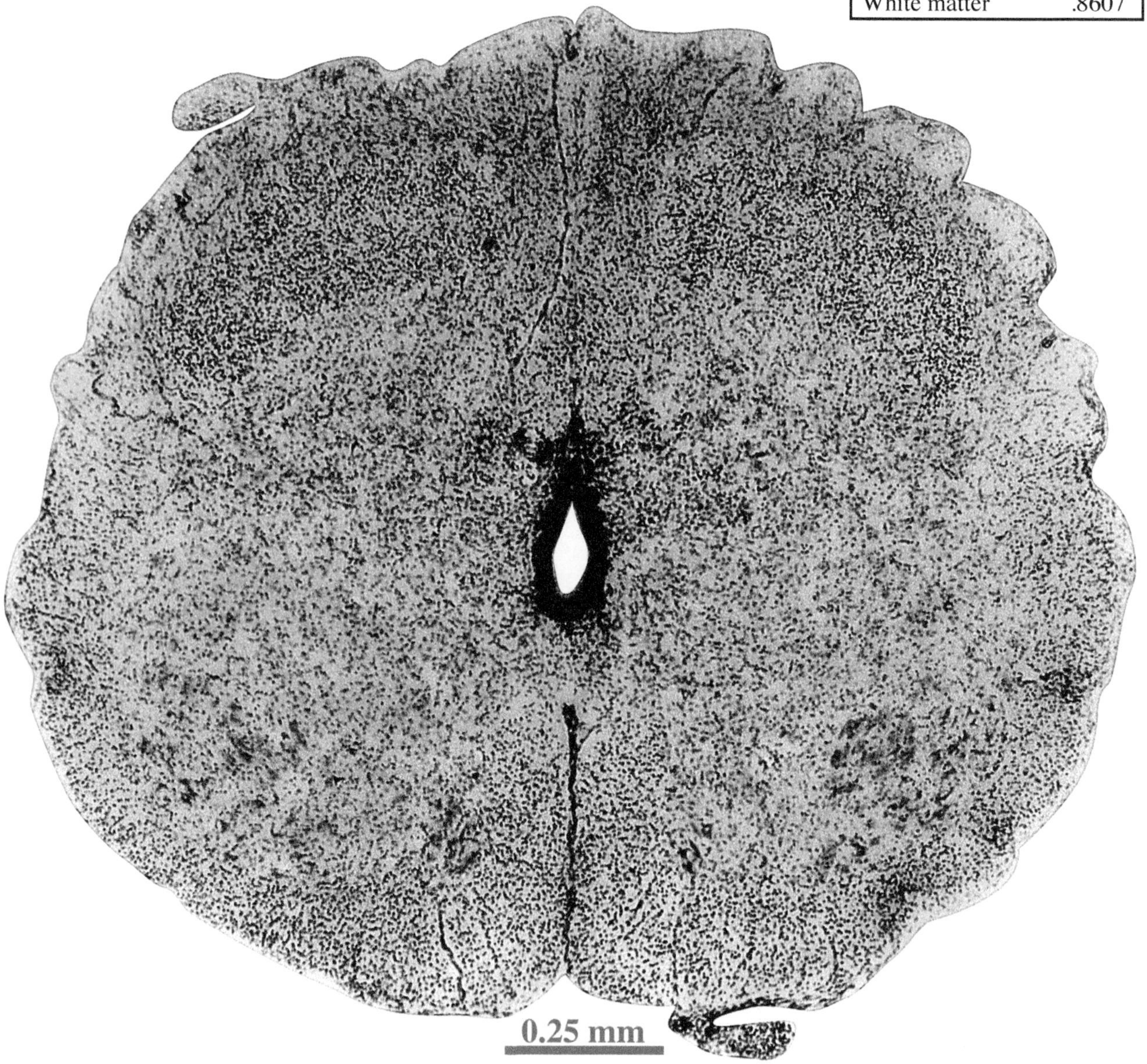

PLATE 6B

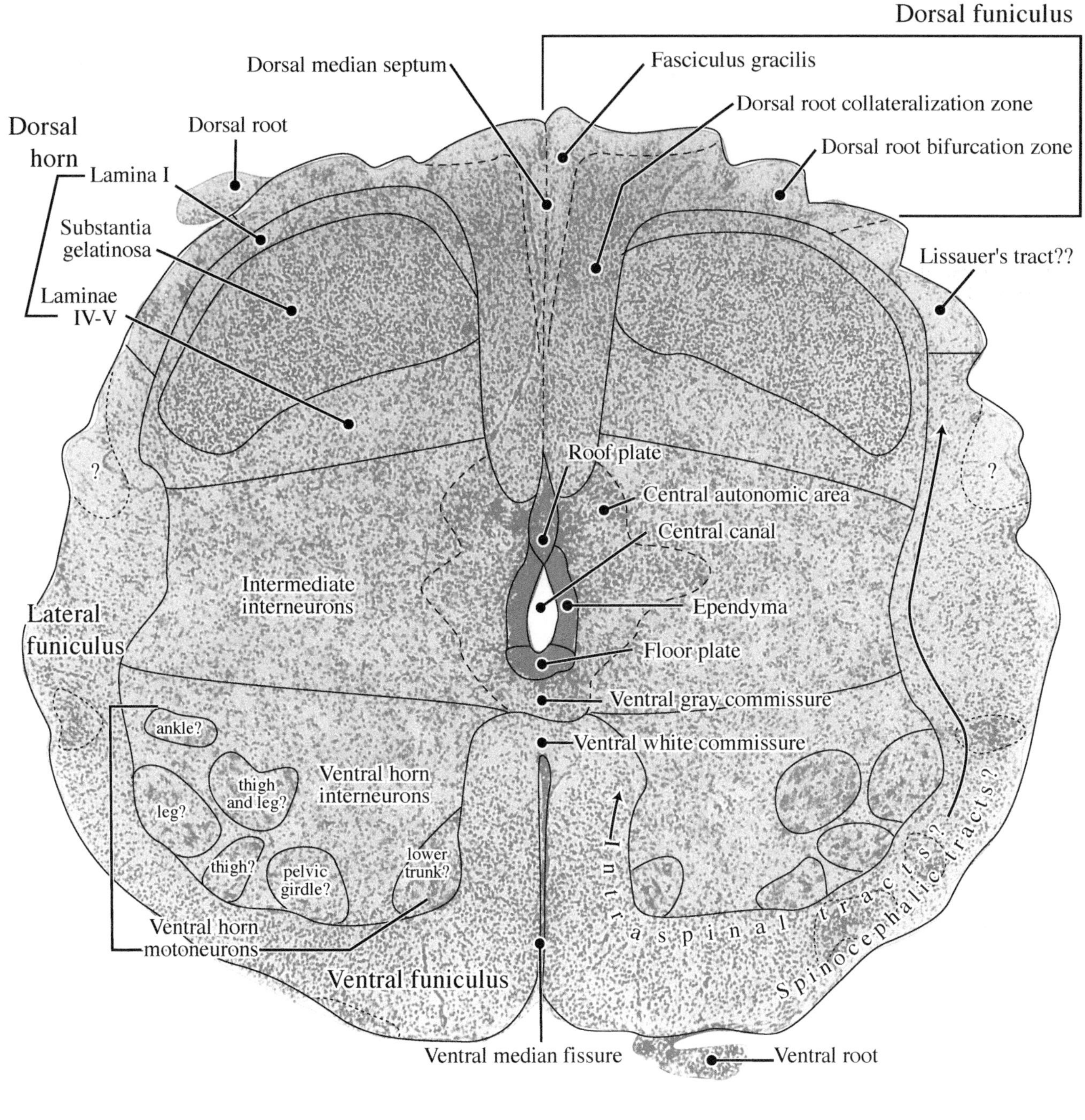
Dorsal funiculus
Dorsal median septum
Fasciculus gracilis
Dorsal root collateralization zone
Dorsal horn
Dorsal root
Dorsal root bifurcation zone
Lamina I
Substantia gelatinosa
Lissauer's tract??
Laminae IV-V
?
?
Roof plate
Central autonomic area
Central canal
Intermediate interneurons
Lateral funiculus
Ependyma
Floor plate
Ventral gray commissure
ankle?
Ventral white commissure
thigh and leg?
Ventral horn interneurons
leg?
Intraspinal tracts??
Spinocephalic tracts?
thigh?
pelvic girdle?
lower trunk?
Ventral horn motoneurons
Ventral funiculus
Ventral median fissure
Ventral root

PLATE 7A

CR 130 mm
GW 19
Y52-61
Lumbar Enlargement
Cell body stain

Areas (mm^2)

Central canal	.0072
Neuroepithelium	.0151
Roof plate	.0170
Floor plate	.0031
Gray matter	1.8235
White matter	.7822

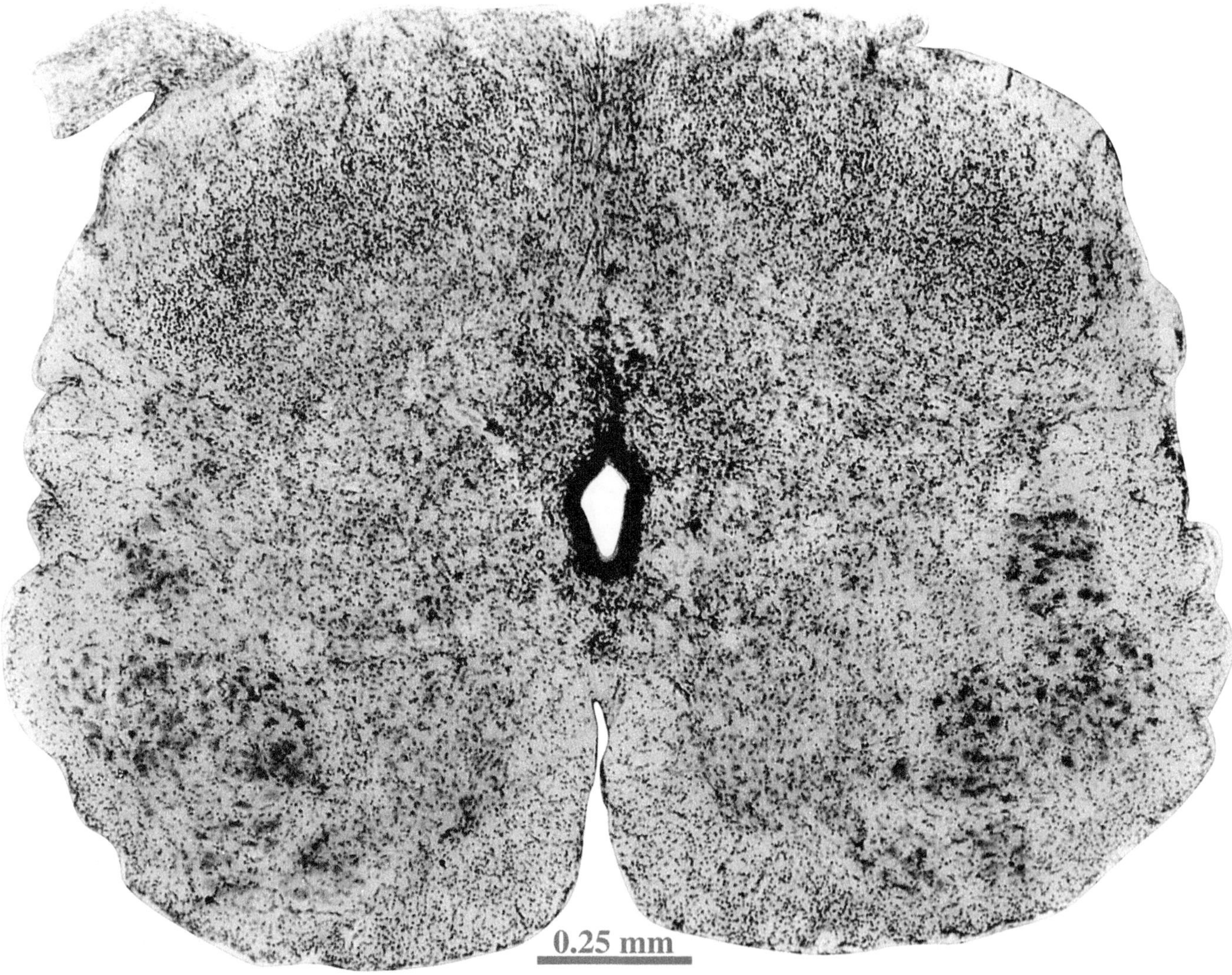

PLATE 7B

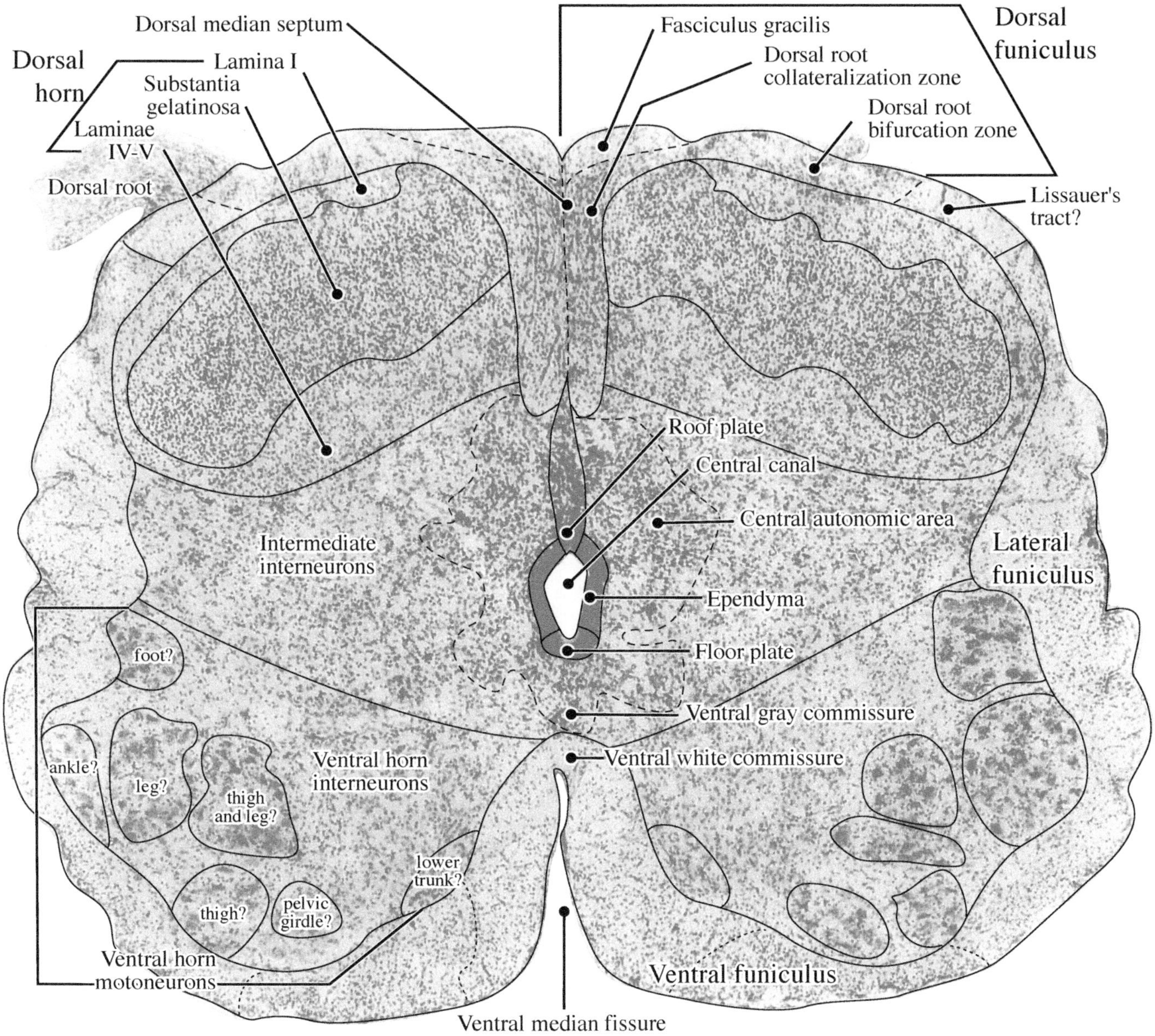
Dorsal median septum
Fasciculus gracilis
Dorsal funiculus
Dorsal horn
Lamina I
Dorsal root collateralization zone
Substantia gelatinosa
Dorsal root bifurcation zone
Laminae IV-V
Dorsal root
Lissauer's tract?
Roof plate
Central canal
Central autonomic area
Intermediate interneurons
Lateral funiculus
Ependyma
Floor plate
foot?
Ventral gray commissure
Ventral white commissure
ankle?
leg?
thigh and leg?
Ventral horn interneurons
lower trunk?
thigh?
pelvic girdle?
Ventral horn motoneurons
Ventral funiculus
Ventral median fissure

PLATE 8A

CR 130 mm
GW 19
Y52-61
Lumbosacral
Cell body stain

Areas (mm^2)

Central canal	.0027
Neuroepithelium	.0187
Roof plate	.0201
Floor plate	.0019
Gray matter	1.4461
White matter	.5269

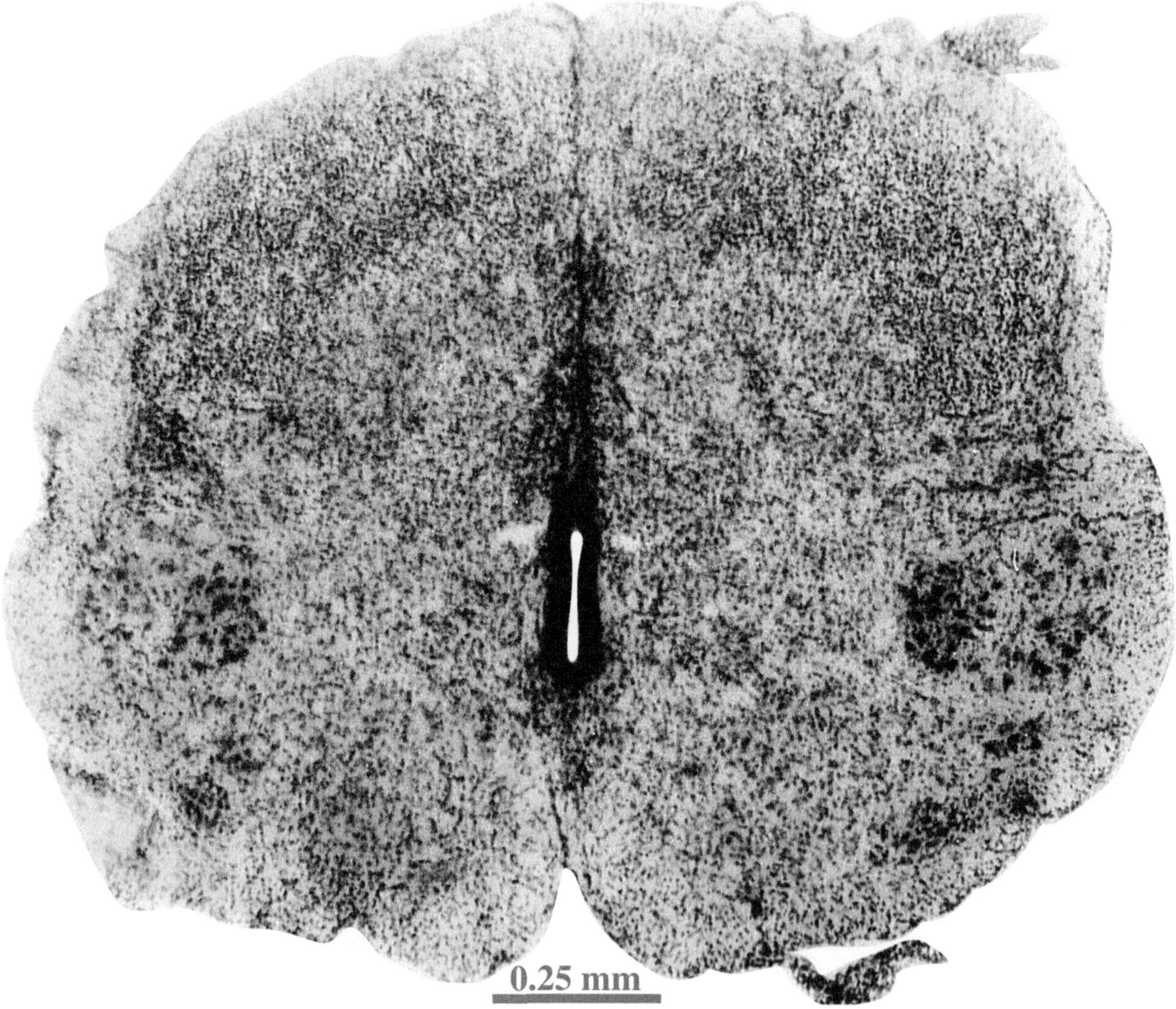

PLATE 8B

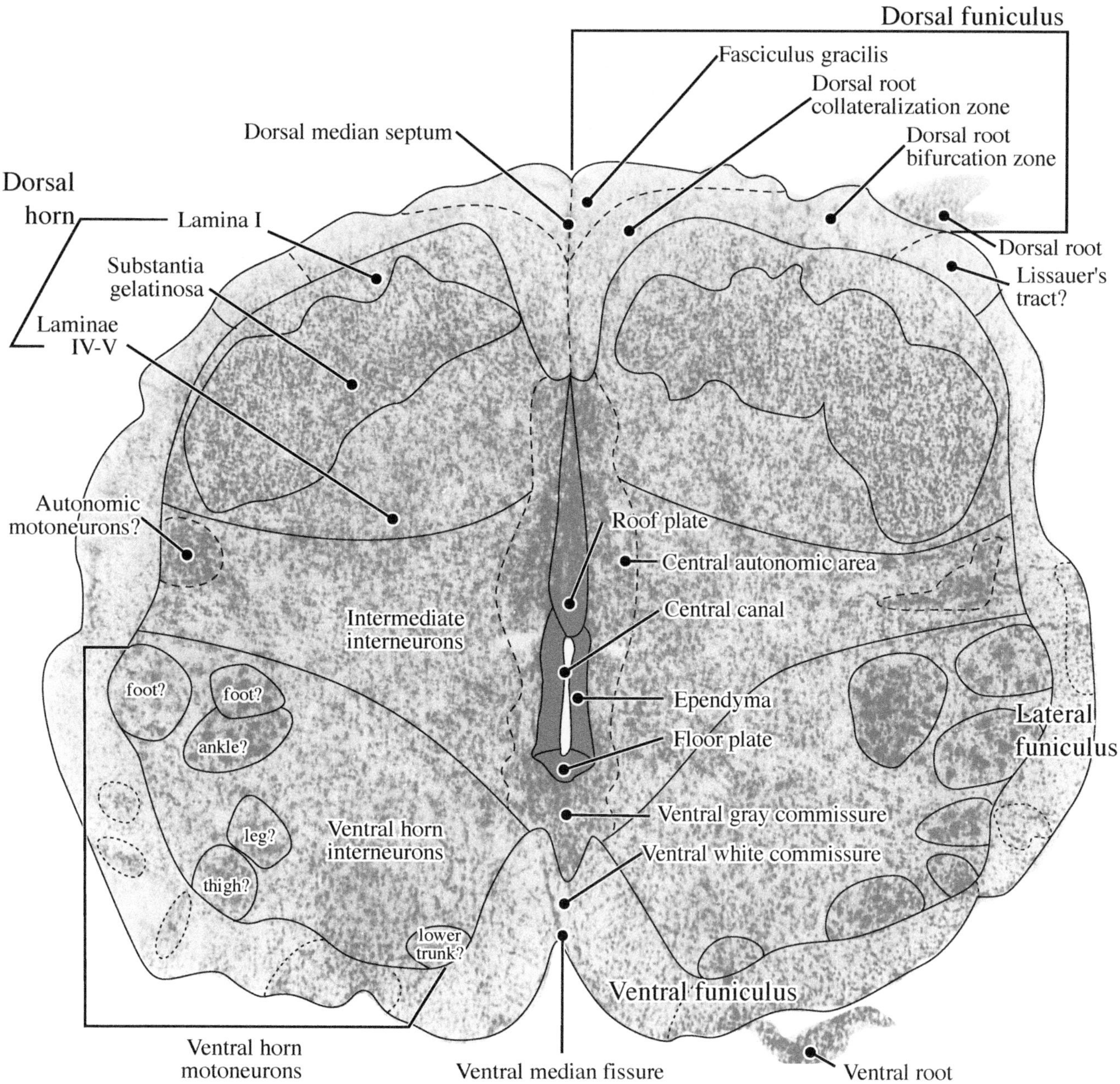
Dorsal funiculus
Fasciculus gracilis
Dorsal root
collateralization zone
Dorsal median septum
Dorsal root
bifurcation zone
Dorsal
horn
Lamina I
Dorsal root
Substantia
gelatinosa
Lissauer's
tract?
Laminae
IV-V
Autonomic
motoneurons?
Roof plate
Central autonomic area
Central canal
Intermediate
interneurons
foot?
foot?
Ependyma
ankle?
Lateral
funiculus
Floor plate
Ventral gray commissure
leg?
Ventral horn
interneurons
Ventral white commissure
thigh?
lower
trunk?
Ventral funiculus
Ventral horn
motoneurons
Ventral median fissure
Ventral root

PART III: Y60-61 CR 210 mm (GW 26)

Plate 9 is a survey of matched myelin-stained and cell-body-stained sections from Y60-61, a specimen in the Yakovlev Collection with a crown-rump length of 210 mm. All sections are shown at the same scale. The boxes enclosing each section list the approximate level and the total area (post-fixation) of the section in square millimeters (mm^2). Full-page normal-contrast photographs of each section are in **Plates 10A–15A.** Low-contrast photographs with superimposed labels and outlines of structural details are in **Plates 10B–15B.** In this specimen, the myelin-stained and cell-body-stained sections were preserved on separate large glass plates without any section numbers. Eighteen myelin-stained and 17 cell-body-stained sections were photographed ranging from upper cervical to thoracic levels. There were no sections preserved at lumbar, sacral, and coccygeal levels. The 35 photographic prints were intuitively arranged in order from upper cervical to thoracic levels, using internal features such as the size of the corticospinal tracts, and the width of the ventral horn. Then, myelin- and cell-body-stained sections were matched. The upper cervical level sections and several middle to lower thoracic level sections were either damaged or had no matches. There were good matches in the region of the cervical enlargement (upper and lower levels are illustrated) and in the upper level of the thoracic cord. The cross-sectional area of a myelin-stained section is smaller than the matching cell-body-stained section in all cases. Evidently, the myelin staining procedure produces greater tissue shrinkage than the cell-body staining procedure.

Y60-61 is one of the youngest specimens to show any myelin-stained areas. Actually, staining for myelin products begins around GW 20 (Y27-60, CR 160 mm, Fig. 6-2 in Altman and Bayer, 2001), but that specimen is very incomplete, and is not shown in this Atlas. Three structures contain solid black stain indicative of myelinating axons (*Column 1* in **Table 2**). Throughout the rest of the white matter in the myelin-stained sections, there is either punctate staining or no staining. Dense punctate stained areas are assumed to have high concentrations of glia that react with the stain prior to production of the myelin sheath, sparse punctate stained areas have low concentrations, and unstained areas have none (*Column 2*, **Table 2**). In the cell-body-stained sections, there are various densities of what is assumed to be proliferating interfascicular glia in the dorsal, lateral, and ventral funiculi (*Column 3*, **Table 2**), but generally, the concentration of proliferating glia is less pronounced than in the GW 19 specimen.

The densities of reactive and proliferating glia within a fiber tract vary independently of each other (compare rows in **Table 2**). These variations are probably due to the different rates of myelination in each fiber tract. In general, glial proliferation precedes reactive gliosis (see the differing concentrations of proliferating glia in the GW 19 specimen, **Plates 1–8**), and that precedes myelination in different fiber tracts (*see* Chapter 6 in Altman and Bayer, 2001).

Table 2: Glia types and concentration in the white matter at GW 26

Name	Myelination	Reactive glia	Proliferating glia
DORSAL ROOT	---	dense	dense
VENTRAL ROOT	advanced	---	sparse
DORSAL FUNICULUS: dorsal root bif. zone	some fibers	---	dense
dorsal root col. zone	---	dense	dense
deep fas. gracilis	---	dense	dense
superficial fas. gracilis	---	sparse	dense
deep fas. cuneatus	---	dense	dense
superficial fas. cuneatus	---	sparse	dense
Lissauer's tract	---	none	very sparse
LATERAL and VENTRAL FUNICULI: lat. corticospinal tract	---	none	very sparse
ven. corticospinal tract	---	none	very sparse
spinocerebellar tracts	---	dense	dense
ven. commissure	some fibers	dense	dense
**intraspinal tracts	---	*gradient	sparse
††spinocephalic tracts	---	†gradient	sparse
med. long. fasciculus	---	dense	sparse
tectospinal tract	---	dense	sparse
vestibulospinal tract	---	sparse	sparse

** Overlaps with the medial longitudinal fasciculus, the tectospinal tract, and the lateral reticulospinal tract.

* The central part of the tract (adjacent to the ventral and lateral parts of the ventral horn) stain more intensely than medial parts or dorsolateral parts.

†† Contains ventral and lateral parts. Overlaps with the vestibulospinal, spinotectal, and spino-olivary tracts.

† Deep parts (adjacent to intraspinal tracts) are more dense than superficial parts (adjacent to the ventrolateral pial membrane).

PLATE 9

CR 210 mm, GW 26, Y60-61

MYELIN STAIN | **CELL BODY STAIN**

Plates 10A, 10B
Upper Cervical Enlargement
Total area:
11.455 mm^2

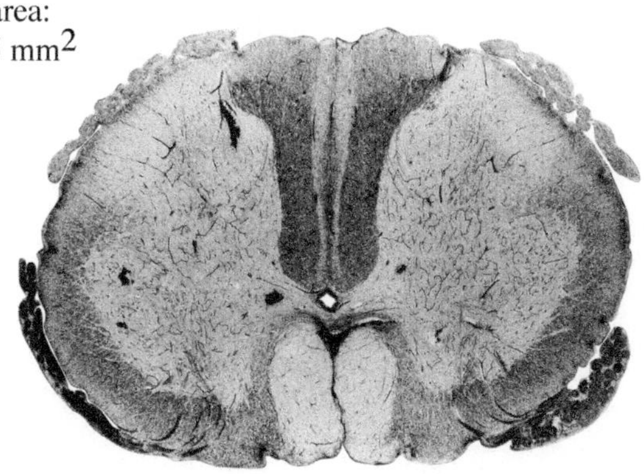

Plates 11A, 11B
Upper Cervical Enlargement
Total area:
12.070 mm^2

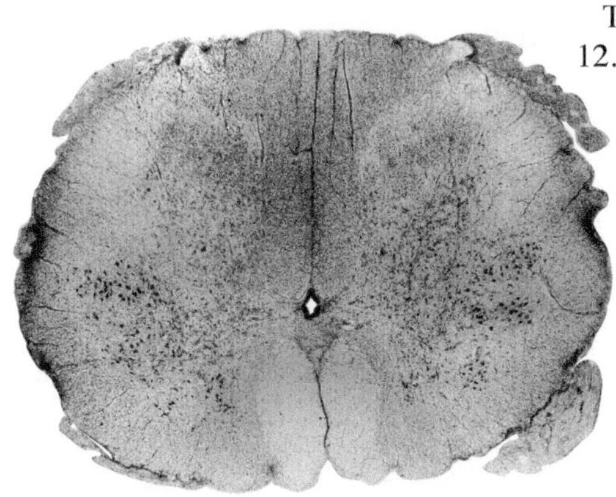

Plates 12A, 12B
Lower Cervical Enlargement
Total area:
11.380 mm^2

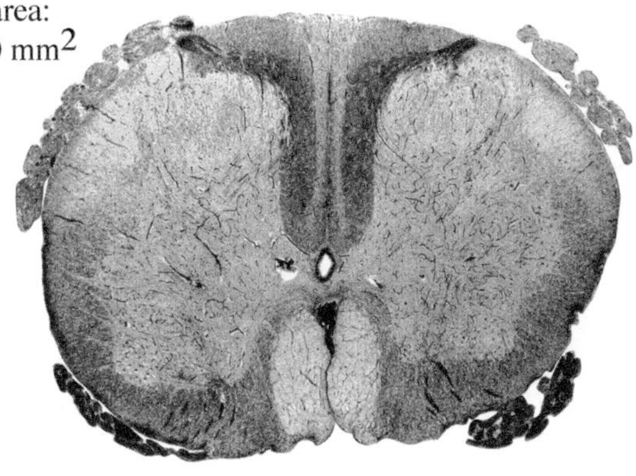

Plates 13A, 13B
Lower Cervical Enlargement
Total area:
12.589 mm^2

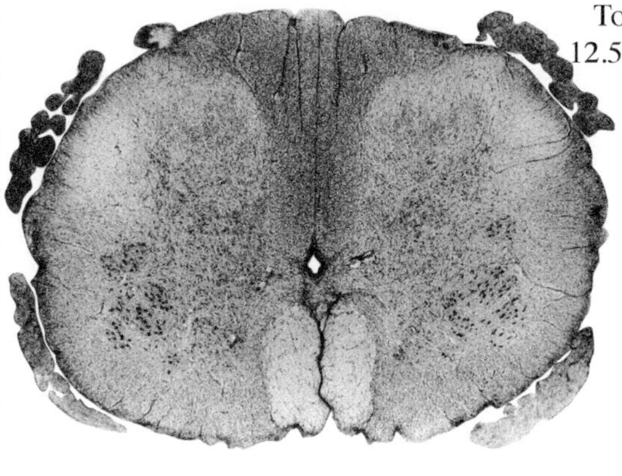

Plates 14A, 14B
Upper Thoracic
Total area:
6.029 mm^2

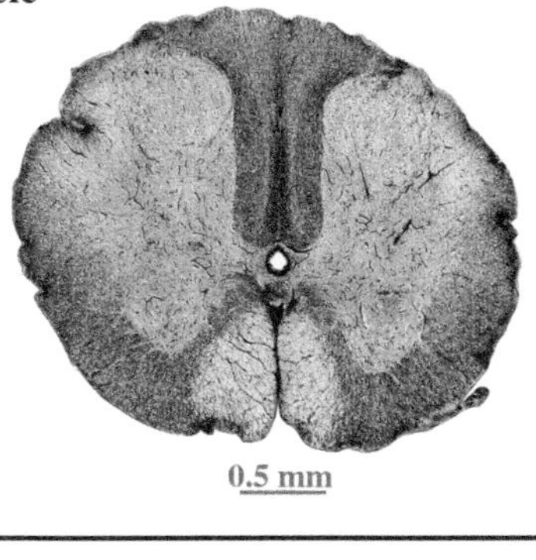

0.5 mm

Plates 15A, 15B
Upper Thoracic
Total area:
6.982 mm^2

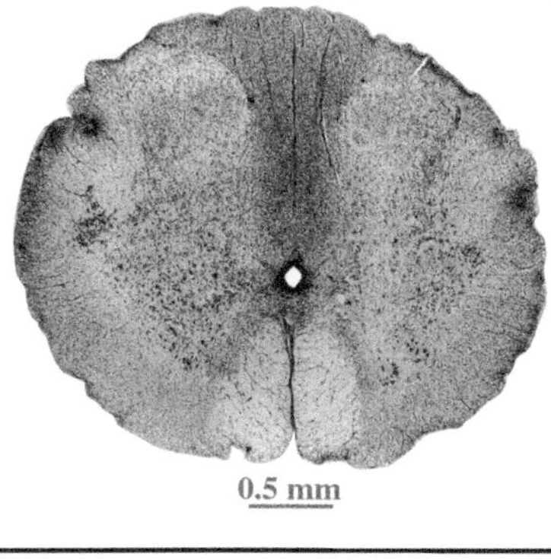

0.5 mm

PLATE 10A

CR 210 mm
GW 26
Y60-61
Upper Cervical Enlargement
Myelin stain

Areas (mm²)

Central canal	.0069
Ependyma	.0163
Gray matter	4.7293
White matter	6.7021

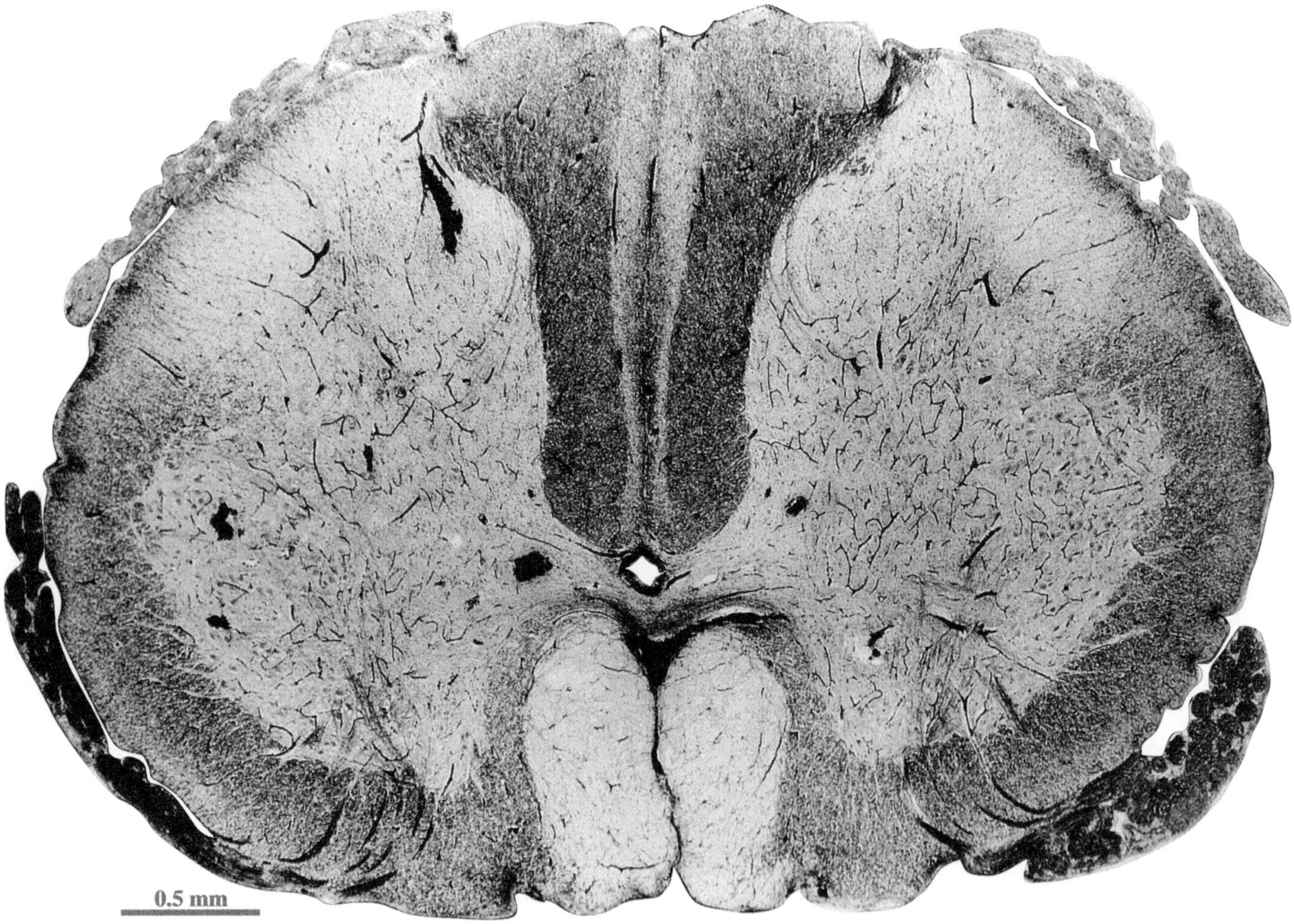

PLATE 10B

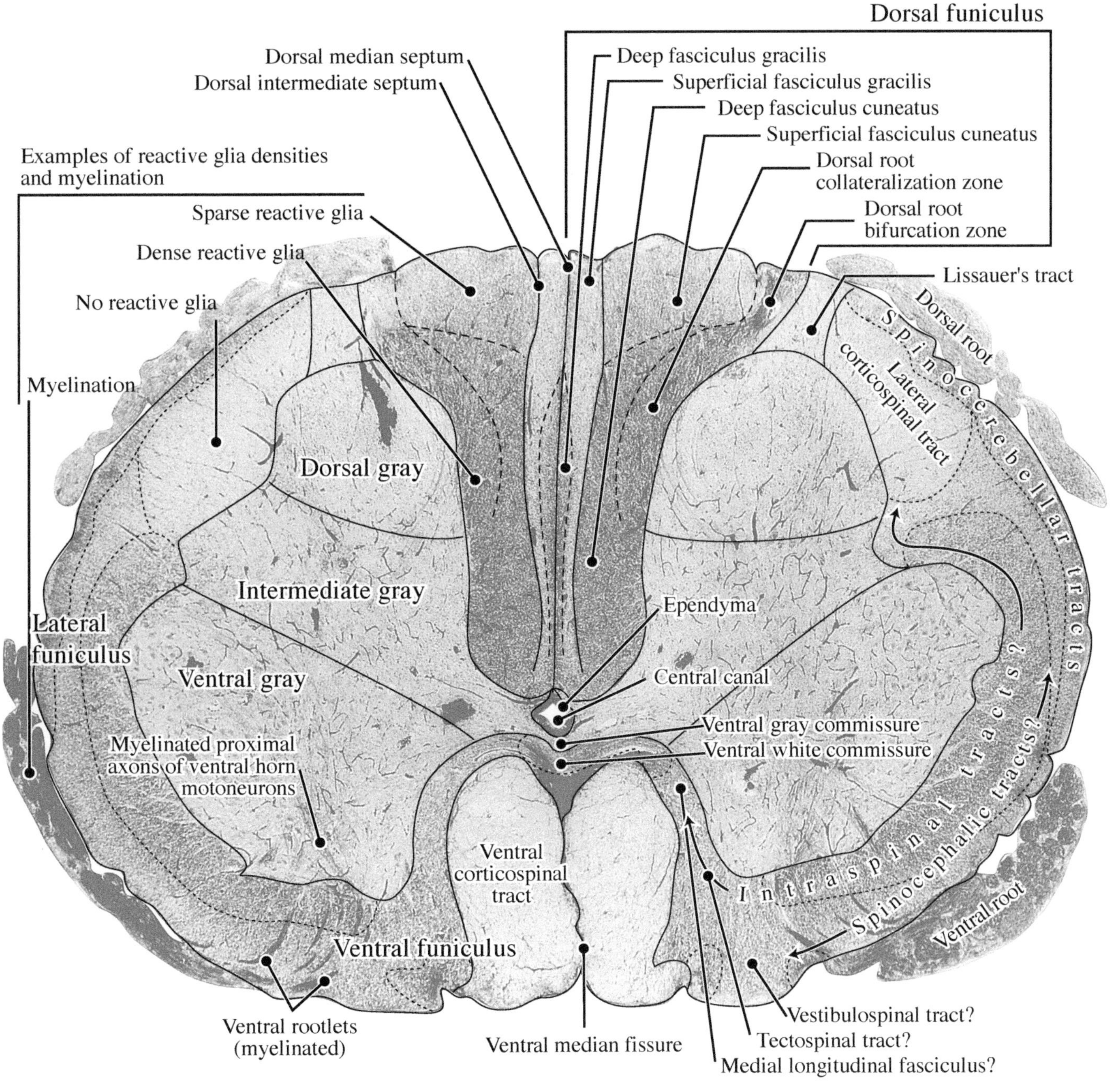

The fine lines in the lateral and ventral funiculi segregate regions of differing densities of reactive glia, not the borders of fiber tracts. These lines may correspond to the borders of some fiber tracts, such as the spinocerebellar and the lateral corticospinal tracts. Other lines run across fiber tracts, such as the dense region adjacent to the ventral and lateral part of the ventral horn that includes the central part of the intraspinal tracts and the deep parts of the spinocephalic tracts. Many of the fiber tracts in the ventral and lateral funiculi overlap and do not have distinct borders. The spinotectal, spino-olivary, spinoreticular, and ventrolateral reticulospinal tracts (all unlabeled) intermingle with the spinocephalic fibers. The medial longitudinal fasciculus and tectospinal tract are interspersed with the medial fibers of the intraspinal tracts; dorsal and lateral fibers of the intraspinal tract are infiltrated by the lateral reticulospinal tract (unlabeled).

PLATE 11A

CR 210 mm
GW 26
Y60-61
Upper Cervical Enlargement
Cell body stain

Areas (mm^2)

Central canal	.0055
Ependyma	.0206
Gray matter	5.0505
White matter	6.9934

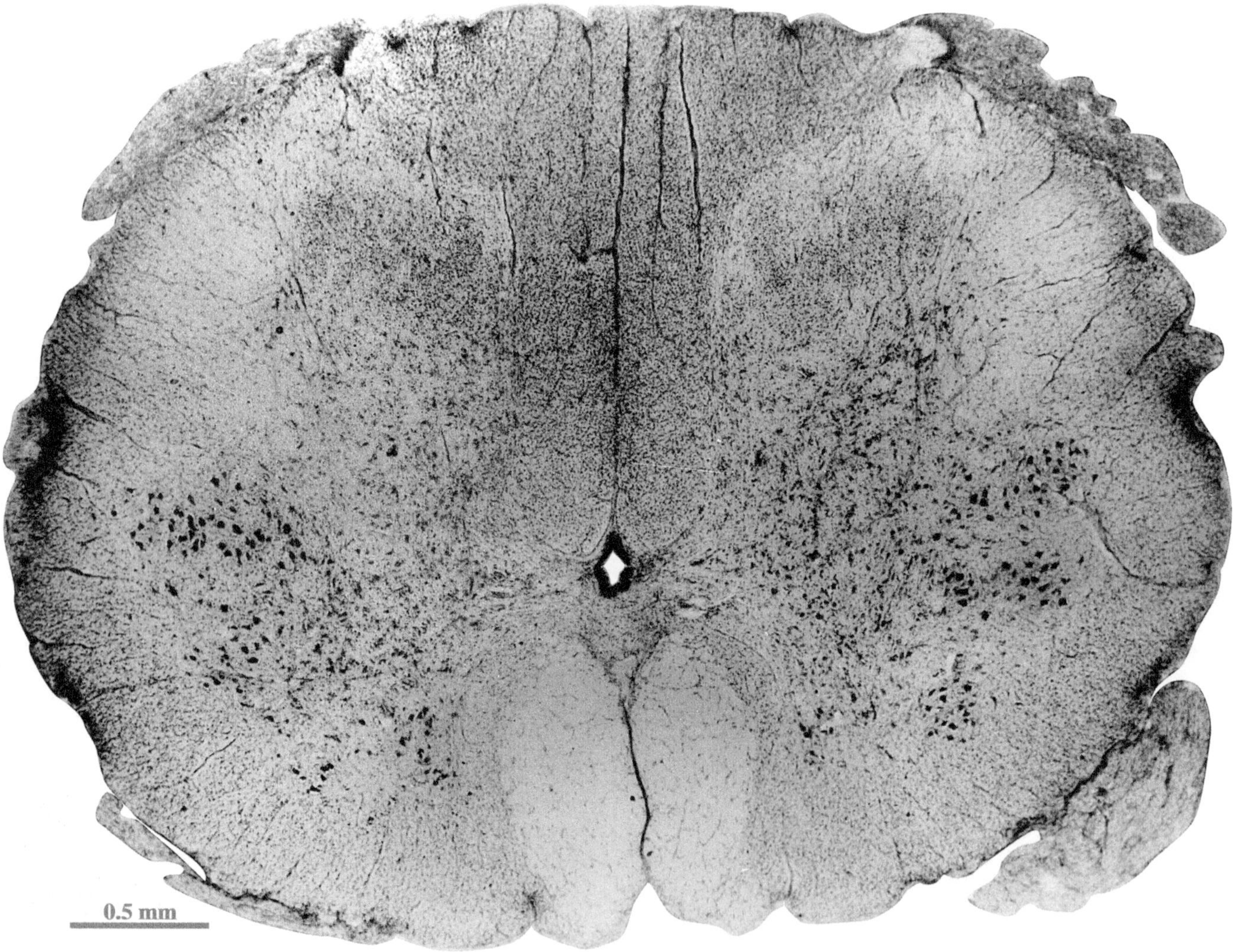

PLATE 11B

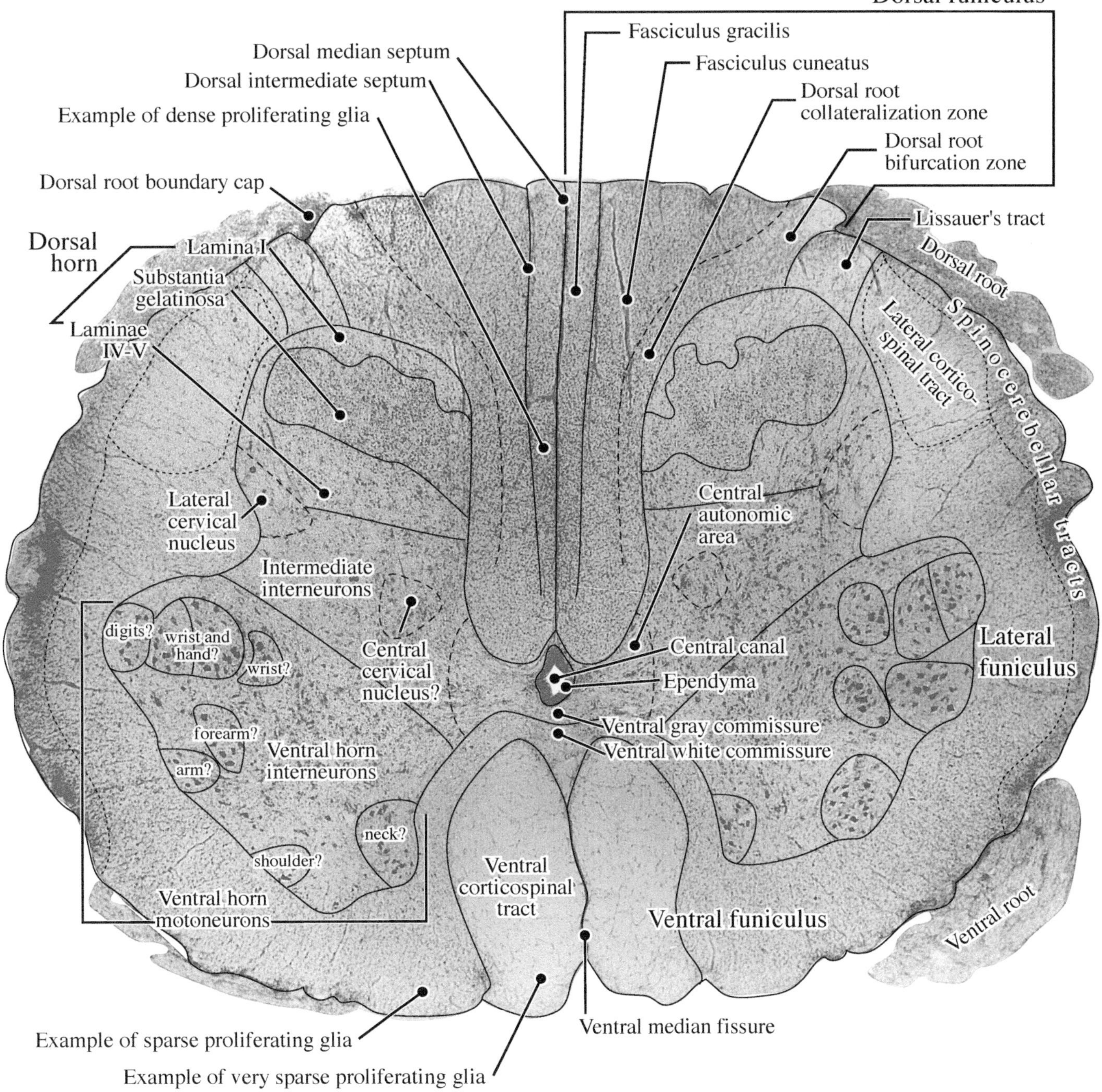

Only the corticospinal and spinocerebellar tracts can be clearly delineated in the ventral and lateral funiculi. The corticospinal tracts stand out as clear areas with very sparse proliferating glia. The spinocerebellar tracts have a denser concentration of proliferating glia. A sparse population of proliferating glia fills the remaining ventral and lateral funiculi and contains several fiber tracts (unlabeled in this section). Refer to the matching myelin-stained section for the approximate locations of the medial longitudinal fasciculus, tectospinal, vestibulospinal, intraspinal, and spinocephalic tracts.

PLATE 12A

CR 210 mm
GW 26
Y60-61
Lower Cervical Enlargement
Myelin stain

Areas (mm^2)

Central canal	.0090
Ependyma	.0196
Gray matter	5.1514
White matter	6.2003

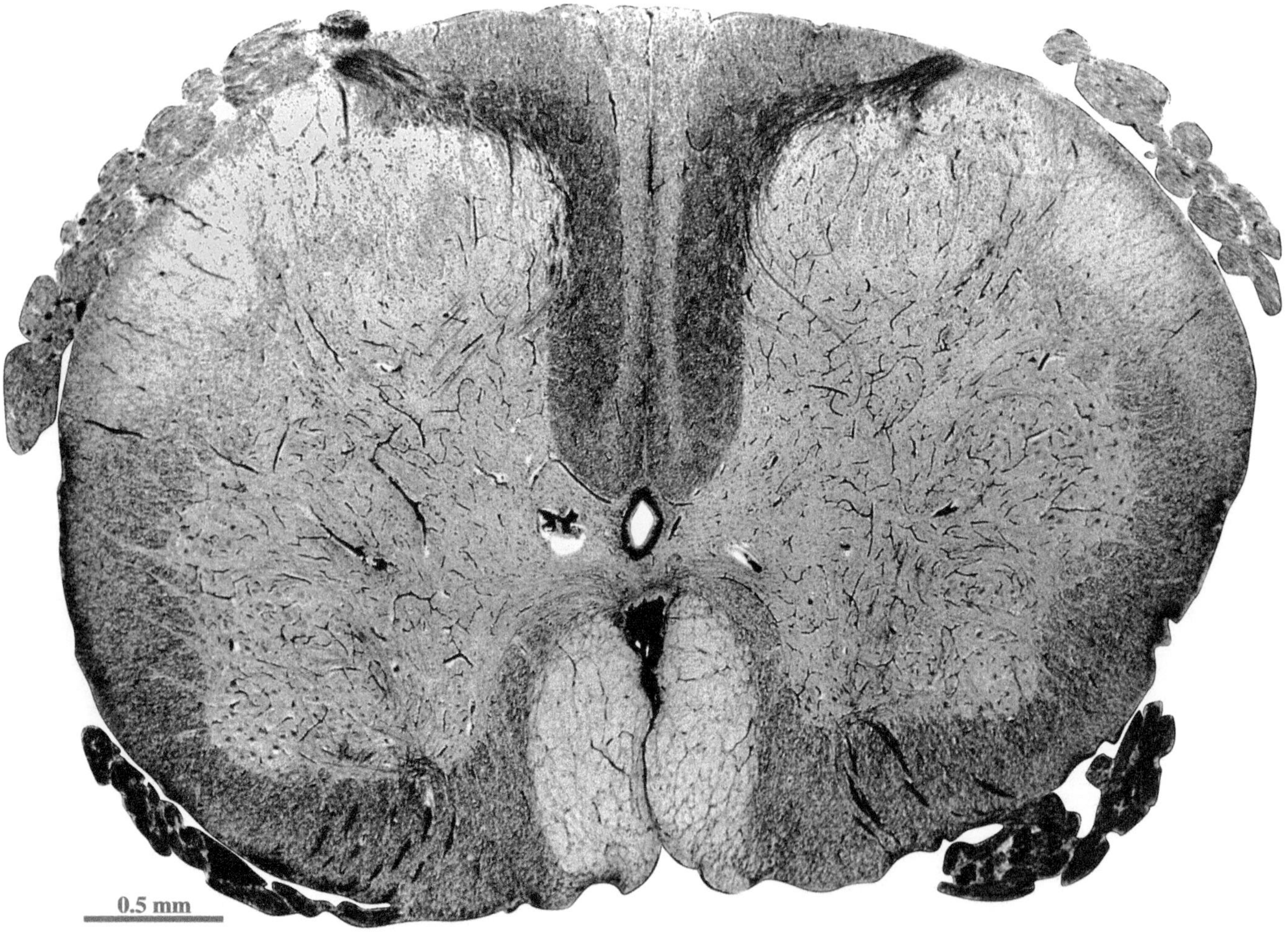

PLATE 12B

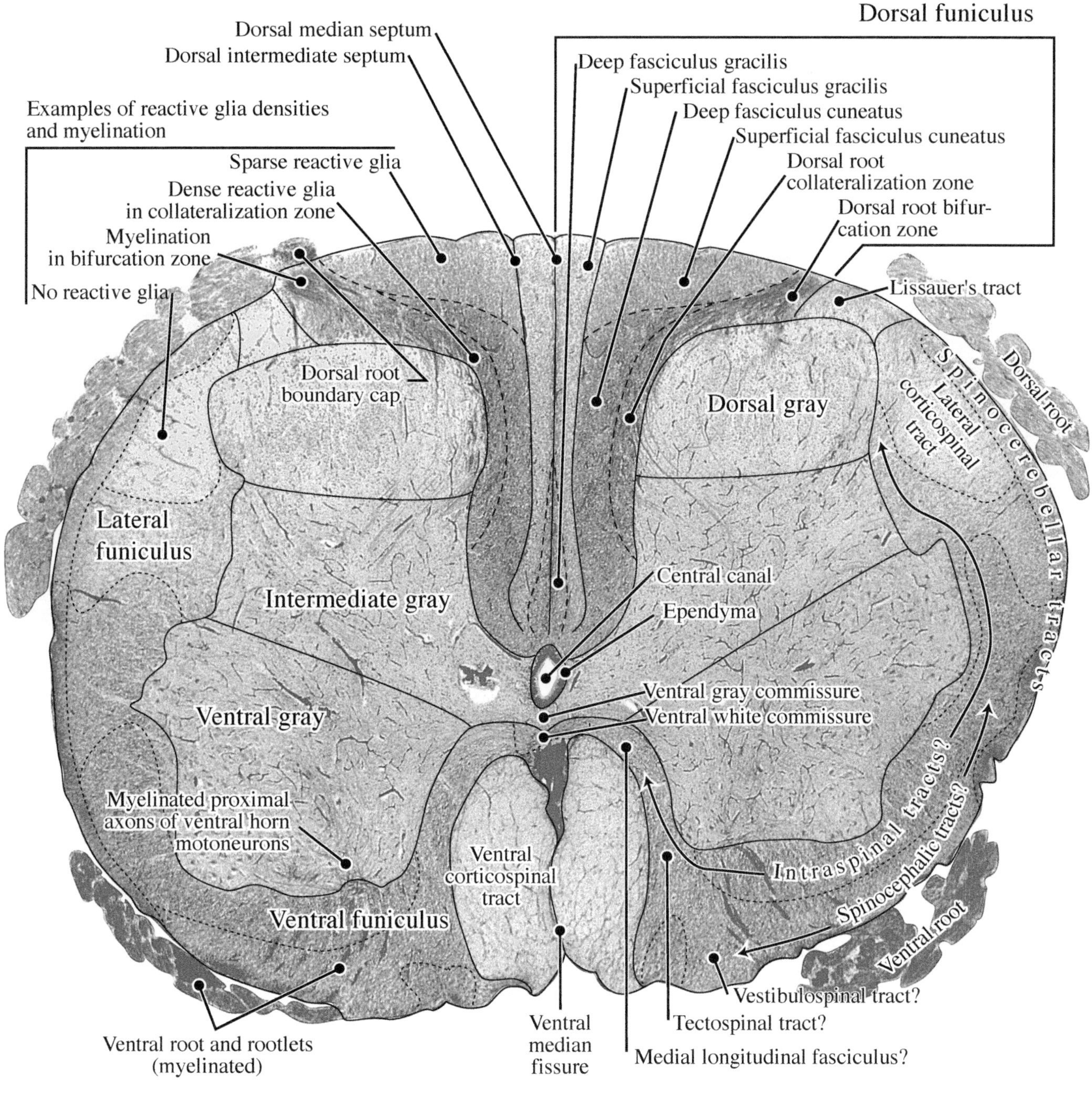
Dorsal funiculus
Dorsal median septum
Dorsal intermediate septum
Deep fasciculus gracilis
Superficial fasciculus gracilis
Deep fasciculus cuneatus
Superficial fasciculus cuneatus
Dorsal root collateralization zone
Dorsal root bifurcation zone
Examples of reactive glia densities and myelination
Sparse reactive glia
Dense reactive glia in collateralization zone
Myelination in bifurcation zone
No reactive glia
Lissauer's tract
Dorsal root boundary cap
Dorsal gray
Spinocerebellar tracts
Lateral corticospinal tract
Dorsal root
Lateral funiculus
Intermediate gray
Central canal
Ependyma
Ventral gray commissure
Ventral white commissure
Ventral gray
Myelinated proximal axons of ventral horn motoneurons
Ventral corticospinal tract
Intraspinal tracts?
Spinocephalic tracts?
Ventral root
Ventral funiculus
Vestibulospinal tract?
Tectospinal tract?
Medial longitudinal fasciculus?
Ventral median fissure
Ventral root and rootlets (myelinated)

PLATE 13A

CR 210 mm
GW 26
Y60-61
Lower Cervical Enlargement
Cell body stain

Areas (mm^2)

Central canal	.0086
Ependyma	.0234
Gray matter	5.3014
White matter	7.2555

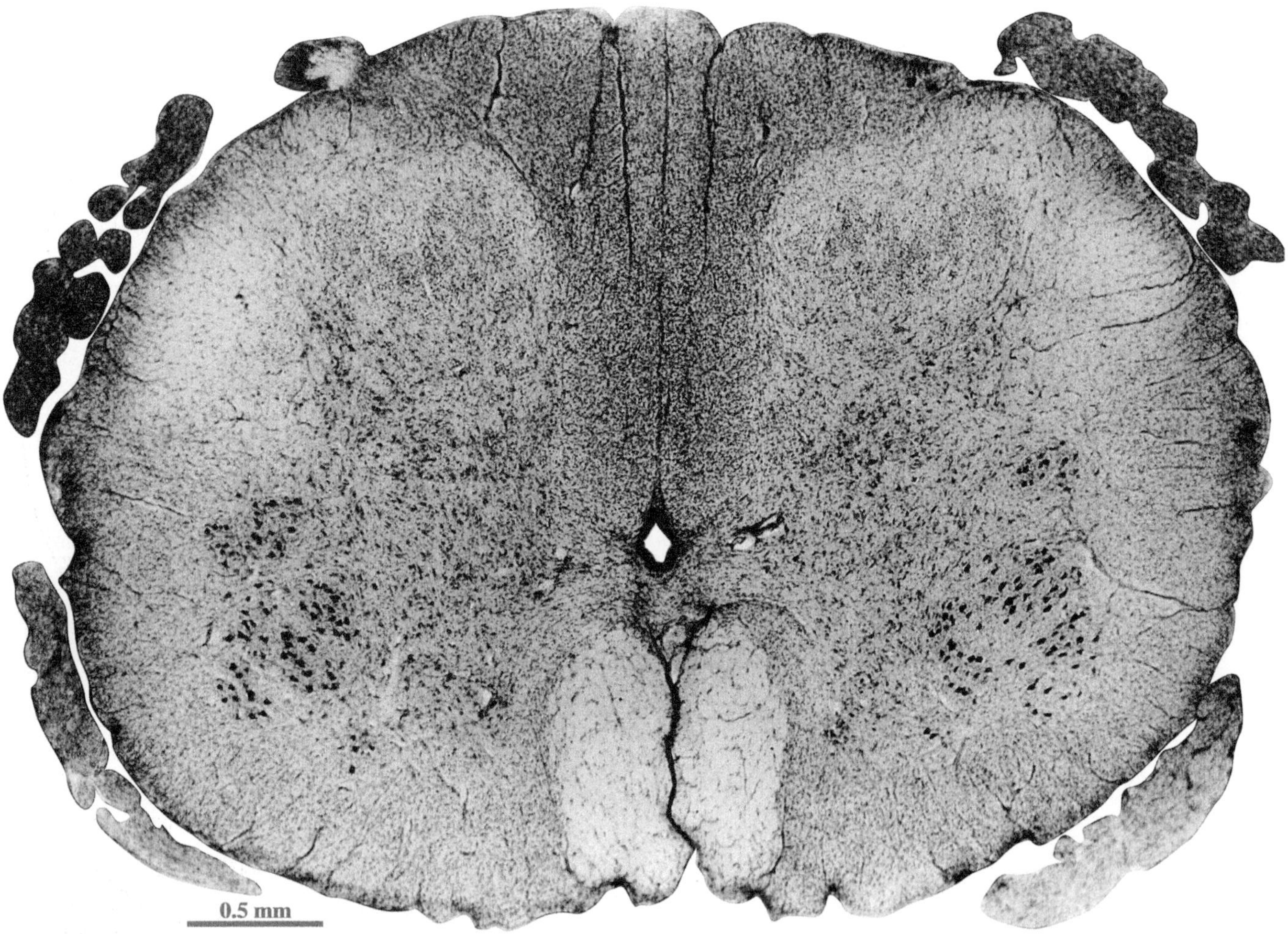

PLATE 13B

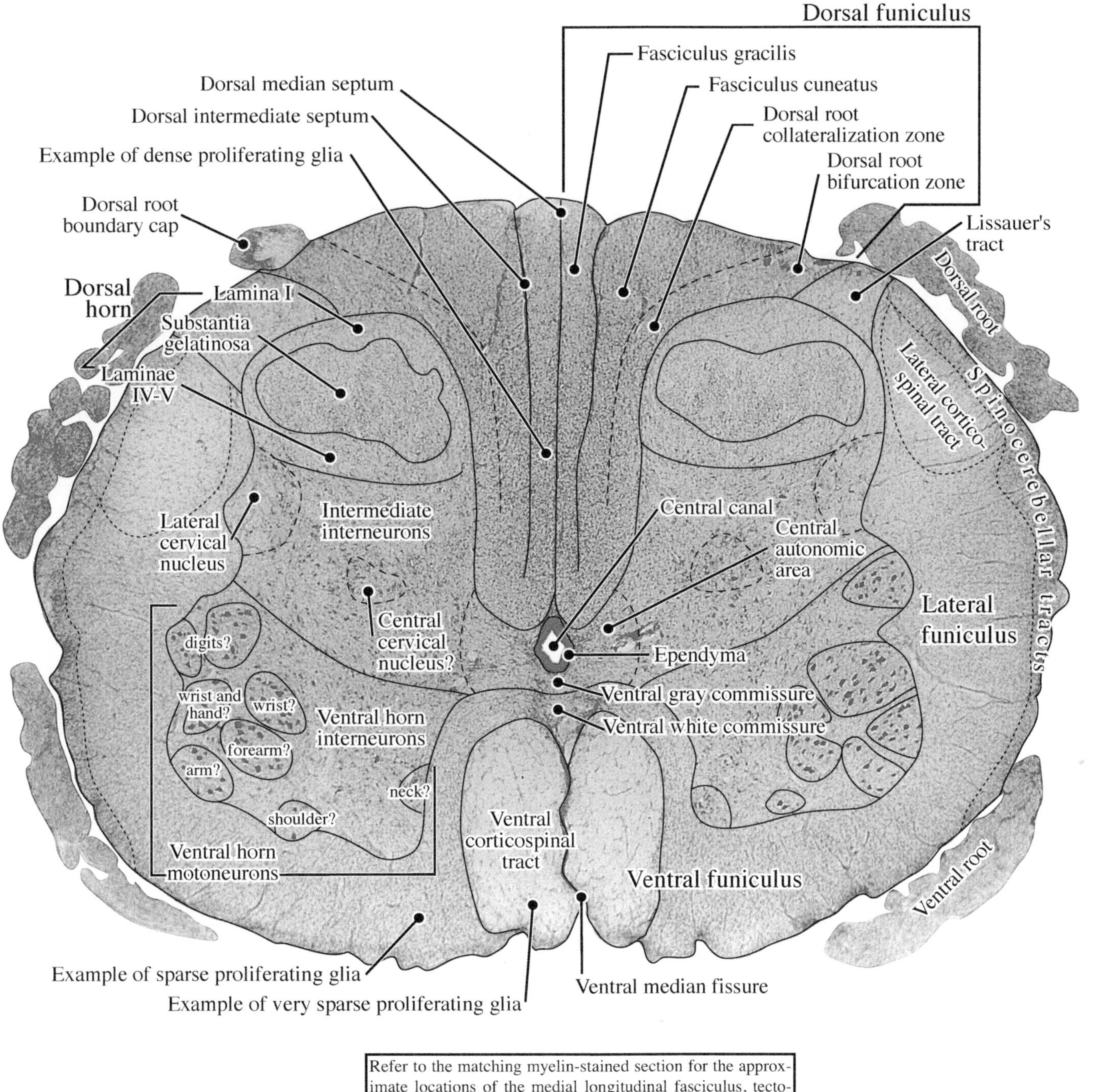

Refer to the matching myelin-stained section for the approximate locations of the medial longitudinal fasciculus, tectospinal, vestibulospinal, intraspinal, and spinocephalic tracts.

PLATE 14A

CR 210 mm
GW 26
Y60-61
Upper Thoracic
Myelin stain

Areas (mm^2)

Central canal	.0071
Ependyma	.0144
Gray matter	2.3030
White matter	3.7045

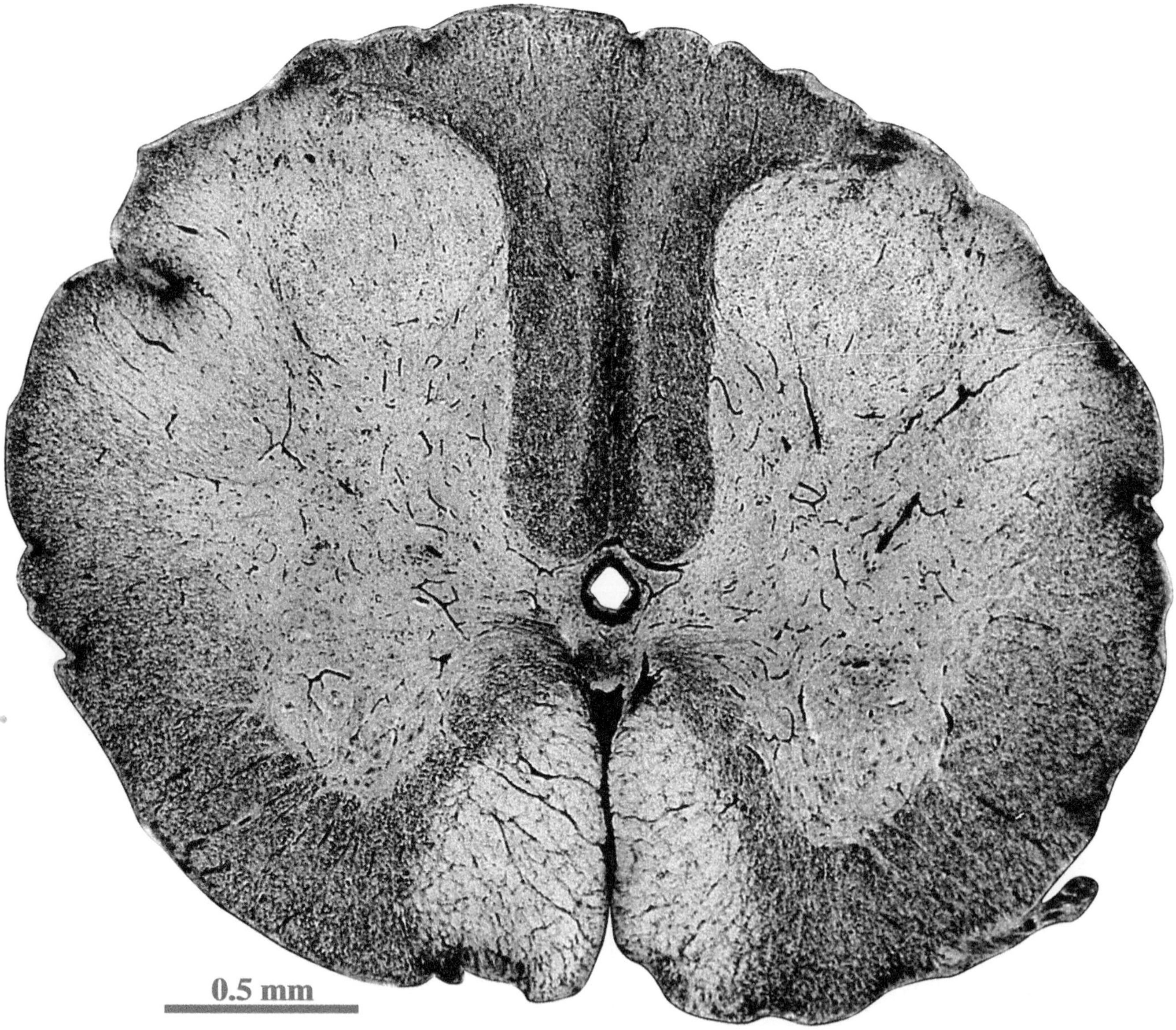

PLATE 14B

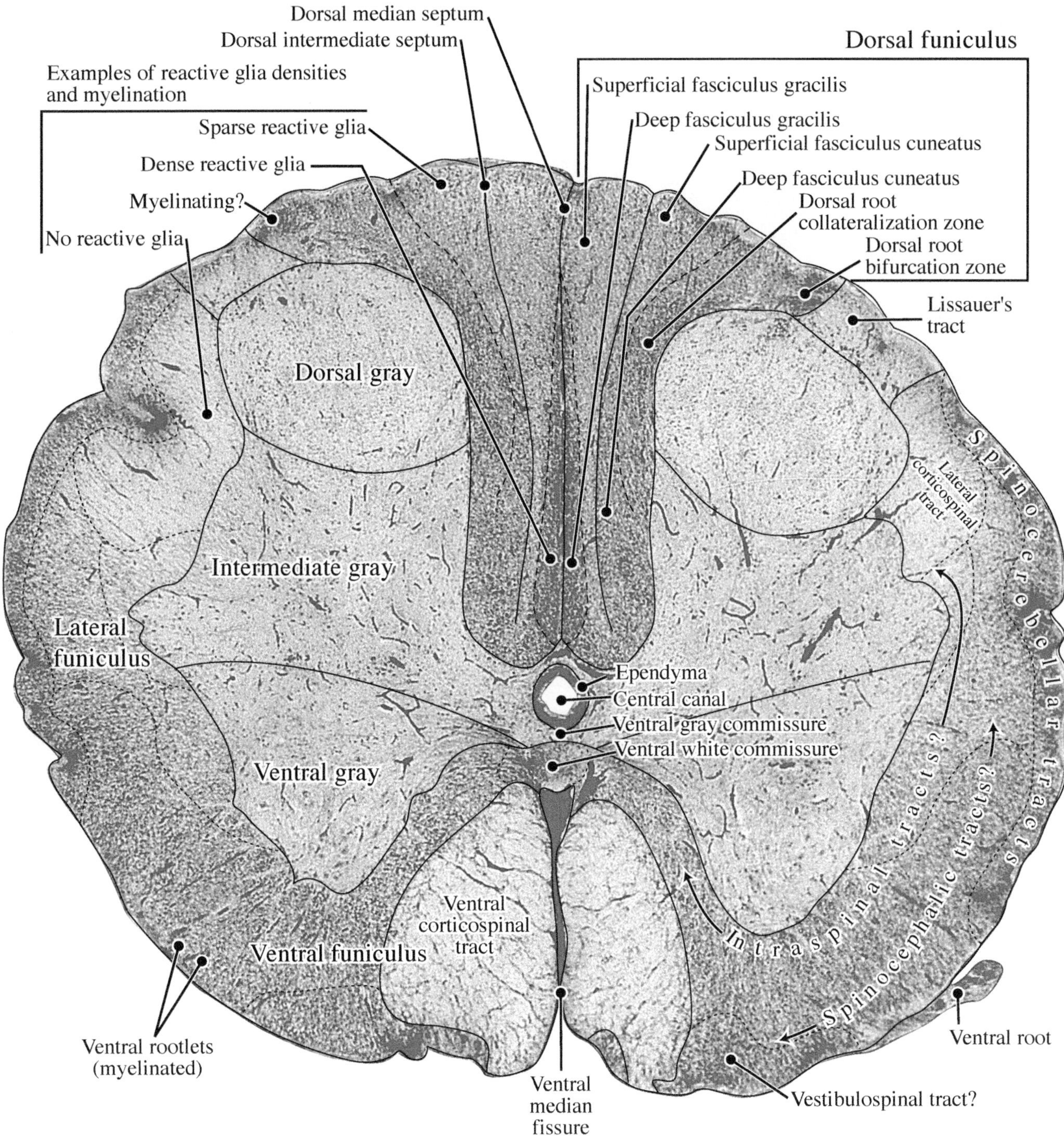
Dorsal median septum
Dorsal intermediate septum
Examples of reactive glia densities and myelination
Sparse reactive glia
Dense reactive glia
Myelinating?
No reactive glia
Dorsal funiculus
Superficial fasciculus gracilis
Deep fasciculus gracilis
Superficial fasciculus cuneatus
Deep fasciculus cuneatus
Dorsal root collateralization zone
Dorsal root bifurcation zone
Lissauer's tract
Dorsal gray
Intermediate gray
Lateral funiculus
Spinocerebellar tracts
Lateral corticospinal tract
Ependyma
Central canal
Ventral gray commissure
Ventral white commissure
Ventral gray
Intraspinal tracts?
Spinocephalic tracts?
Spinocephalic tracts?
Ventral corticospinal tract
Ventral funiculus
Ventral rootlets (myelinated)
Ventral median fissure
Ventral root
Vestibulospinal tract?

PLATE 15A

CR 210 mm
GW 26
Y60-61
Upper Thoracic
Cell body stain

Areas (mm^2)

Central canal	.0083
Ependyma	.0204
Gray matter	2.8518
White matter	4.1012

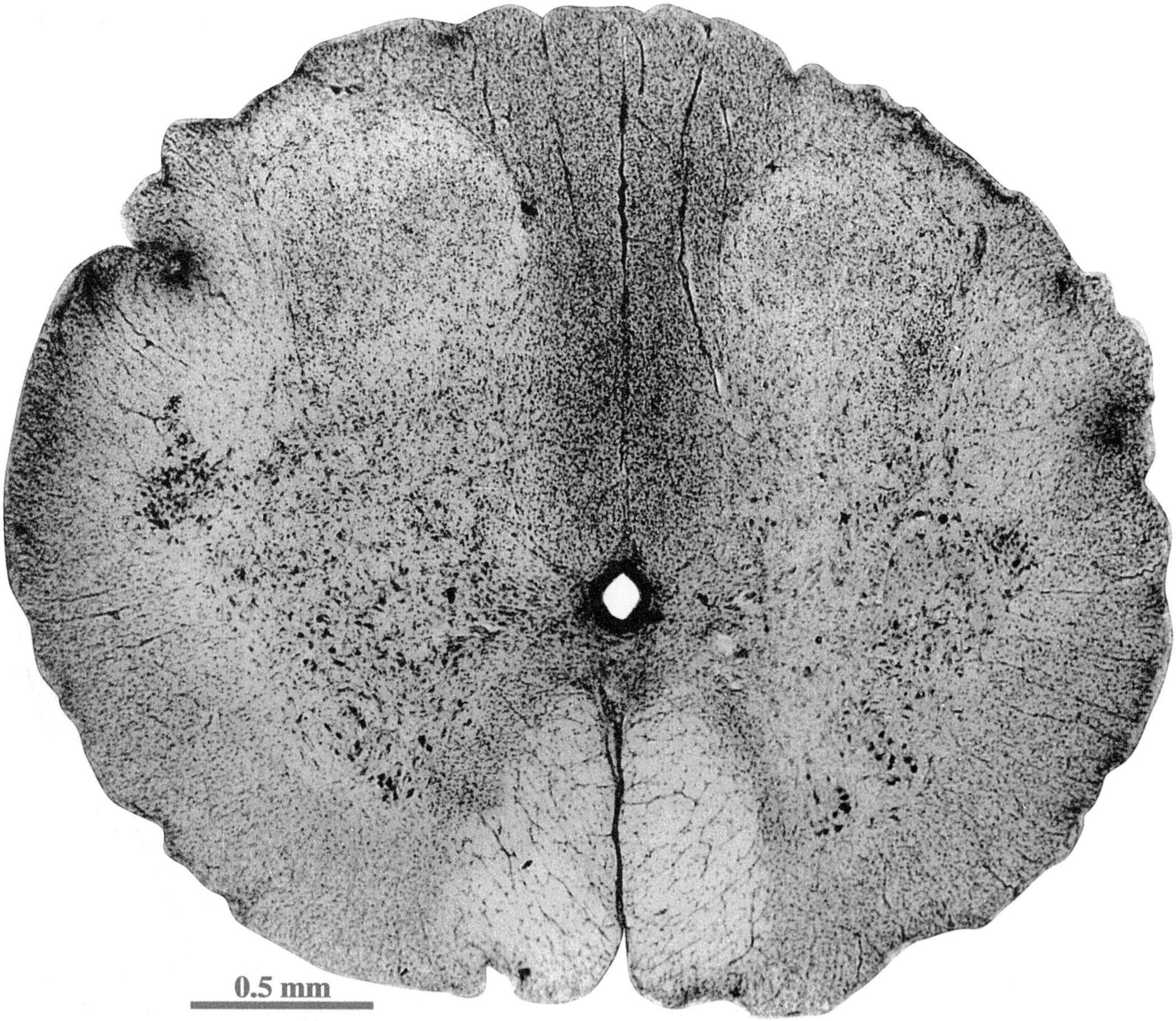

PLATE 15B

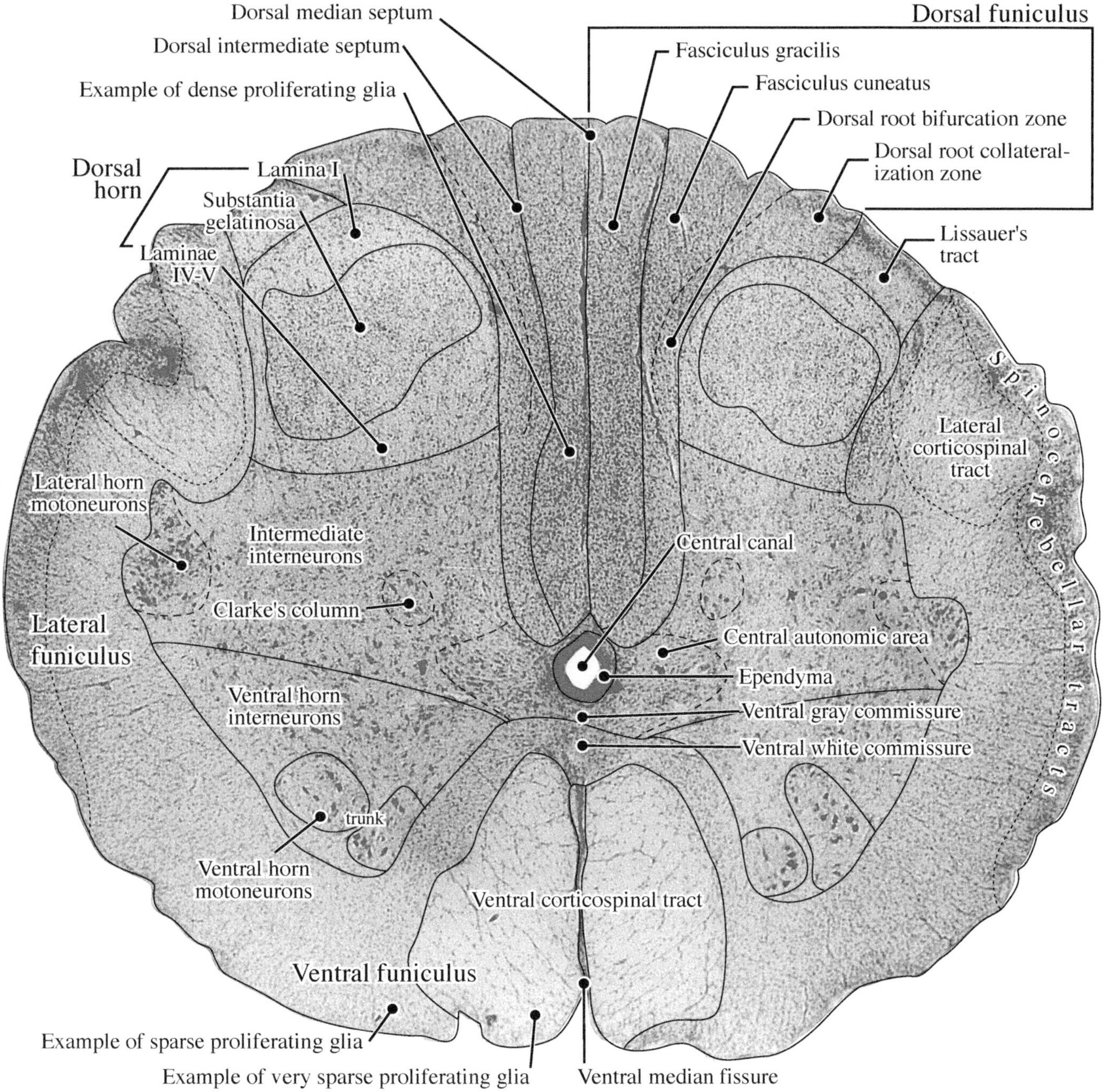

Refer to the matching myelin-stained section for the approximate locations of the vestibulospinal, intraspinal, and spinocephalic tracts.

PART IV: Y162-61 CR 270 mm (GW 31)

Plate 16 is a survey of matched myelin-stained and cell-body-stained sections from Y162-61, a specimen in the Yakovlev Collection. All sections are shown at the same scale. The boxes enclosing each section list the approximate level and the total area of the section in square millimeters (mm^2). Full-page normal-contrast photographs of each section are in **Plates 17A–24A.** Low-contrast photographs with superimposed labels and outlines of structural details are in **Plates 17B–24B.** In this specimen, the myelin-stained and cell-body-stained sections were preserved on separate large glass plates without any section numbers. Thirteen myelin-stained and thirteen cell-body-stained sections were photographed ranging from lower thoracic to sacral/coccygeal levels. There were no sections preserved at cervical, upper thoracic, and middle thoracic levels. The 26 photographic prints were intuitively arranged in order from upper cervical to thoracic, using internal features such as the size of the corticospinal tracts, and the width of the ventral horn. Then, myelin- and cell-body-stained sections were matched.

As in the previous specimen, the cross-sectional area of a myelin-stained section is smaller than the matching cell-body-stained section in all cases. Evidently, the myelin-staining procedure produces greater tissue shrinkage than the cell-body-staining procedure. The lumbar enlargement is definitely expanding relative to regions above and below. Using the myelin-stained section areas for comparison, the lumbar enlargement is 90% larger than the lower thoracic level, 41% larger than the upper lumbar level, and 278% larger than the sacral/coccygeal level.

Myelination in this specimen is more advanced than in the previous specimen, even though the two are only 5 weeks apart, and we are looking at a more immature region of the spinal cord. Dense staining indicative of true myelination is seen in the ventral commissure, the ventral rootlets, the intraspinal tracts, the dorsal root bifurcation zone, the dorsal root collateralization zone, and deep regions of the fasciculus gracilis (*Column 2*, **Table 3**). Myelinated fibers from the dorsal root collateralization zone penetrate the gray matter, and there is a light dusting of reactive glia in the subgelatinosal plexus in the dorsal gray as fine collateral axons prepare for later myelination (*see* Fig. 6-39 in Altman and Bayer 2001). Clumps of myelinated axons are in the lateral part of the intermediate gray and the lateral neck region of the dorsal gray (the reticulated area, labeled in Figs. 9-25 through 9-31 in Altman and Bayer, 2001). The lateral corticospinal tract and Lissauer's tract contain only nonreactive glia or very sparse reactive glia. The dorsal root bifurcation zone is a complex area that not only contains heavily myelinated fibers, but also unmyelinated fibers. The heavily myelinated axons are from large sensory neurons in the dorsal root ganglia, while the unmyelinated axons are most likely those from small ganglion cells that form Lissauer's tract. The remaining fiber tracts contain varying densities of reactive glia (see *column 3* in **Table 3**). In most cases, labels of the fiber tracts include our assessment of the concentration of reactive glia. In the cell-body-stained sections, there is a different density of proliferating glia in various fiber tracts. It is lowest in the lateral and ventral corticospinal tracts and highest in the dorsal funiculus (except Lissauer's tract) and spinocerebellar tracts (*Column 3*, **Table 3**).

In the cell-body-stained sections, columns of motoneurons continue to be prominent and show more segregation in the ventral horn. The dorsal horn has fairly well-defined clusters of small neurons in the substantia gelatinosa. The accumulation of lateral horn motoneurons is very obvious at the thoracic level. Clarke's column is also prominent in this specimen.

Table 3: Glia types and concentration in the white matter at GW 31

Name	Myelination	Reactive glia	Proliferating glia
DORSAL FUNICULUS:			
dorsal root bif. zone	*many fibers	---	sparse
dorsal root col. zone	many fibers	---	very dense
deep fas. gracilis	many fibers	---	very dense
superficial fas. gracilis	some fibers	dense	dense
Lissauer's tract	---	none	very sparse
LATERAL and VENTRAL FUNICULI:			
lat. corticospinal tract	---	very sparse	very sparse
ven. corticospinal tract	---	sparse	very sparse
rubrospinal tract	some fibers	dense	sparse
spinocerebellar tracts	some fibers	dense	dense
lat. reticulospinal tract	some fibers	dense	sparse
intraspinal tracts	some fibers	dense	sparse
spinocephalic tracts	some fibers	dense	sparse
vestibulospinal tract	some fibers	dense	sparse
ven. commissure	many fibers	---	sparse

* intermingled in a bed of nonreactive glia (associated with Lissauer's tract fibers?)

PLATE 16

CR 270 mm, GW 31, Y162-61

MYELIN STAIN | **CELL BODY STAIN**

Plates 17A, 17B
Lower Thoracic
Total area:
5.1272 mm^2

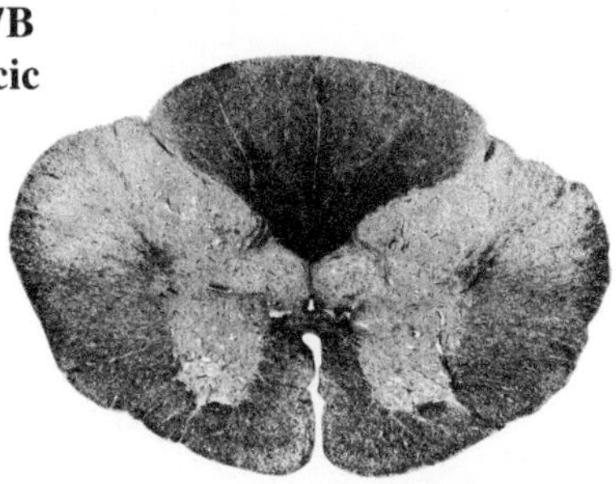

Plates 18A, 18B
Lower Thoracic
Total area:
5.2380 mm^2

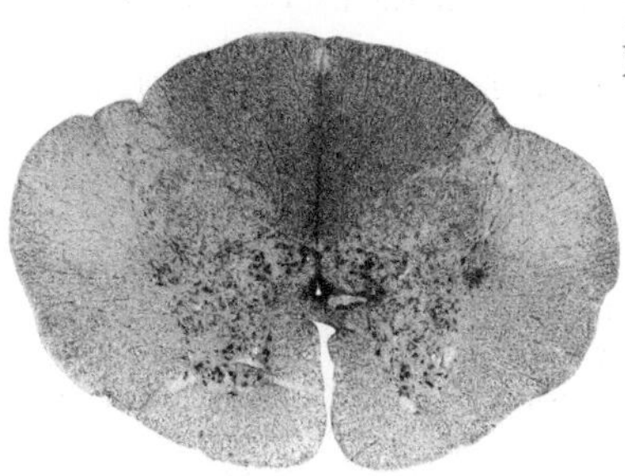

Plates 19A, 19B
Upper Lumbar
Total area:
6.9121 mm^2

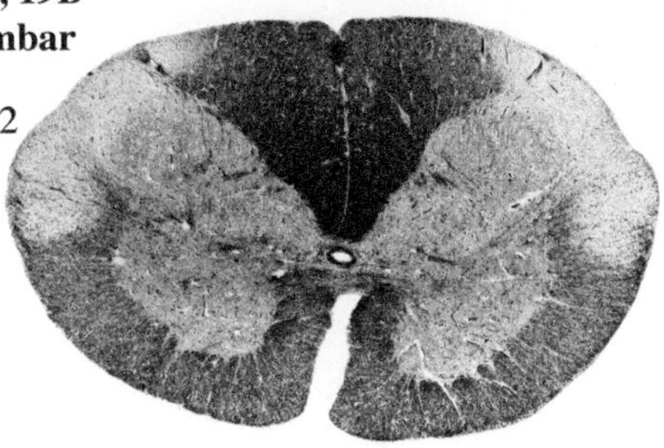

Plates 20A, 20B
Upper Lumbar
Total area:
7.1380 mm^2

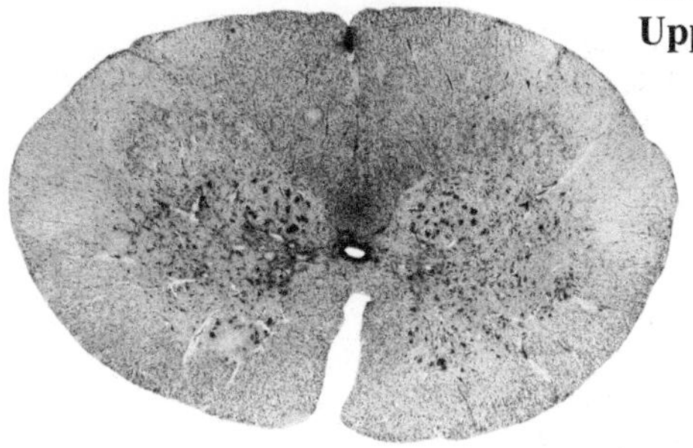

Plates 21A, 21B
Lumbar Enlargement
Total area:
9.7553 mm^2

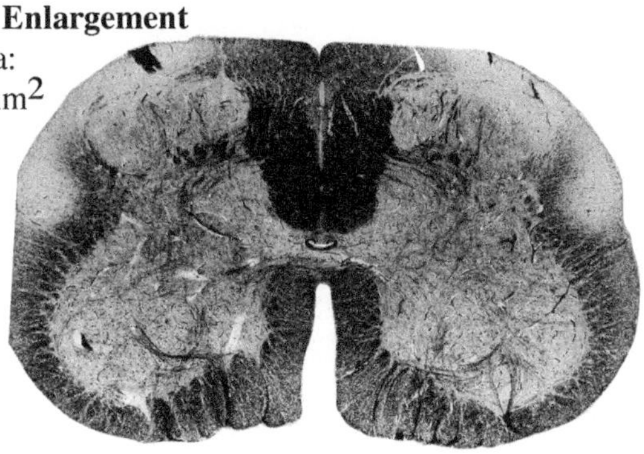

Plates 22A, 229B
Lumbar Enlargement
Total area:
10.918 mm^2

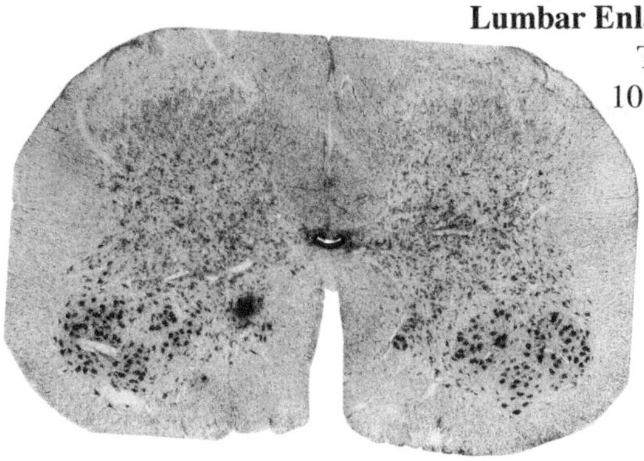

Plates 23A, 23B
Sacral/Coccygeal
Total area:
3.5153 mm^2

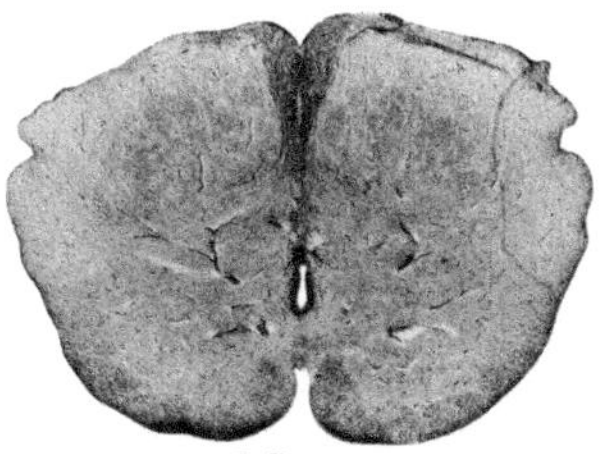

0.5 mm

Plates 24A, 24B
Sacral/Coccygeal
Total area:
3.5751 mm^2

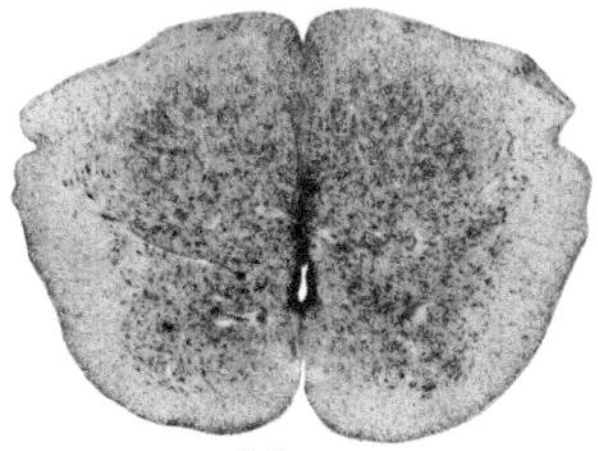

0.5 mm

PLATE 17A

CR 270 mm
GW 31
Y162-61
Lower Thoracic
Myelin stain

Areas (mm^2)

Central canal	.0012
Ependyma	.0057
Gray matter	1.4171
White matter	3.7032

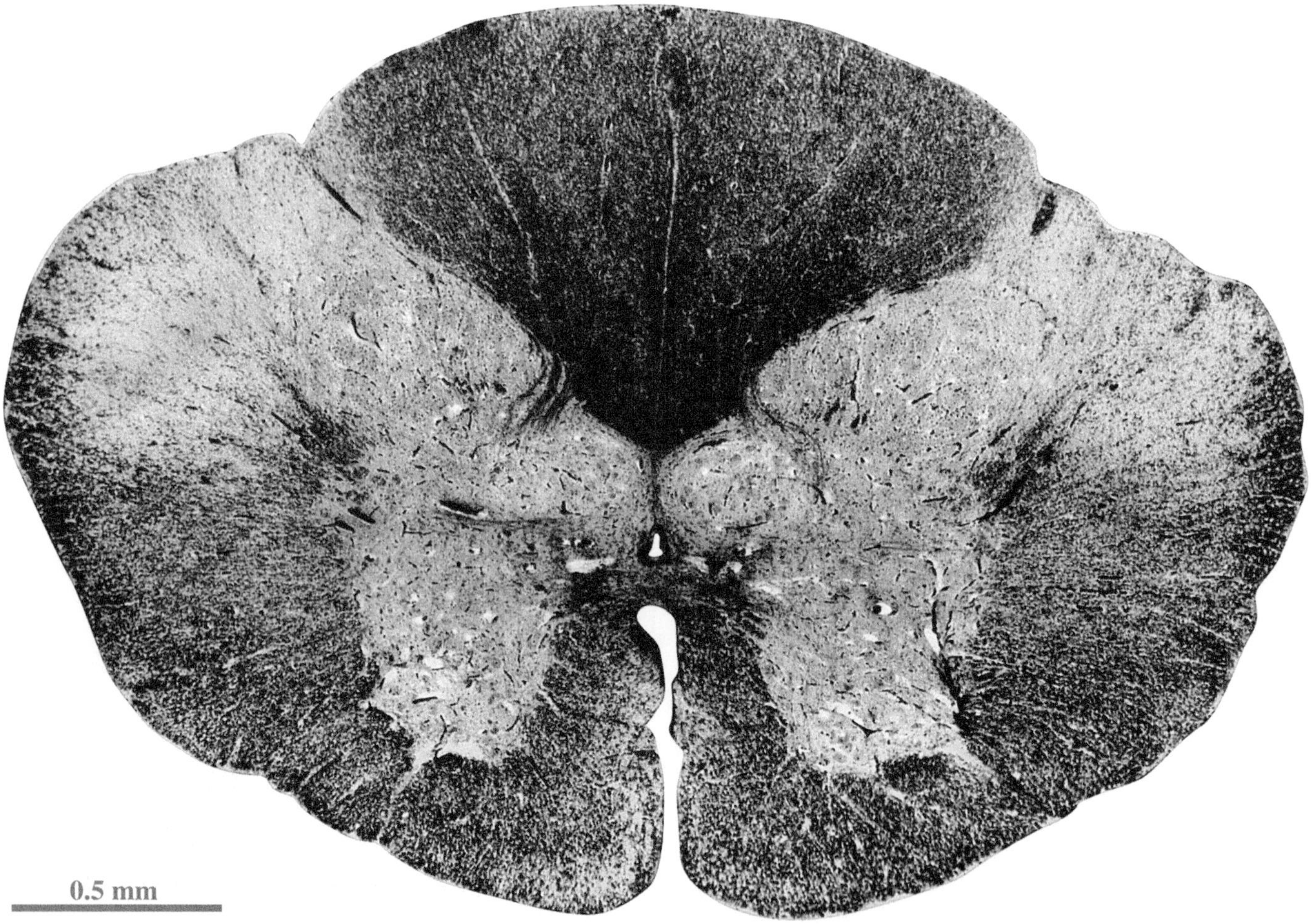

PLATE 17B

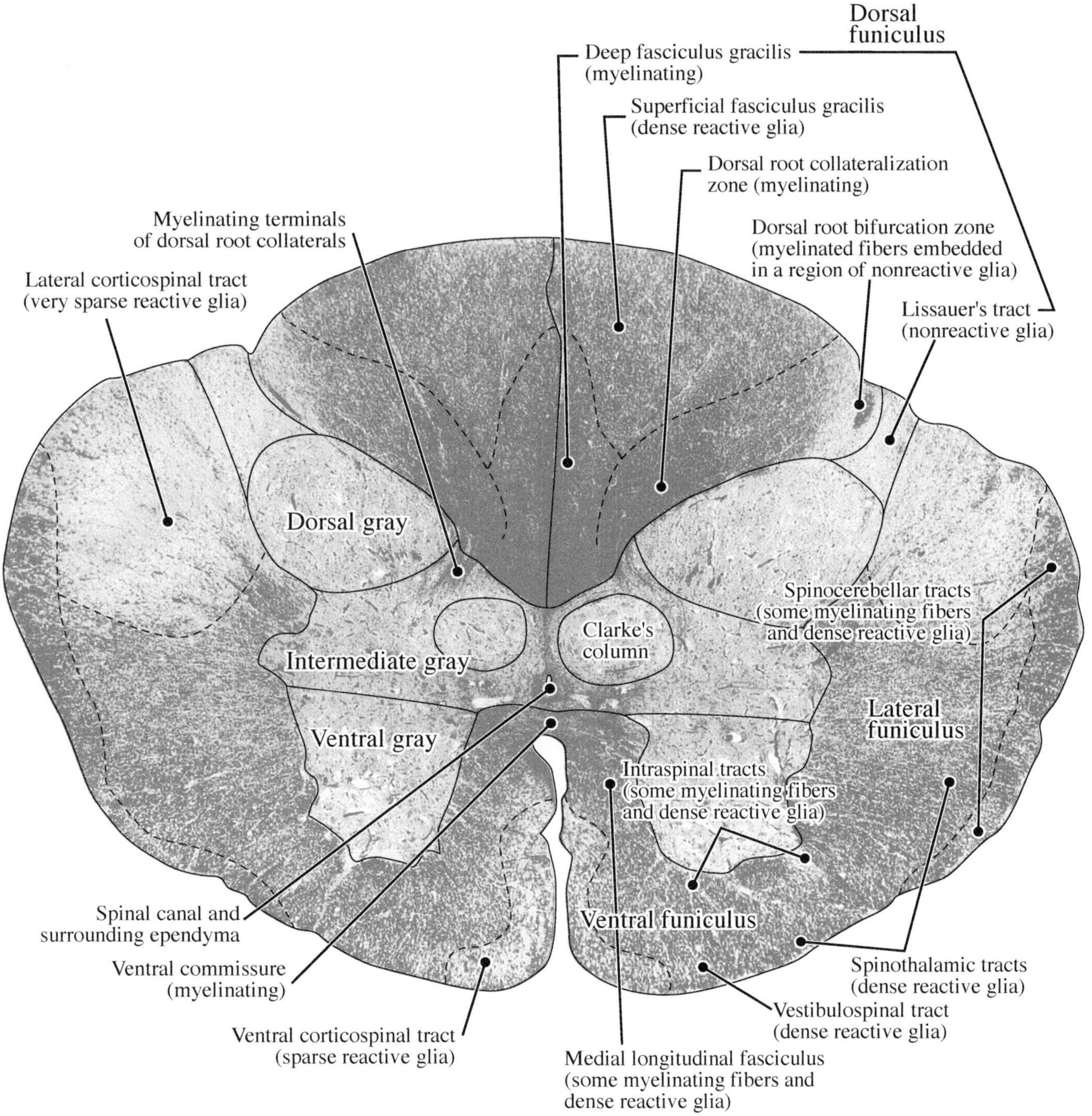
Dorsal funiculus
Deep fasciculus gracilis (myelinating)
Superficial fasciculus gracilis (dense reactive glia)
Dorsal root collateralization zone (myelinating)
Myelinating terminals of dorsal root collaterals
Dorsal root bifurcation zone (myelinated fibers embedded in a region of nonreactive glia)
Lateral corticospinal tract (very sparse reactive glia)
Lissauer's tract (nonreactive glia)
Dorsal gray
Spinocerebellar tracts (some myelinating fibers and dense reactive glia)
Clarke's column
Intermediate gray
Lateral funiculus
Ventral gray
Intraspinal tracts (some myelinating fibers and dense reactive glia)
Ventral funiculus
Spinal canal and surrounding ependyma
Ventral commissure (myelinating)
Spinothalamic tracts (dense reactive glia)
Vestibulospinal tract (dense reactive glia)
Ventral corticospinal tract (sparse reactive glia)
Medial longitudinal fasciculus (some myelinating fibers and dense reactive glia)

PLATE 18A

CR 270 mm
GW 31
Y162-61
Lower Thoracic
Cell body stain

Areas (mm²)

Central canal	.0011
Ependyma	.0087
Gray matter	1.5228
White matter	3.7054

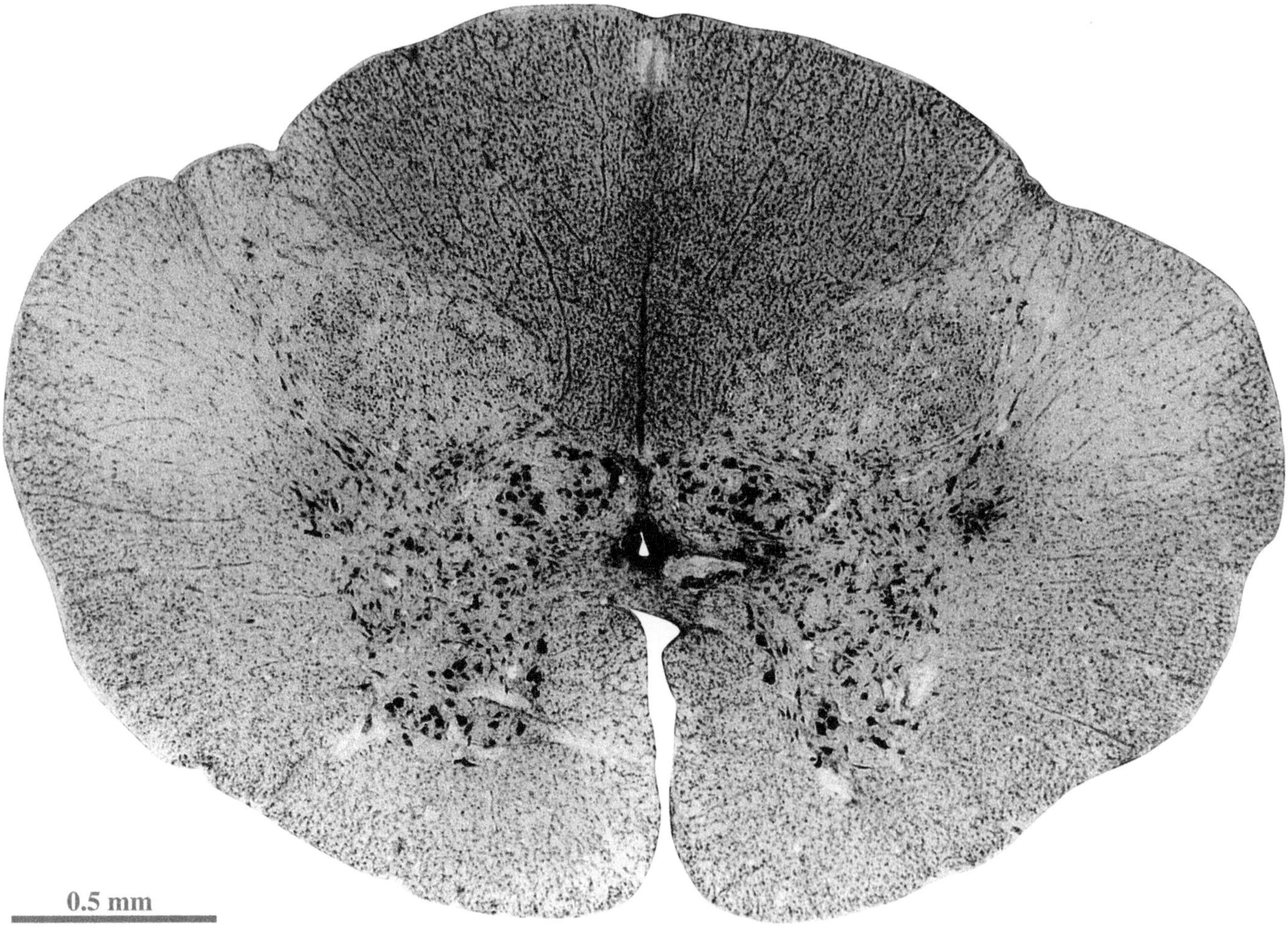

PLATE 18B

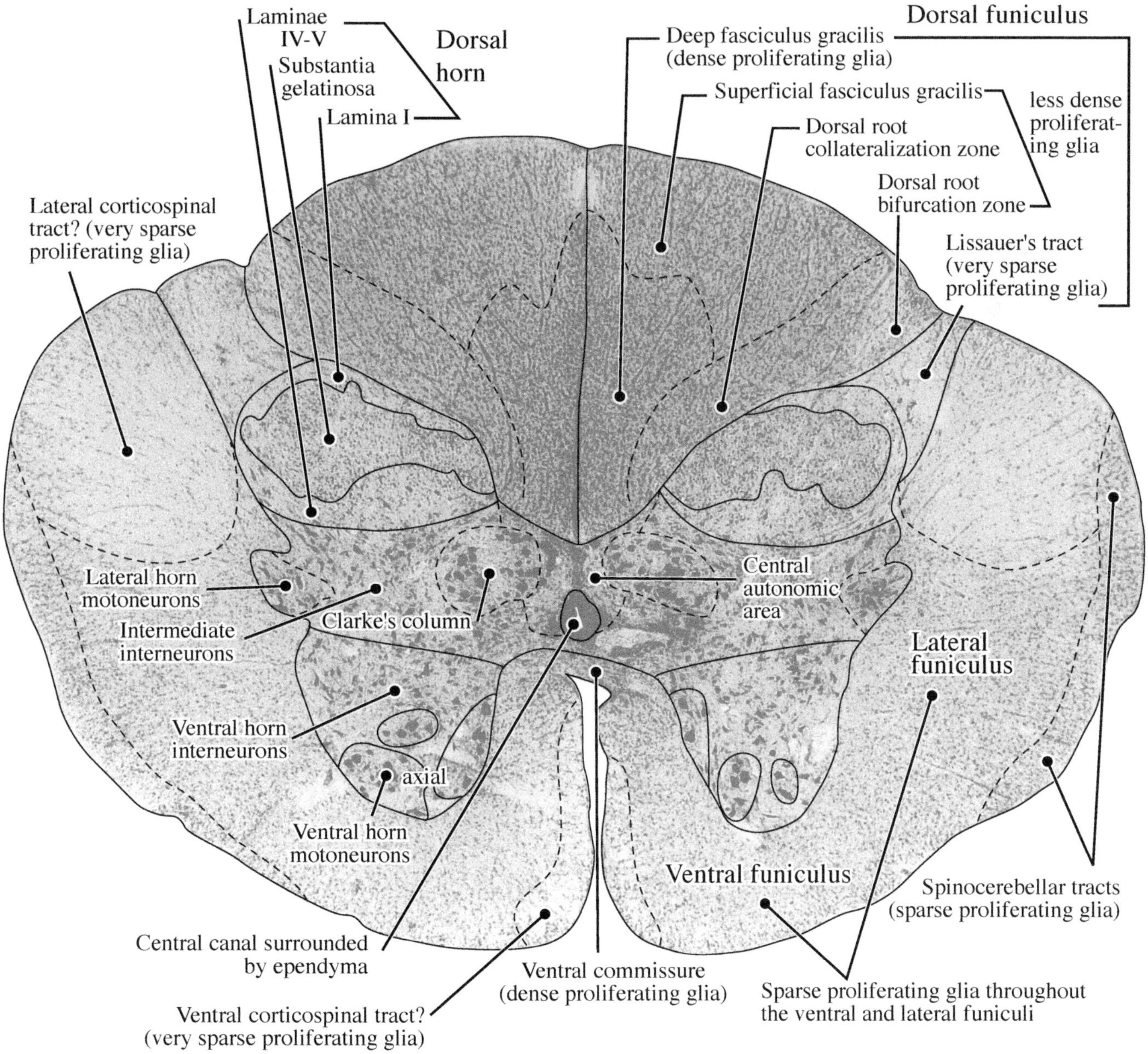
Laminae
IV-V
Substantia
gelatinosa
Lamina I
Dorsal
horn
Deep fasciculus gracilis
(dense proliferating glia)
Dorsal funiculus
Superficial fasciculus gracilis
Dorsal root
collateralization zone
Dorsal root
bifurcation zone
less dense
proliferat-
ing glia
Lissauer's tract
(very sparse
proliferating glia)
Lateral corticospinal
tract? (very sparse
proliferating glia)
Lateral horn
motoneurons
Intermediate
interneurons
Clarke's column
Central
autonomic
area
Lateral
funiculus
Ventral horn
interneurons
axial
Ventral horn
motoneurons
Ventral funiculus
Spinocerebellar tracts
(sparse proliferating glia)
Central canal surrounded
by ependyma
Ventral commissure
(dense proliferating glia)
Sparse proliferating glia throughout
the ventral and lateral funiculi
Ventral corticospinal tract?
(very sparse proliferating glia)

PLATE 19A

CR 270 mm
GW 31
Y162-61
Upper Lumbar
Myelin stain

Areas (mm²)

Central canal	.0038
Ependyma	.0110
Gray matter	2.5641
White matter	4.3333

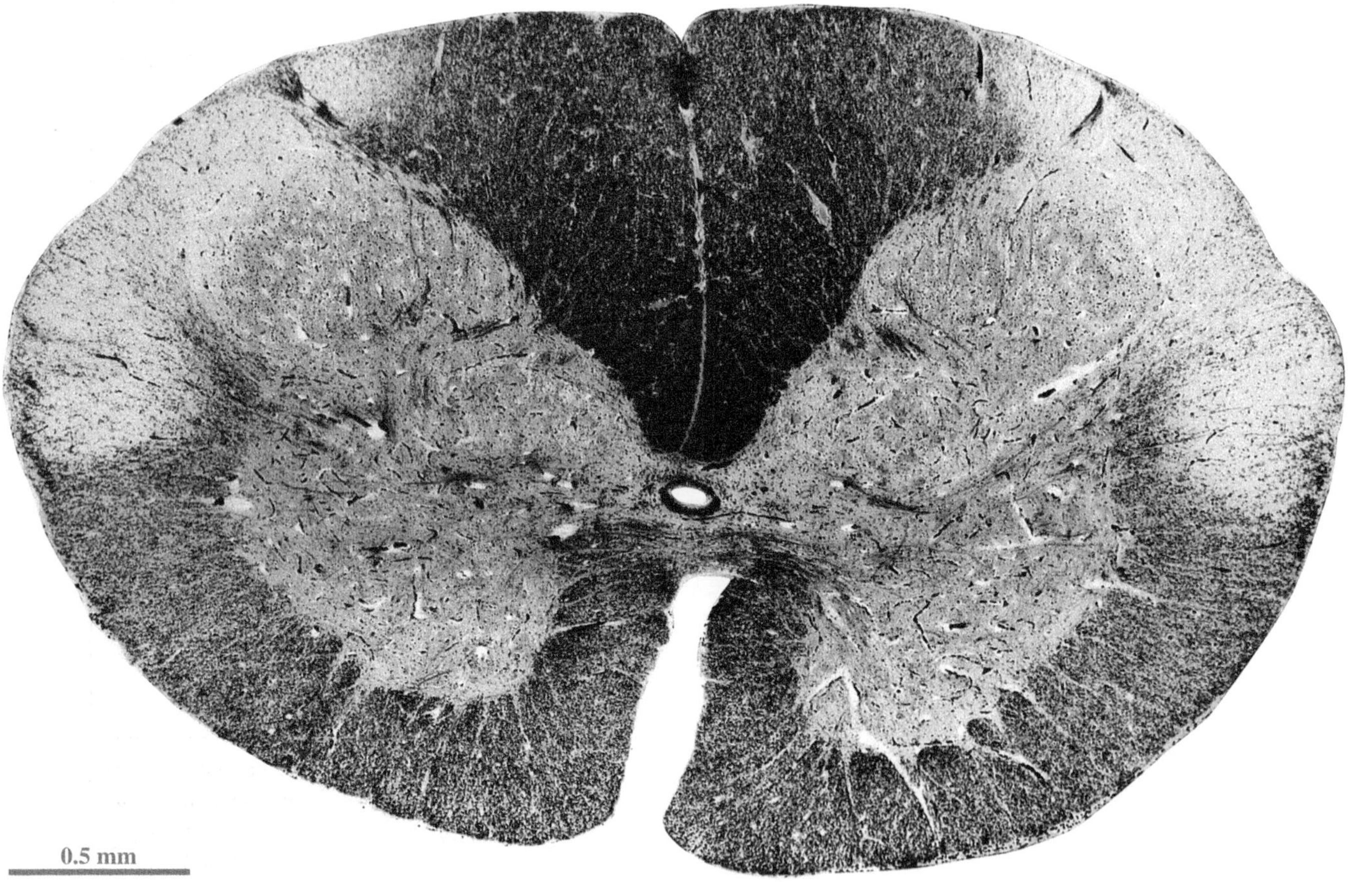

PLATE 19B

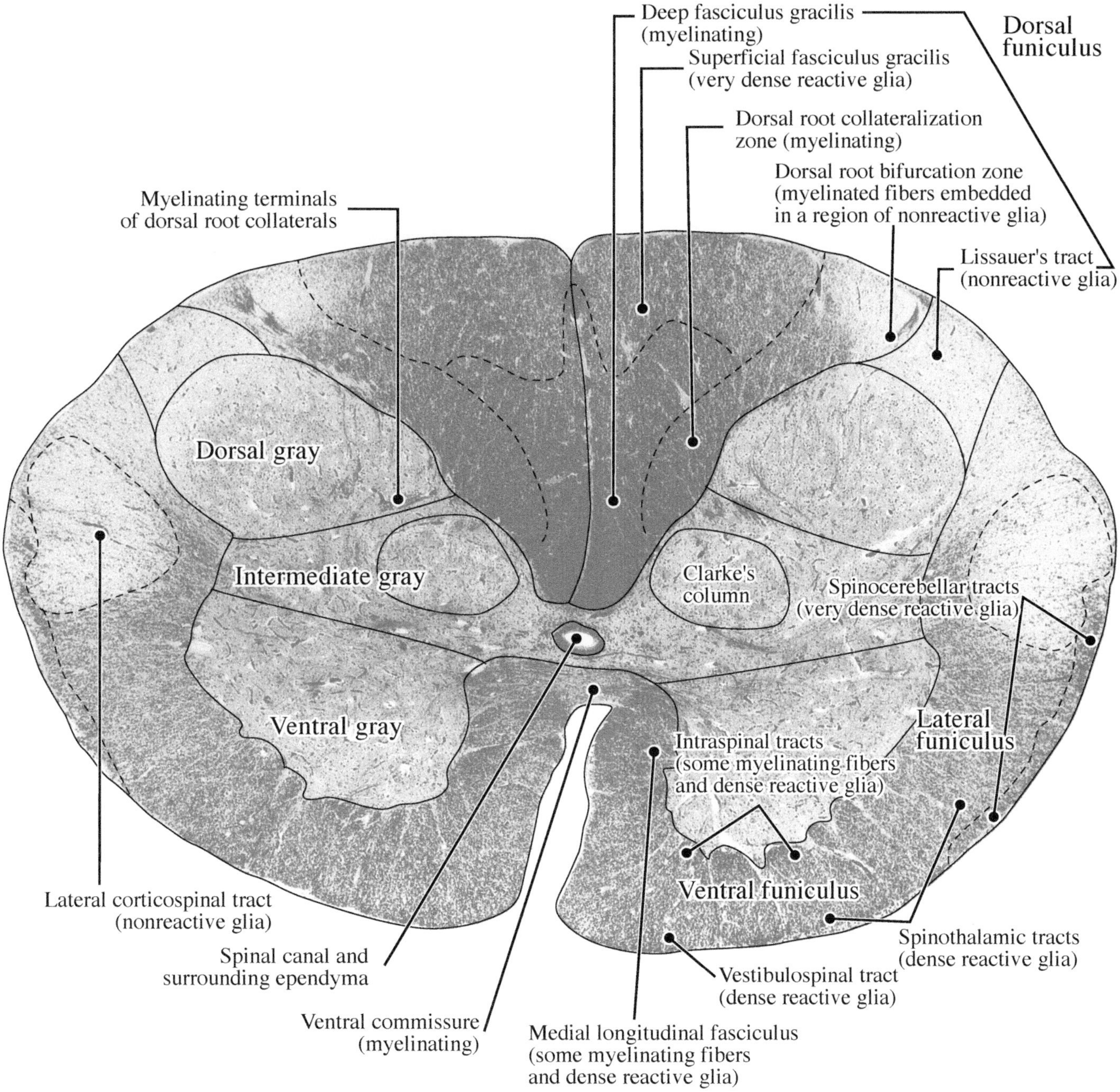
Deep fasciculus gracilis
(myelinating)
Superficial fasciculus gracilis
(very dense reactive glia)
Dorsal root collateralization
zone (myelinating)
Dorsal root bifurcation zone
(myelinated fibers embedded
in a region of nonreactive glia)
Dorsal
funiculus
Lissauer's tract
(nonreactive glia)
Myelinating terminals
of dorsal root collaterals
Dorsal gray
Intermediate gray
Clarke's
column
Spinocerebellar tracts
(very dense reactive glia)
Ventral gray
Lateral
funiculus
Intraspinal tracts
(some myelinating fibers
and dense reactive glia)
Ventral funiculus
Lateral corticospinal tract
(nonreactive glia)
Spinal canal and
surrounding ependyma
Spinothalamic tracts
(dense reactive glia)
Vestibulospinal tract
(dense reactive glia)
Ventral commissure
(myelinating)
Medial longitudinal fasciculus
(some myelinating fibers
and dense reactive glia)

PLATE 20A

CR 270 mm
GW 31
Y162-61
Upper Lumbar
Cell body stain

Areas (mm^2)

Central canal	.0050
Ependyma	.0135
Gray matter	2.7017
White matter	4.4178

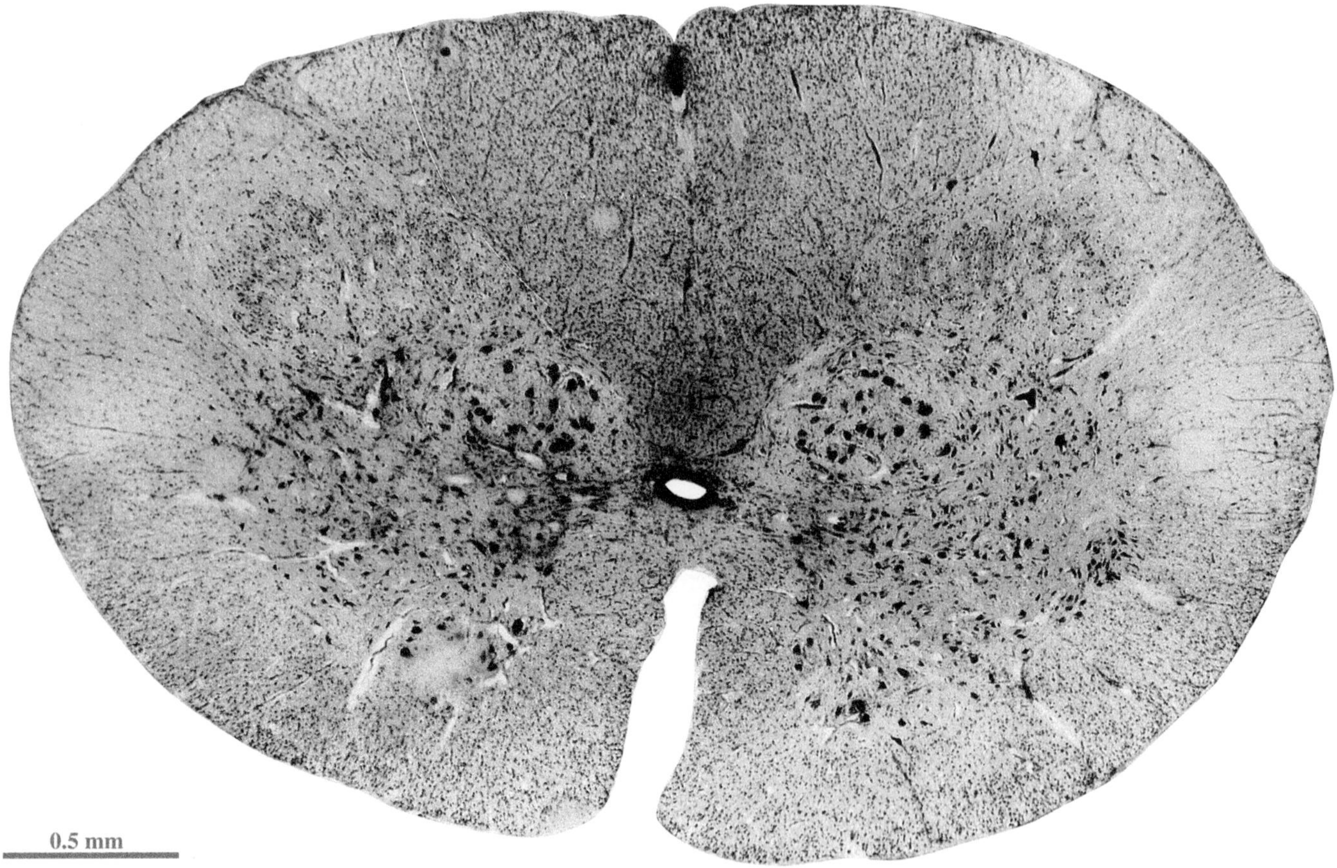

0.5 mm

PLATE 20B

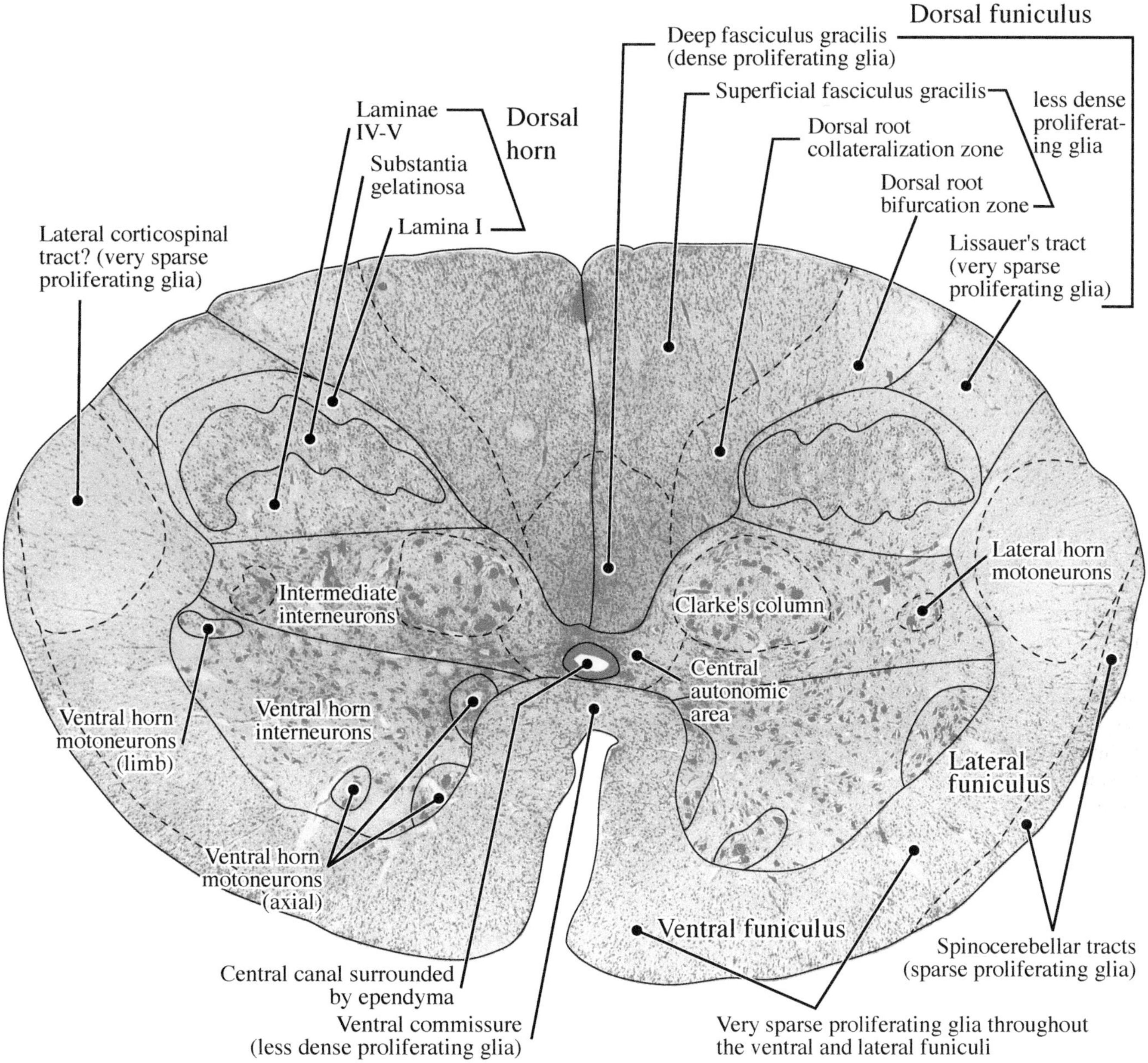
Dorsal funiculus
Deep fasciculus gracilis (dense proliferating glia)
Superficial fasciculus gracilis
less dense proliferating glia
Dorsal root collateralization zone
Dorsal root bifurcation zone
Lissauer's tract (very sparse proliferating glia)
Laminae IV-V
Dorsal horn
Substantia gelatinosa
Lamina I
Lateral corticospinal tract? (very sparse proliferating glia)
Intermediate interneurons
Clarke's column
Lateral horn motoneurons
Central autonomic area
Ventral horn motoneurons (limb)
Ventral horn interneurons
Lateral funiculus
Ventral horn motoneurons (axial)
Ventral funiculus
Spinocerebellar tracts (sparse proliferating glia)
Central canal surrounded by ependyma
Ventral commissure (less dense proliferating glia)
Very sparse proliferating glia throughout the ventral and lateral funiculi

PLATE 21A

CR 270 mm
GW 31
Y162-61
Lumbar Enlargement
Myelin stain

Areas (mm^2)

Central canal	.0030
Ependyma	.0085
Gray matter	4.9128
White matter	4.8309

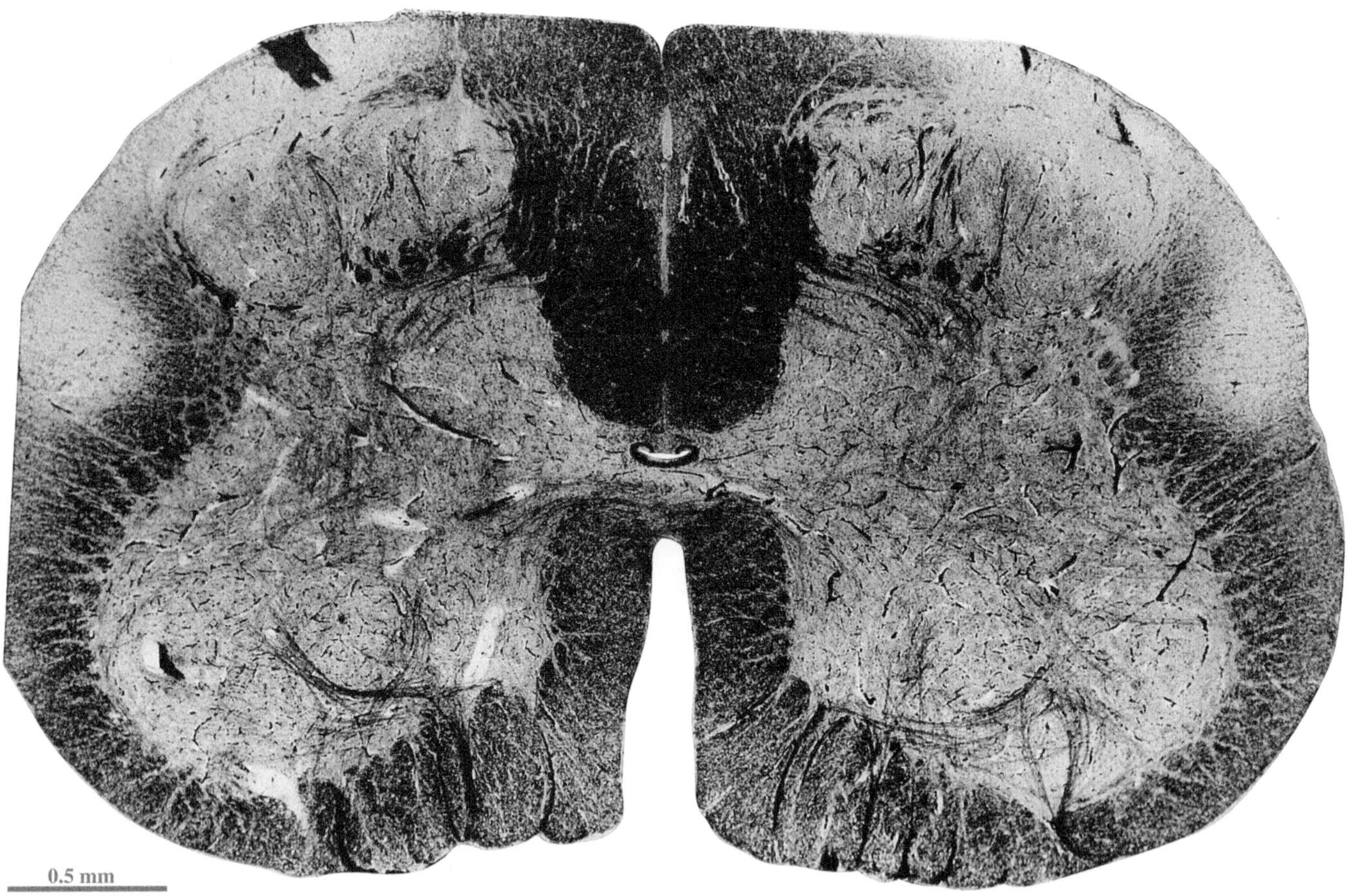

PLATE 21B

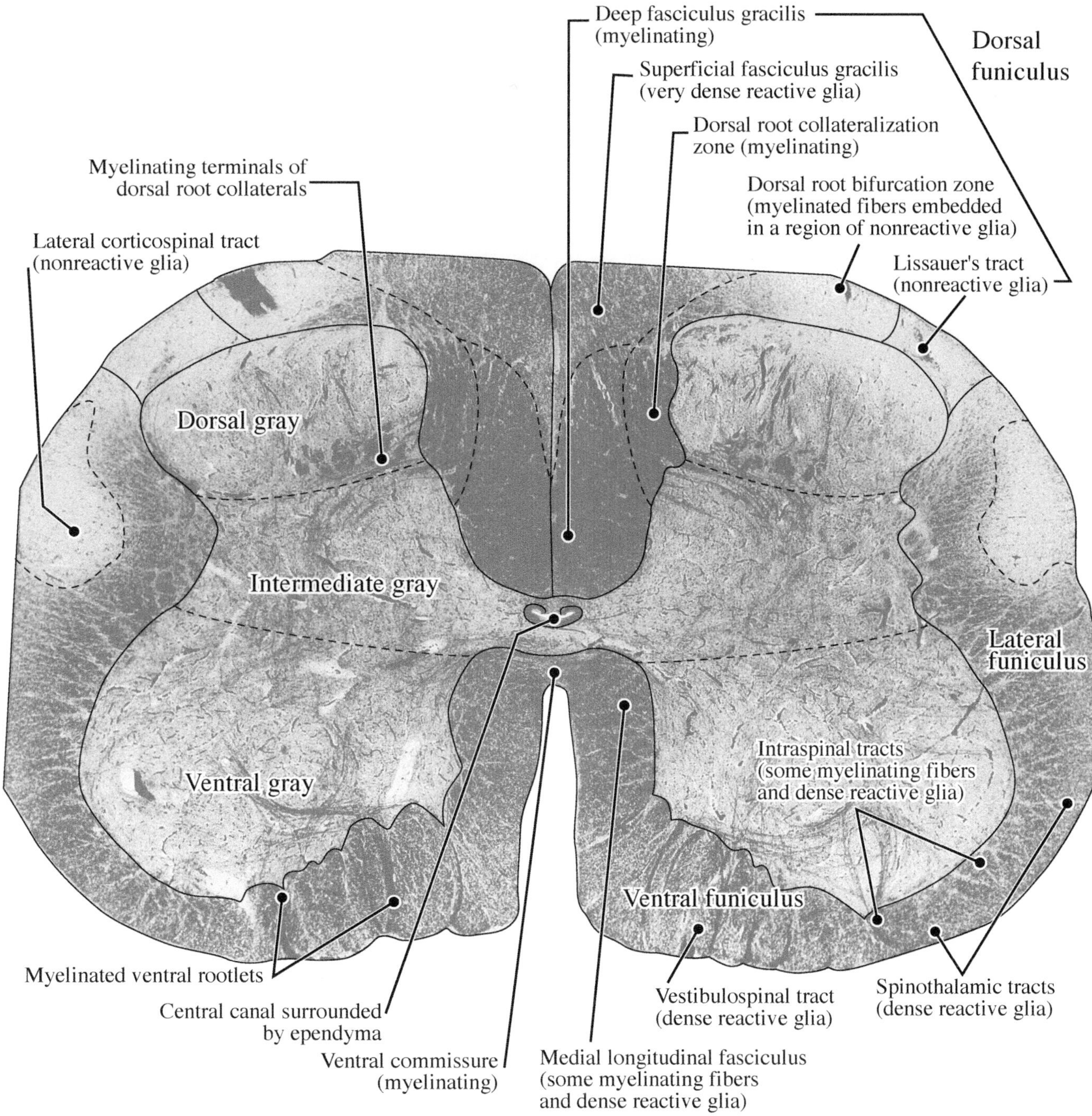
Deep fasciculus gracilis
(myelinating)
Dorsal
funiculus
Superficial fasciculus gracilis
(very dense reactive glia)
Dorsal root collateralization
zone (myelinating)
Myelinating terminals of
dorsal root collaterals
Dorsal root bifurcation zone
(myelinated fibers embedded
in a region of nonreactive glia)
Lateral corticospinal tract
(nonreactive glia)
Lissauer's tract
(nonreactive glia)
Dorsal gray
Intermediate gray
Lateral
funiculus
Intraspinal tracts
(some myelinating fibers
and dense reactive glia)
Ventral gray
Ventral funiculus
Myelinated ventral rootlets
Central canal surrounded
by ependyma
Ventral commissure
(myelinating)
Medial longitudinal fasciculus
(some myelinating fibers
and dense reactive glia)
Vestibulospinal tract
(dense reactive glia)
Spinothalamic tracts
(dense reactive glia)

PLATE 22A

CR 270 mm
GW 31
Y162-61
Lumbar Enlargement
Cell body stain

Areas (mm²)

Central canal	.0028
Ependyma	.0154
Gray matter	5.7639
White matter	5.1354

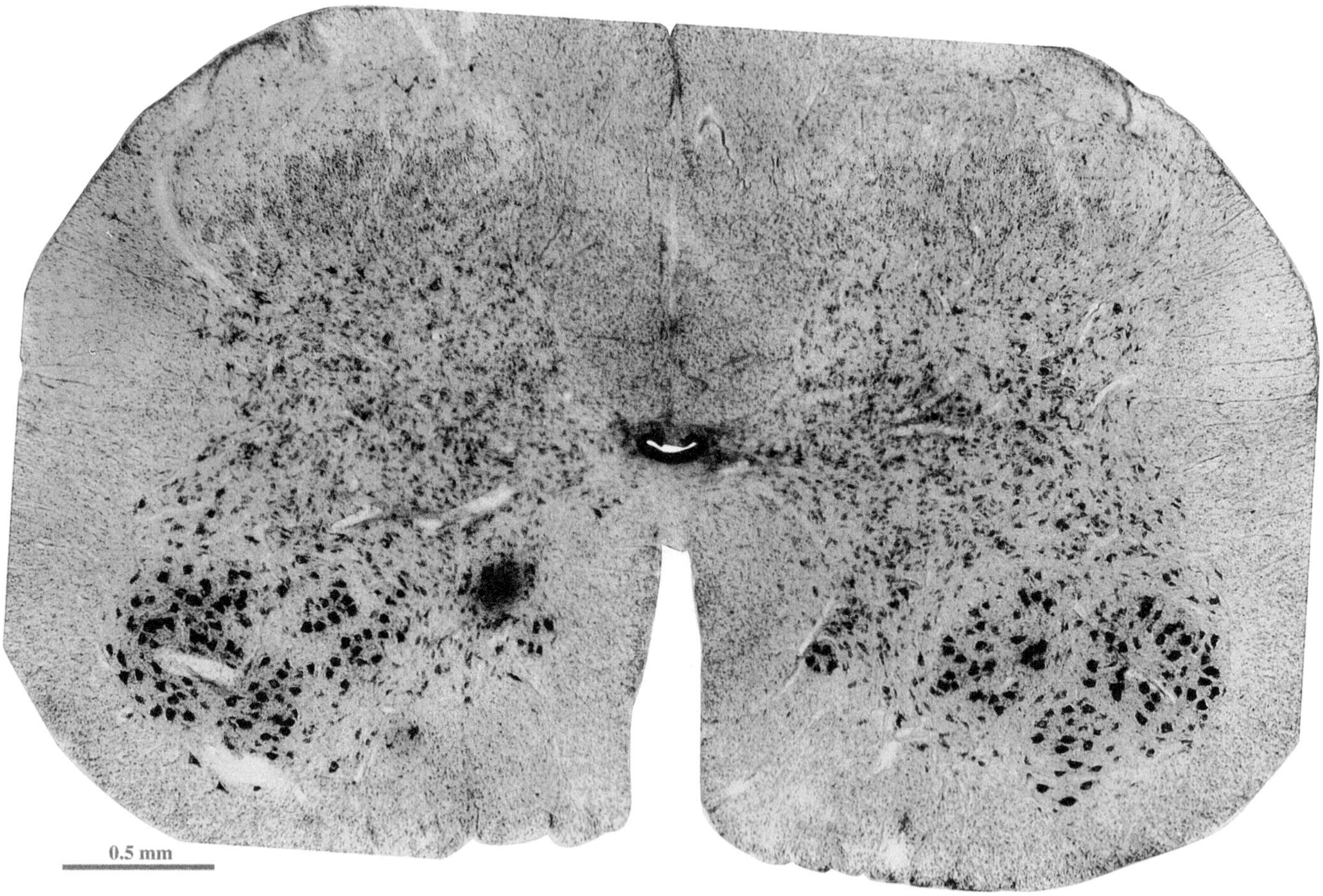

PLATE 22B

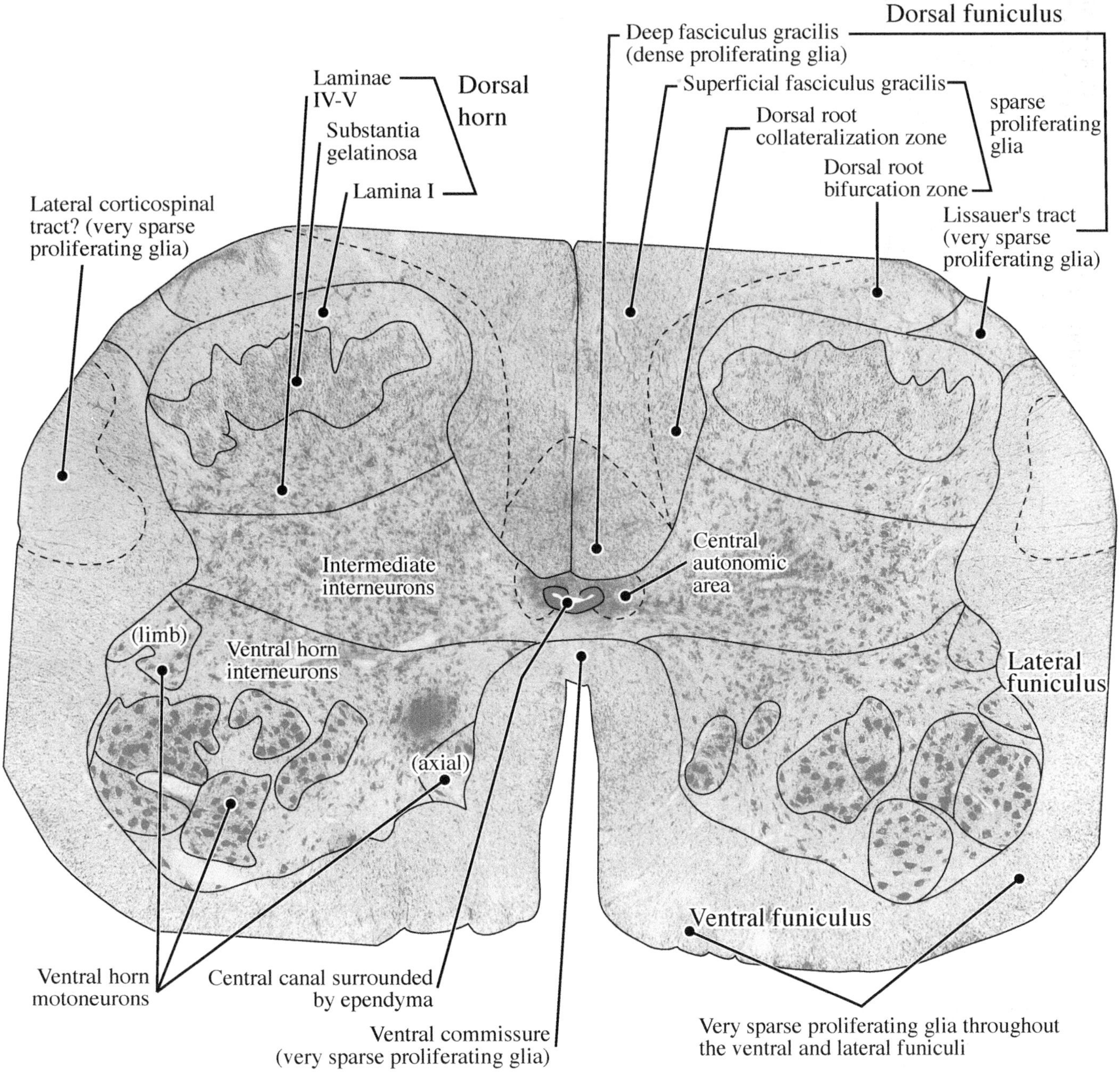
Dorsal funiculus
Deep fasciculus gracilis
(dense proliferating glia)
Superficial fasciculus gracilis
Dorsal root
collateralization zone
Dorsal root
bifurcation zone
sparse
proliferating
glia
Lissauer's tract
(very sparse
proliferating glia)
Laminae
IV-V
Substantia
gelatinosa
Lamina I
Dorsal
horn
Lateral corticospinal
tract? (very sparse
proliferating glia)
Intermediate
interneurons
Central
autonomic
area
(limb)
Ventral horn
interneurons
(axial)
Lateral
funiculus
Ventral funiculus
Ventral horn
motoneurons
Central canal surrounded
by ependyma
Ventral commissure
(very sparse proliferating glia)
Very sparse proliferating glia throughout
the ventral and lateral funiculi

PLATE 23A

CR 270 mm
GW 31
Y162-61
Sacral/Coccygeal
Myelin stain

Areas (mm^2)

Central canal	.0053
Ependyma	.0184
Gray matter	2.1433
White matter	1.3482

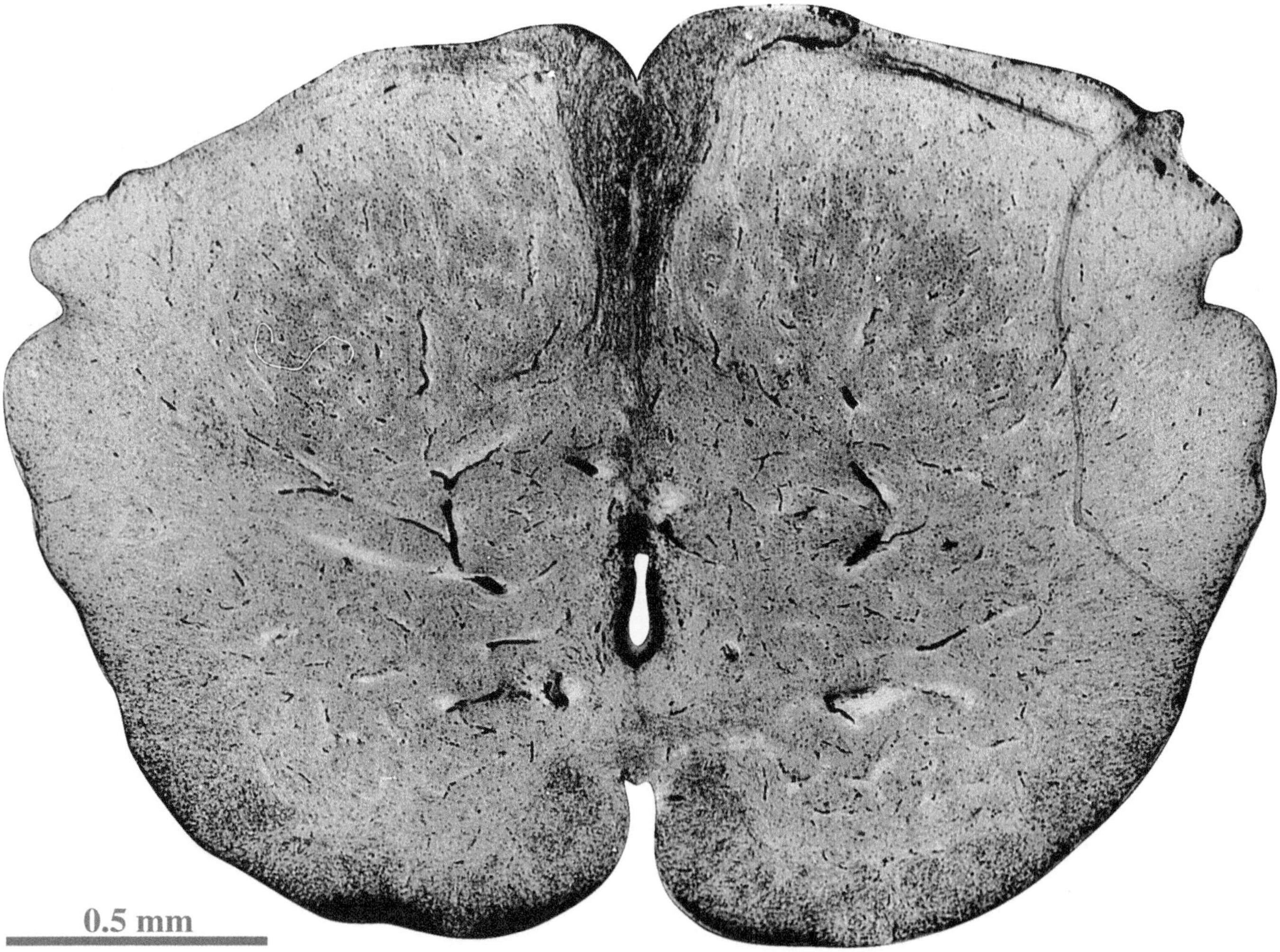

PLATE 23B

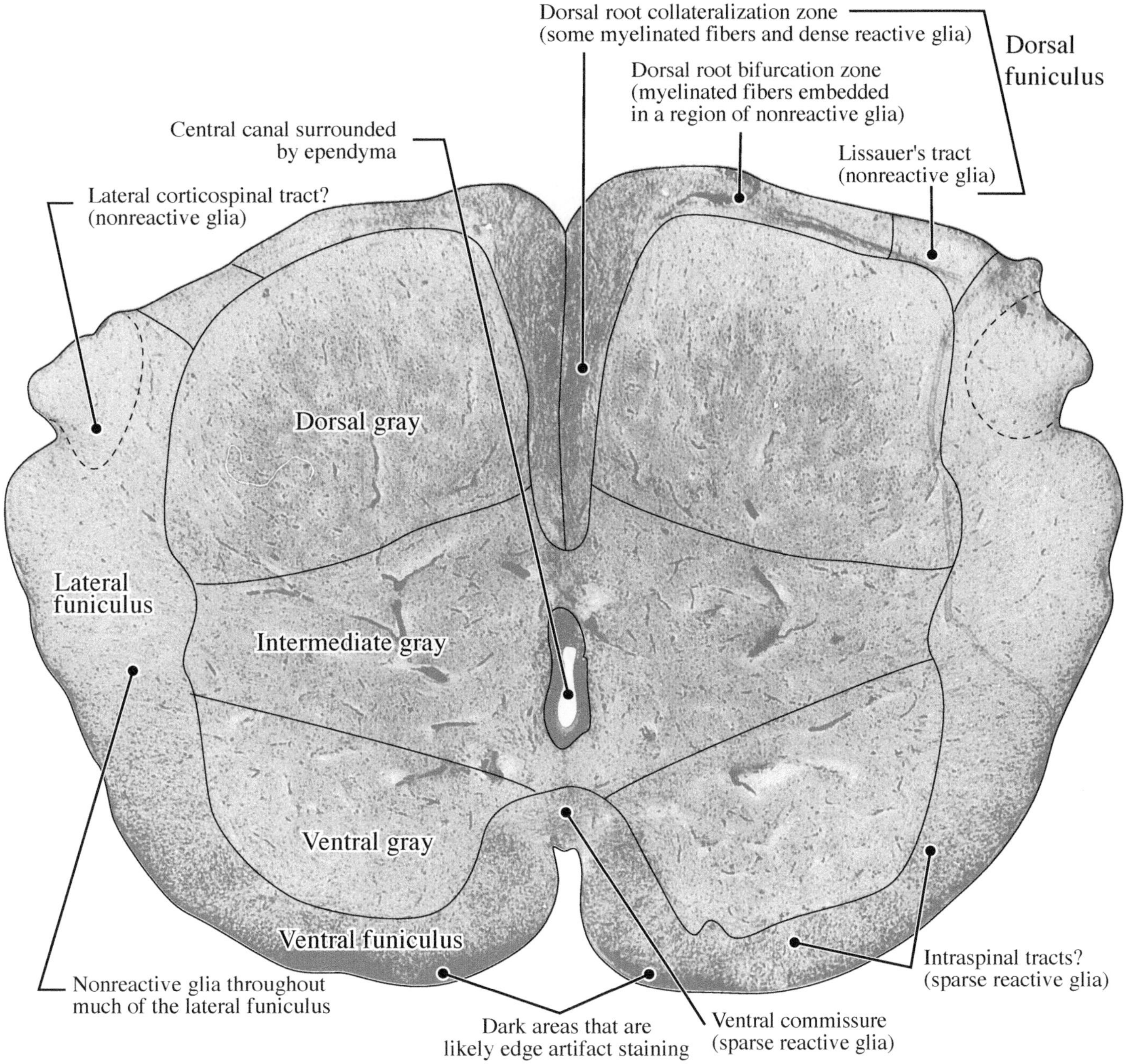
Dorsal root collateralization zone
(some myelinated fibers and dense reactive glia)
Dorsal funiculus
Dorsal root bifurcation zone
(myelinated fibers embedded
in a region of nonreactive glia)
Central canal surrounded
by ependyma
Lissauer's tract
(nonreactive glia)
Lateral corticospinal tract?
(nonreactive glia)
Dorsal gray
Lateral funiculus
Intermediate gray
Ventral gray
Ventral funiculus
Intraspinal tracts?
(sparse reactive glia)
Nonreactive glia throughout
much of the lateral funiculus
Dark areas that are
likely edge artifact staining
Ventral commissure
(sparse reactive glia)

PLATE 24A

CR 270 mm
GW 31
Y162-61
Sacral/Coccygeal
Cell body stain

Areas (mm^2)

Central canal	.0046
Ependyma	.0253
Gray matter	2.2183
White matter	1.3268

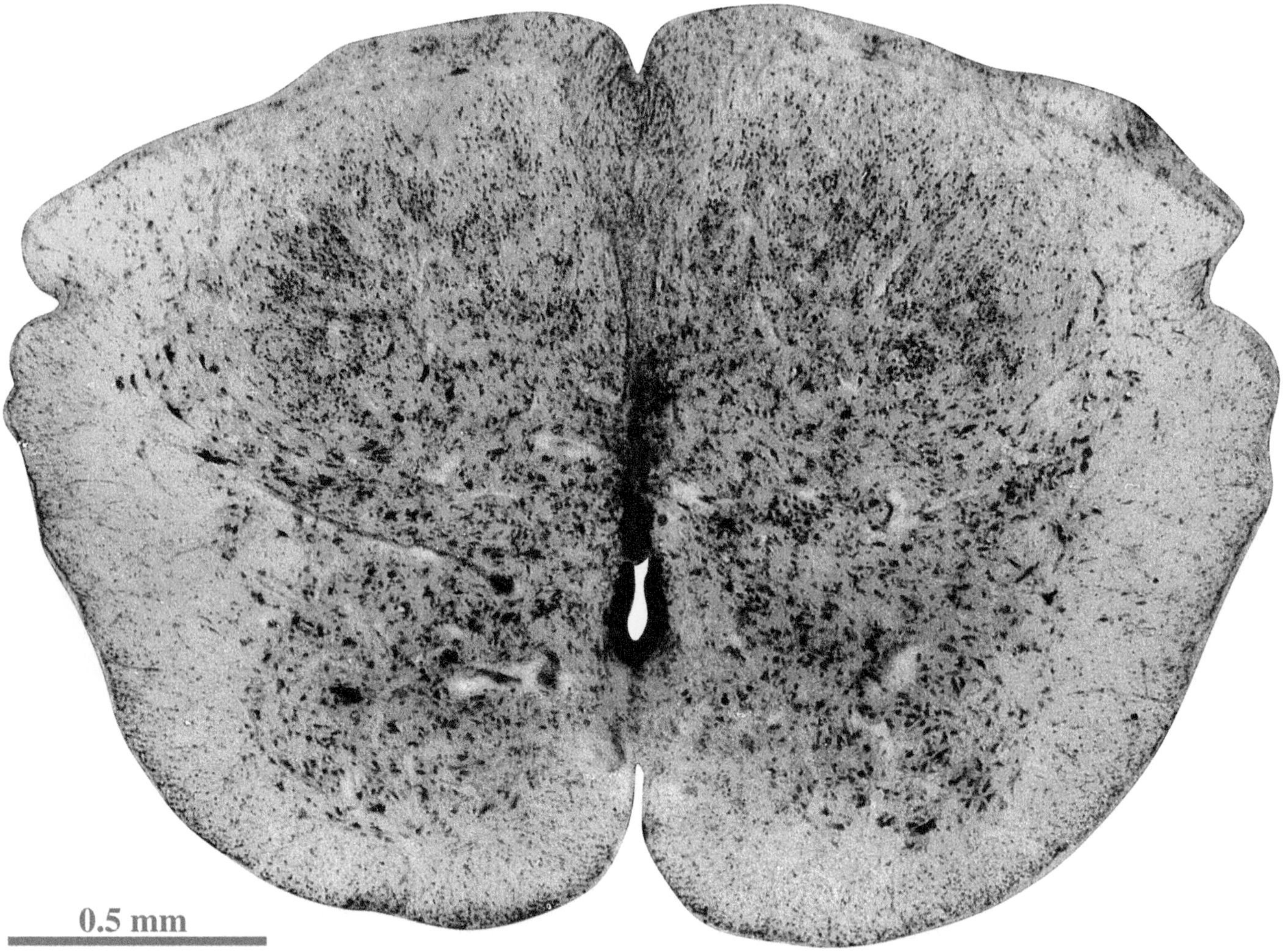

PLATE 24B

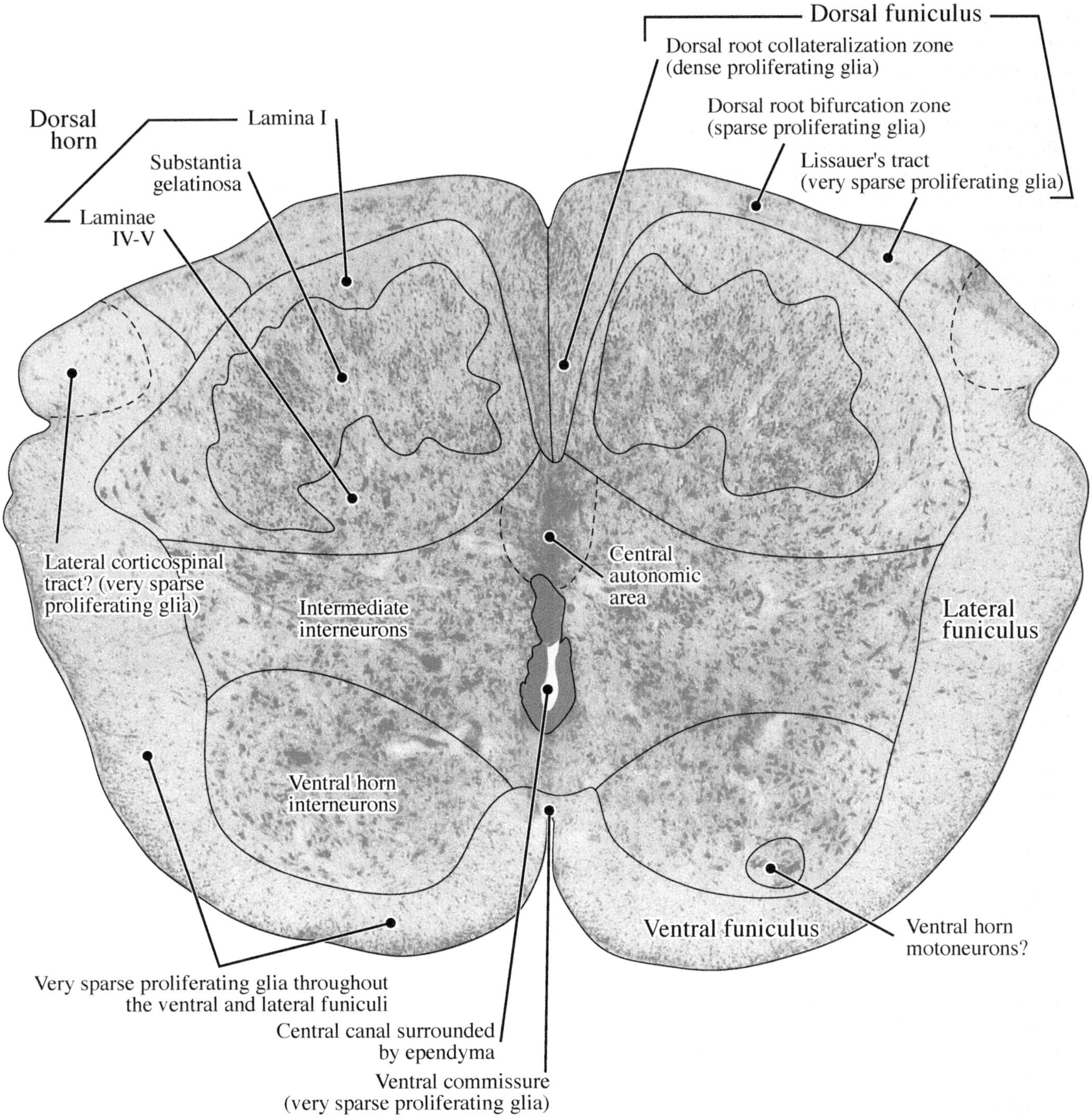
Dorsal funiculus
Dorsal root collateralization zone (dense proliferating glia)
Dorsal root bifurcation zone (sparse proliferating glia)
Lissauer's tract (very sparse proliferating glia)
Dorsal horn
Lamina I
Substantia gelatinosa
Laminae IV-V
Lateral corticospinal tract? (very sparse proliferating glia)
Intermediate interneurons
Central autonomic area
Lateral funiculus
Ventral horn interneurons
Ventral funiculus
Ventral horn motoneurons?
Very sparse proliferating glia throughout the ventral and lateral funiculi
Central canal surrounded by ependyma
Ventral commissure (very sparse proliferating glia)

PART V: Y117-61 CR 310 mm (GW 37)

Plate 25 is a survey of matched myelin-stained and cell-body-stained sections from Y117-61, a specimen in the Yakovlev Collection. All sections are shown at the same scale. The boxes enclosing each section list the approximate level and the total area of the section in square millimeters (mm^2). Full-page normal-contrast photographs of each section are in **Plates 26A–41A.** Low-contrast photographs with superimposed labels and outlines of structural details are in **Plates 26B–41B.** This specimen has the the most complete set of spinal cord sections in the Yakovlev Collection because all levels are preserved. Twenty myelin-stained sections and nineteen cell-body-stained sections were photographed ranging from upper cervical to sacral/coccygeal levels. In this specimen, the myelin-stained and cell-body-stained sections were preserved on separate large glass plates without any section numbers. The 39 photographic prints were intuitively arranged in order from upper cervical to sacral/coccygeal levels, using internal features such as the size of the corticospinal tracts, and the width of the ventral horn. Then, myelin- and cell-body-stained sections were matched and 8 different levels were analyzed.

As in the previous specimens, the cross-sectional area of a myelin-stained section is smaller than the matching cell-body-stained section in all cases, except the upper thoracic level, where the areas are the same. The myelin staining procedure consistently produces greater tissue shrinkage than the cell-body staining procedure. Using the total areas of the myelin-stained sections, the overall size differences between levels indicate the following comparisons: In this specimen the cervical enlargement is the level with the largest cross-sectional area, being larger than the lumbar enlargement by 30%. The middle thoracic level has the smallest cross-sectional area and is even 6% smaller than the sacral/coccygeal level.

Myelination continues to advance in this specimen (**Table 4**). Dense staining indicative of either advanced or beginning myelination is seen throughout the ventral funiculus and the lateral funiculus except the ventral and lateral corticospinal tracts, the sacral/coccygeal level excepted. All parts of the dorsal funiculus are either myelinated or are myelinating, except Lissauer's tract and the areas in the dorsal root bifurcation zone, which most likely contains incoming axons that will join Lissauer's tract. In most cases, labels of the fiber tracts include our assessment of the progression of myelination. The most densely stained area is considered myelinated, while the next most densely stained area is considered to be myelinating. It is interesting to note that the white matter in the cell-body-stained sections show a uniform density of glia, and various fiber tracts are difficult to delineate. But again, the lateral and ventral corticospinal tracts generally stand out as having slightly lower concentrations of glia. With the exception of the corticospinal tracts and the dorsal funiculus, other fiber tracts are not identified in the cell-body-stained sections.

As in the previous specimens, the labels in the gray matter of the myelin-stained sections are limited to just the dorsal horn, ventral horn, and intermediate gray. In the cell-body-stained sections, the structures in the gray matter are becoming more clearly defined, especially the ventral horn motoneurons, lateral horn motoneurons, Clarke's column, and the substantia gelatinosa.

Table 4: Glia types and concentration in the white matter at GW 37

Name	Myelination	Reactive glia	Proliferating glia
DORSAL ROOT	myelinated	---	dense
VENTRAL ROOT	myelinated	---	---
DORSAL FUNICULUS: dorsal root bif. zone	*many fibers	---	sparse
dorsal root col. zone	myelinated	---	†sparse
deep fas. gracilis	myelinated	---	sparse
superficial fas. gracilis	many fibers	---	sparse
deep fas. cuneatus	myelinated	---	sparse
superficial fas. cuneatus	many fibers	---	sparse
Lissauer's tract	---	none	sparse
LATERAL and VENTRAL FUNICULI: lat. corticospinal tract	---	very sparse	very sparse
ven. corticospinal tract	---	very sparse	very sparse
lat. reticulospinal tract	many fibers	---	sparse
spinocerebellar tracts	many fibers	---	sparse
ven. commissure	myelinated	---	sparse
intraspinal tract	many fibers	---	sparse
spinocephalic tract	some fibers	very dense	sparse
med. long. fasciculus	many fibers	---	sparse
vestibulospinal tract	---	dense	sparse

* intermingled in a bed of nonreactive glia (associated with Lissauer's tract fibers?)
† dense at the sacral/coccygeal level

PLATE 25

CR 310 mm, GW 37, Y117-61

MYELIN STAIN	CELL BODY STAIN
Plates 26A, 26B **Upper Cervical** Total area: 15.138 mm^2 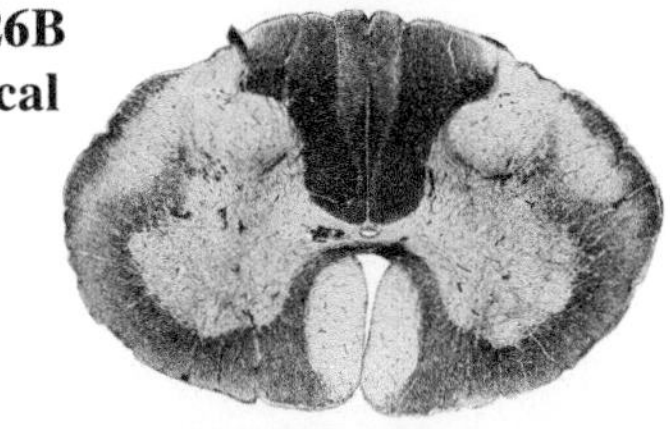	**Plates 27A, 27B** **Upper Cervical** Total area: 15.896 mm^2 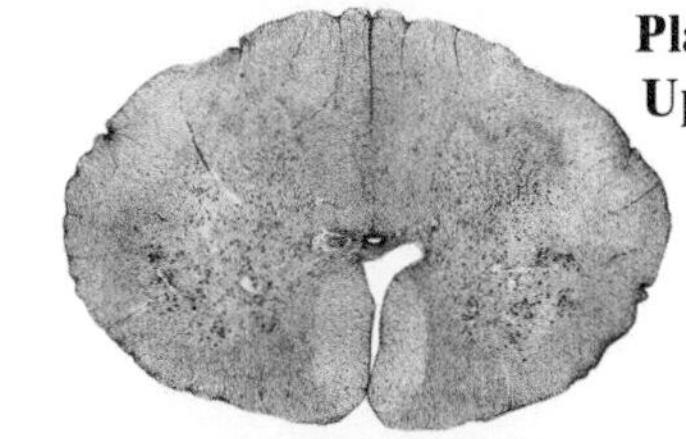
Plates 28A, 28B **Cervical Enlargement** Total area: 16.046 mm^2 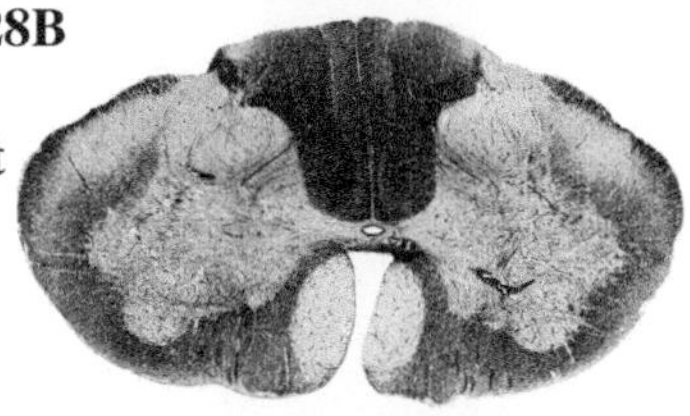	**Plates 29A, 29B** **Cervical Enlargement** Total area: 17.036 mm^2 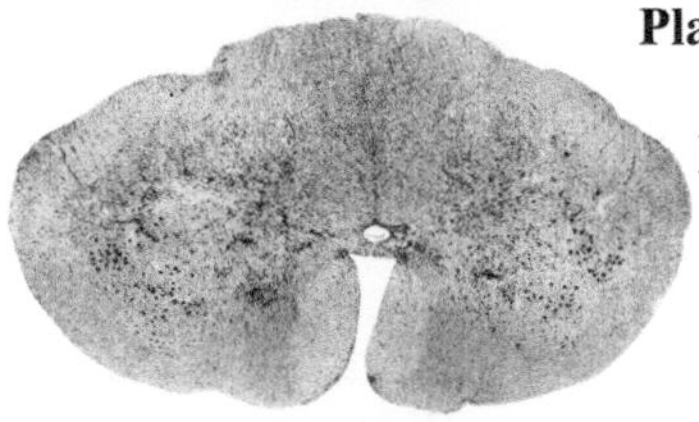
Plates 30A, 30B **Upper Thoracic** Total area: 8.4267 mm^2 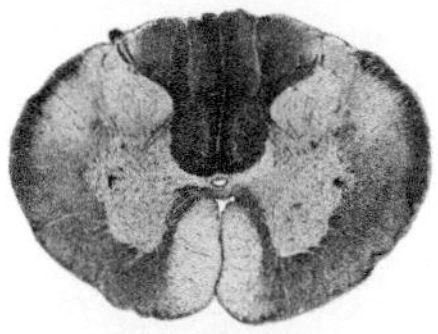	**Plates 31A, 31B** **Upper Thoracic** Total area: 8.4266 mm^2 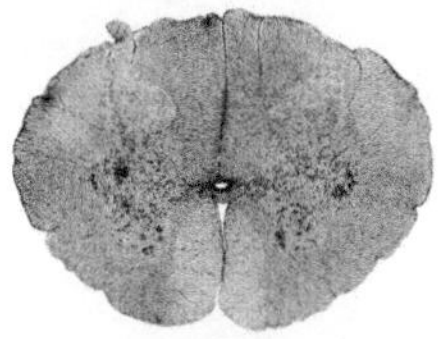
Plates 32A, 32B **Middle Thoracic** Total area: 6.1539 mm^2 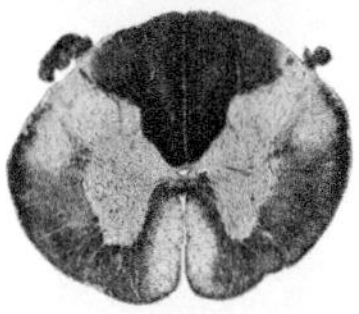	**Plates 33A, 33B** **Middle Thoracic** Total area: 6.7020 mm^2 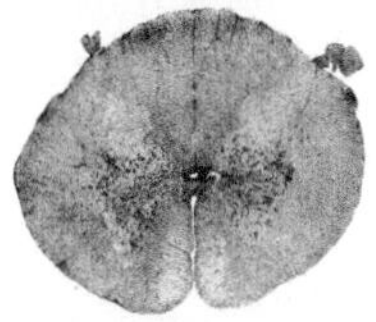
Plates 34A, 34B **Lower Thoracic** Total area: 7.2005 mm^2 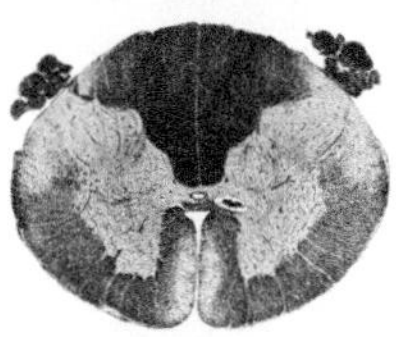	**Plates 35A, 35B** **Lower Thoracic** Total area: 7.9554 mm^2 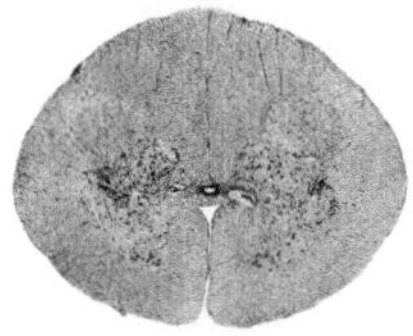
Plates 36A, 36B **Upper Lumbar** Total area: 11.530 mm^2 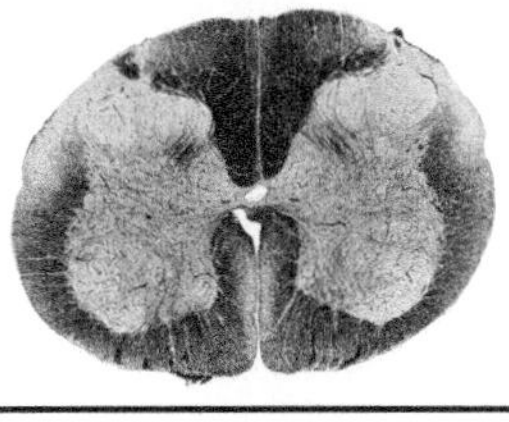	**Plates 37A, 37B** **Upper Lumbar** Total area: 12.297 mm^2 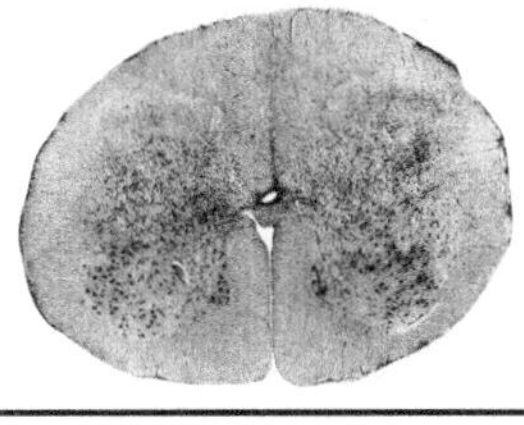
Plates 38A, 38B **Lumbar Enlargement** Total area: 12.297 mm^2 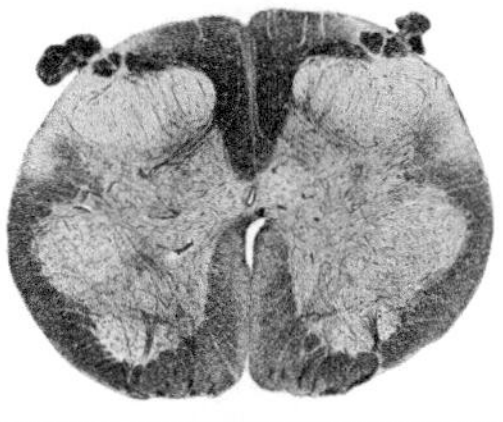	**Plates 39A, 39B** **Lumbar Enlargement** Total area: 12.872 mm^2 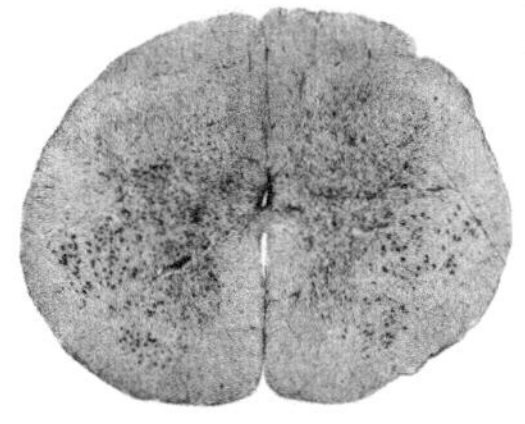
Plates 40A, 40B **Sacral/Coccygeal** Total area: 6.5068 mm^2 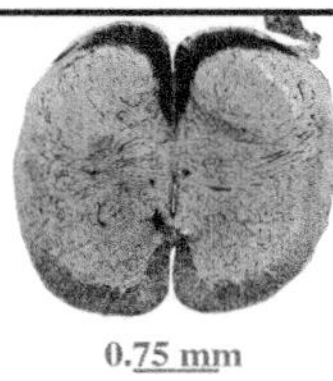 0.75 mm	**Plates 41A, 41B** **Sacral/Coccygeal** Total area: 6.7829 mm^2 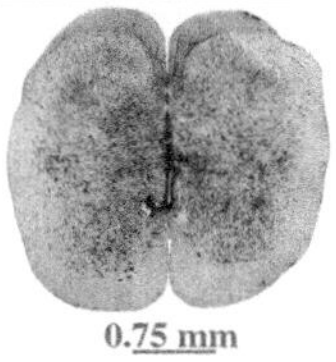 0.75 mm

PLATE 26A

CR 310 mm
GW 37
Y117-61
Upper Cervical
Myelin stain

Areas (mm^2)

Central canal	.0044
Ependyma	.0102
Gray matter	5.2519
White matter	9.8713

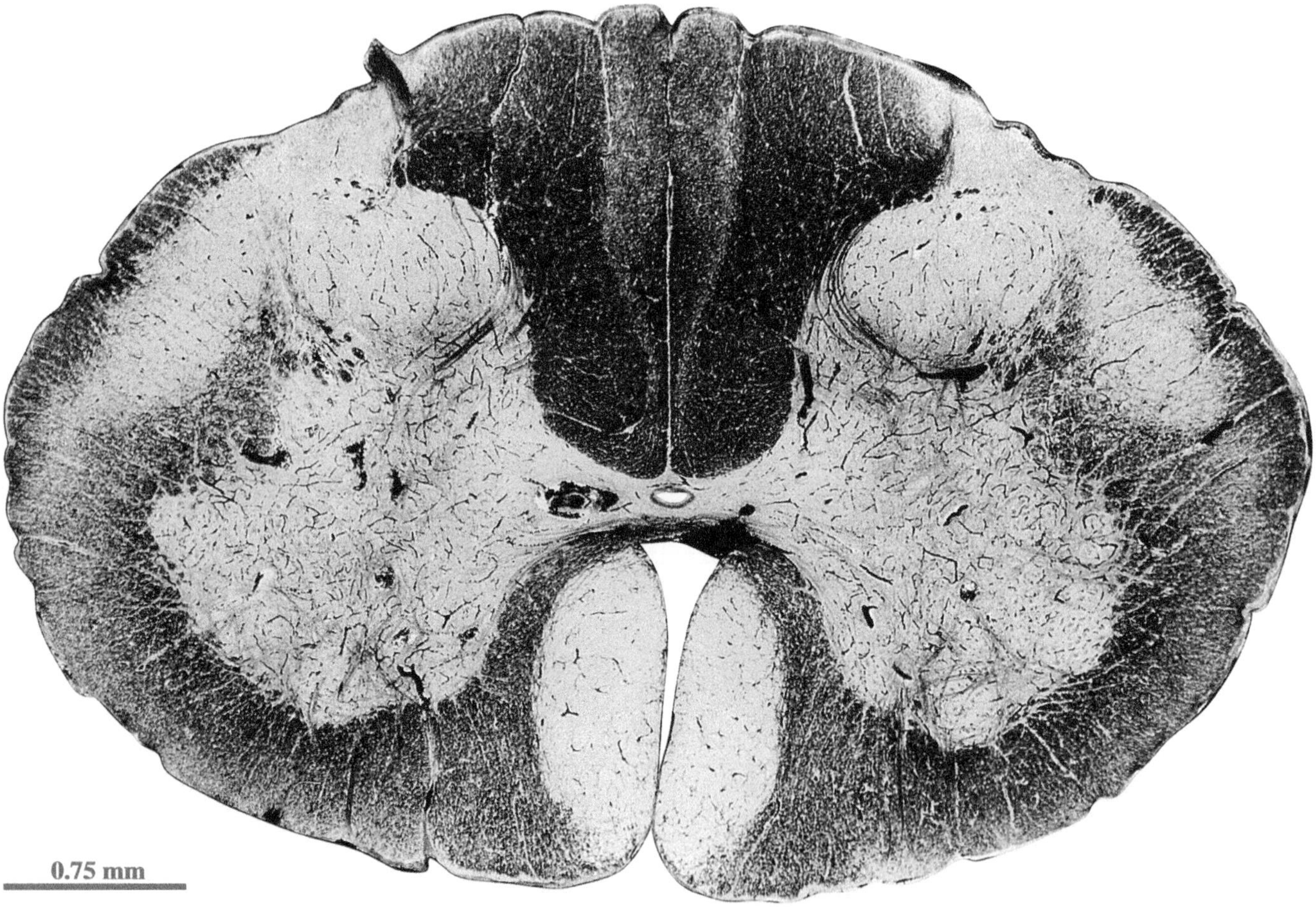

0.75 mm

PLATE 26B

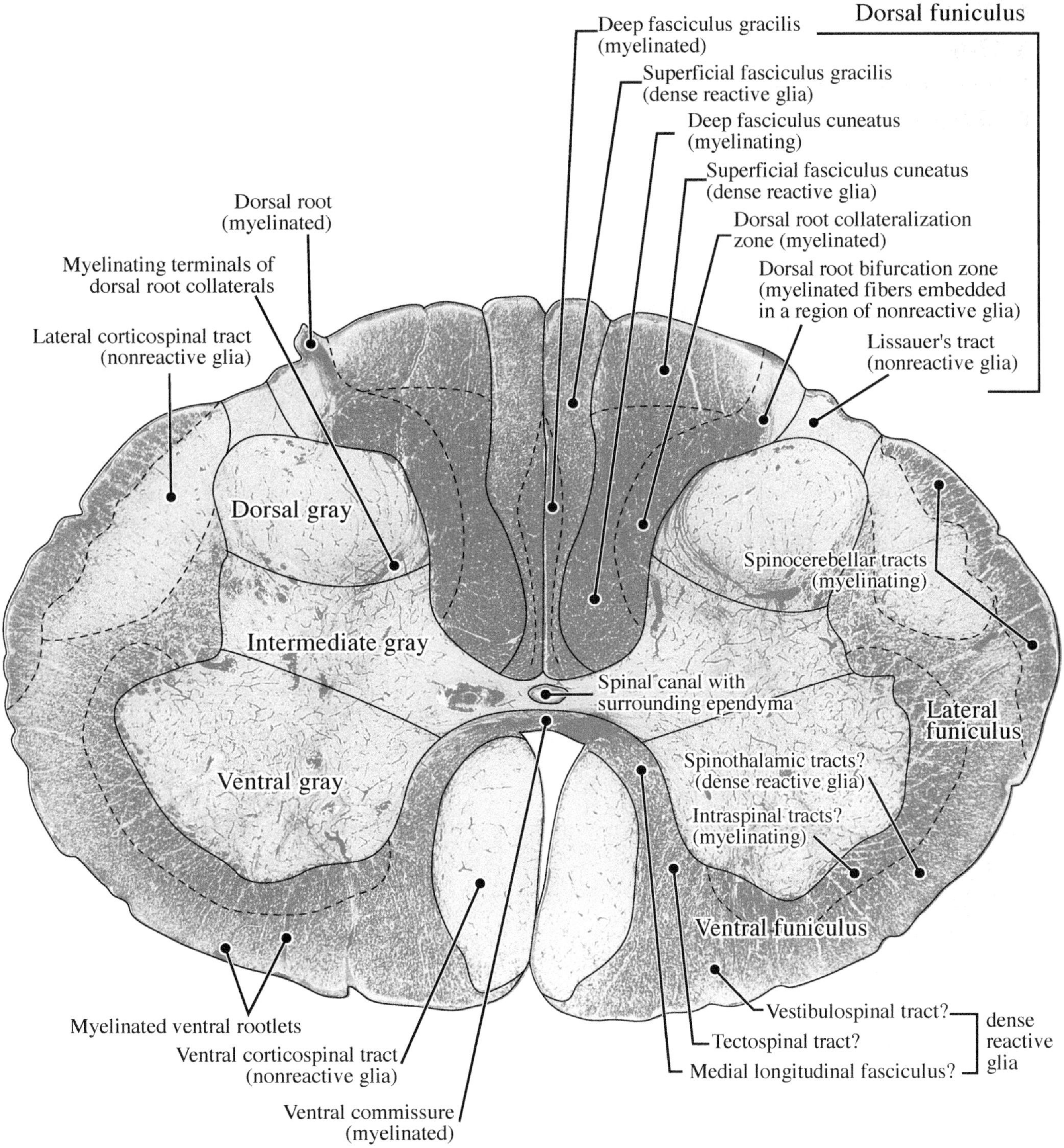
Dorsal funiculus
Deep fasciculus gracilis (myelinated)
Superficial fasciculus gracilis (dense reactive glia)
Deep fasciculus cuneatus (myelinating)
Superficial fasciculus cuneatus (dense reactive glia)
Dorsal root collateralization zone (myelinated)
Dorsal root bifurcation zone (myelinated fibers embedded in a region of nonreactive glia)
Lissauer's tract (nonreactive glia)
Dorsal root (myelinated)
Myelinating terminals of dorsal root collaterals
Lateral corticospinal tract (nonreactive glia)
Dorsal gray
Intermediate gray
Spinocerebellar tracts (myelinating)
Spinal canal with surrounding ependyma
Lateral funiculus
Ventral gray
Spinothalamic tracts? (dense reactive glia)
Intraspinal tracts? (myelinating)
Ventral funiculus
Myelinated ventral rootlets
Ventral corticospinal tract (nonreactive glia)
Ventral commissure (myelinated)
Vestibulospinal tract?
Tectospinal tract?
Medial longitudinal fasciculus?
dense reactive glia

PLATE 27A

CR 310 mm
GW 37
Y117-61
Upper Cervical
Cell body stain

Areas (mm²)

Central canal	.0038
Ependyma	.0175
Gray matter	5.8125
White matter	10.0620

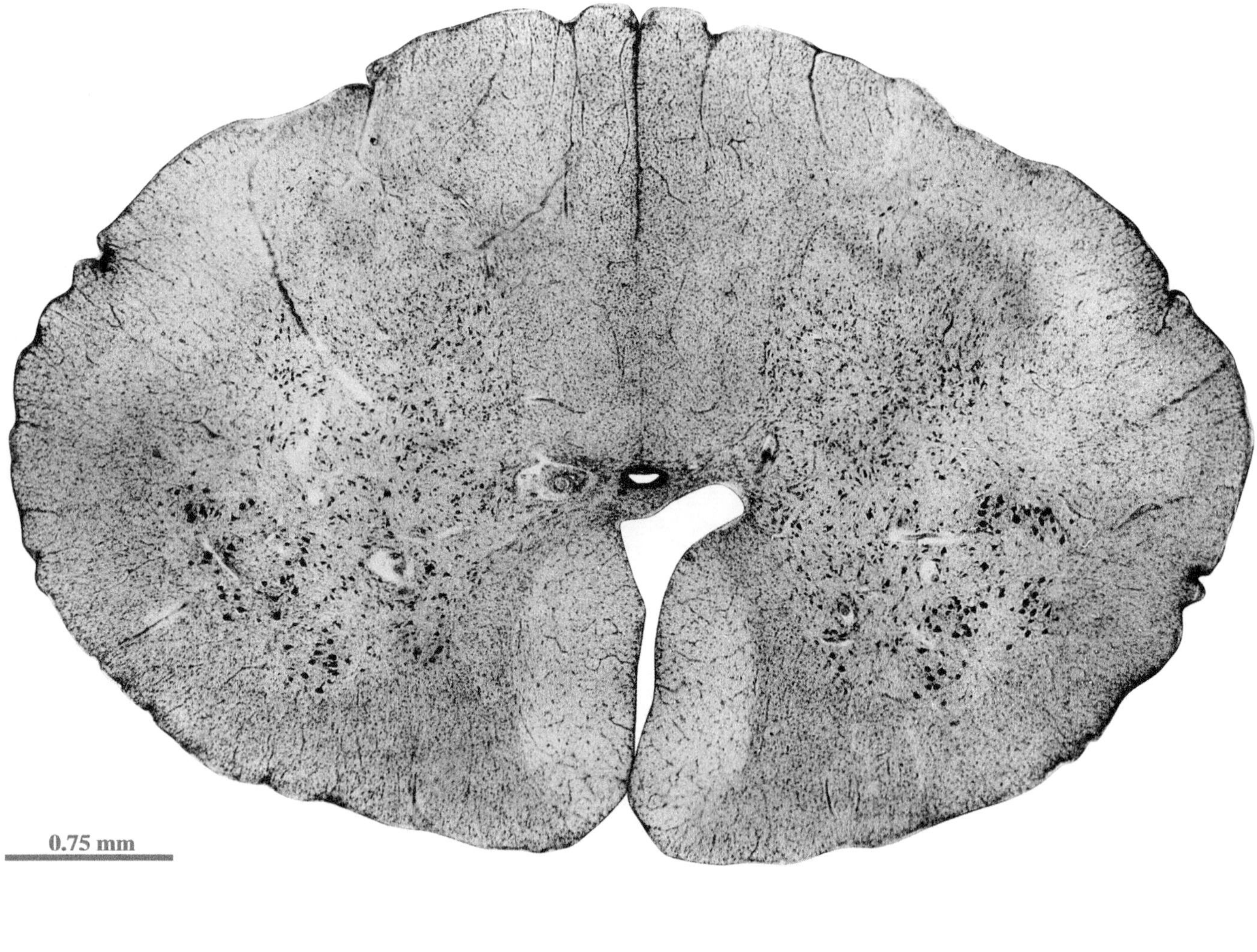

PLATE 27B

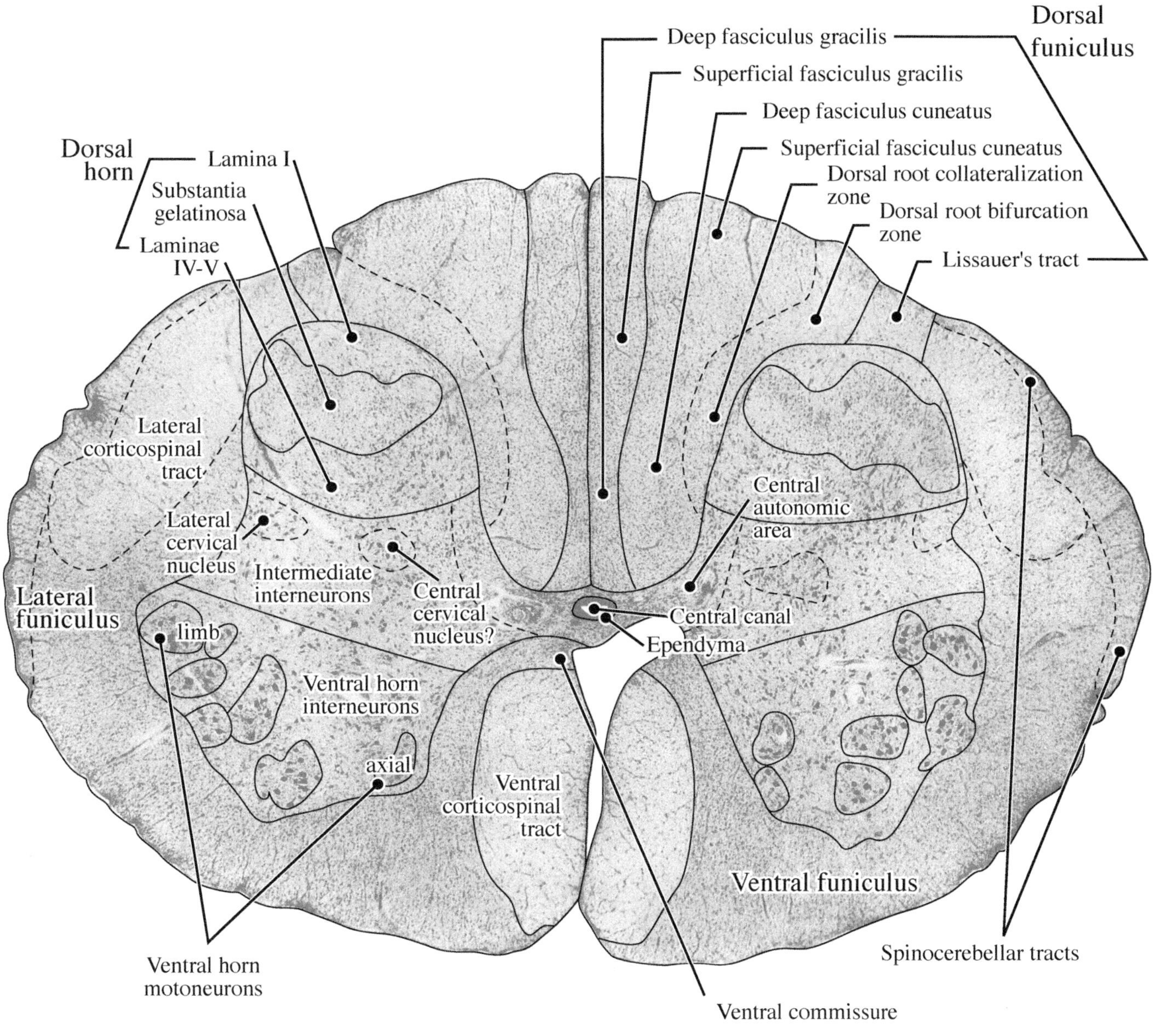
Deep fasciculus gracilis
Dorsal funiculus
Superficial fasciculus gracilis
Deep fasciculus cuneatus
Superficial fasciculus cuneatus
Dorsal root collateralization zone
Dorsal root bifurcation zone
Lissauer's tract
Dorsal horn
Lamina I
Substantia gelatinosa
Laminae IV-V
Lateral corticospinal tract
Lateral cervical nucleus
Intermediate interneurons
Central cervical nucleus?
Central autonomic area
Central canal
Ependyma
Lateral funiculus
limb
Ventral horn interneurons
axial
Ventral corticospinal tract
Ventral funiculus
Ventral horn motoneurons
Ventral commissure
Spinocerebellar tracts

PLATE 28A

CR 310 mm
GW 37
Y117-61
Cervical Enlargement
Myelin stain

Areas (mm²)

Central canal	.0073
Ependyma	.0137
Gray matter	6.0566
White matter	9.9683

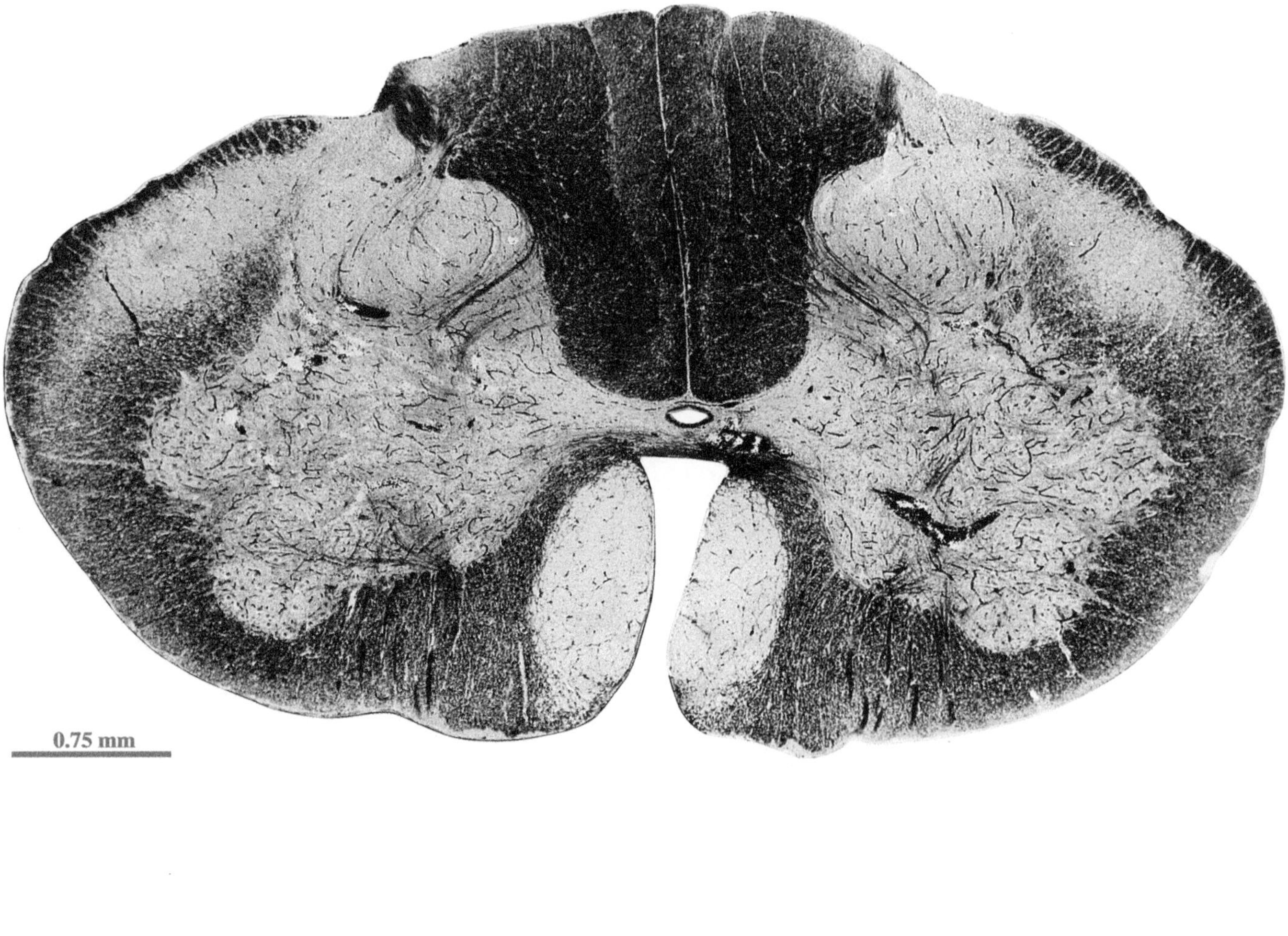

PLATE 28B

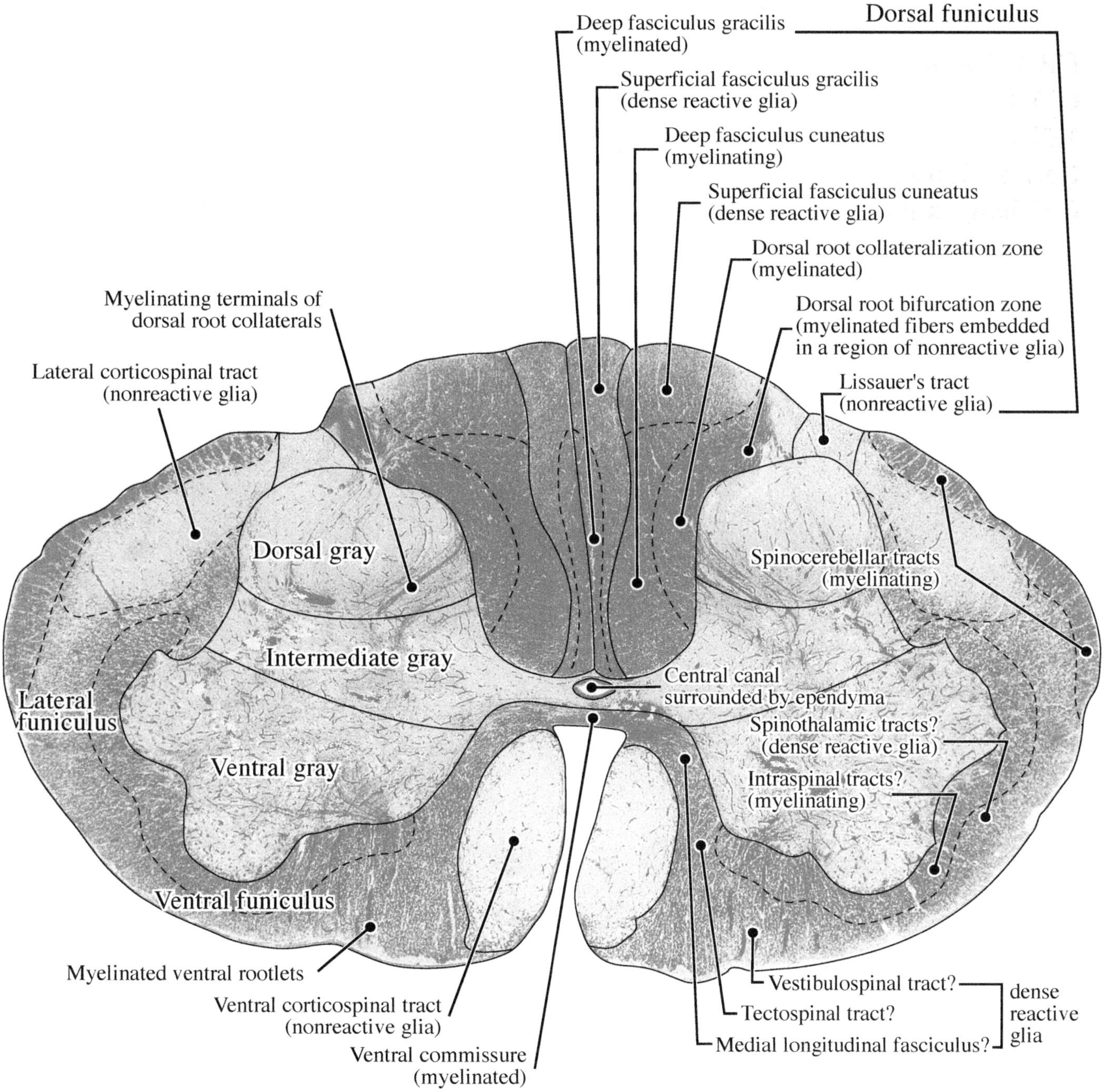
Dorsal funiculus
Deep fasciculus gracilis (myelinated)
Superficial fasciculus gracilis (dense reactive glia)
Deep fasciculus cuneatus (myelinating)
Superficial fasciculus cuneatus (dense reactive glia)
Dorsal root collateralization zone (myelinated)
Dorsal root bifurcation zone (myelinated fibers embedded in a region of nonreactive glia)
Lissauer's tract (nonreactive glia)
Myelinating terminals of dorsal root collaterals
Lateral corticospinal tract (nonreactive glia)
Dorsal gray
Spinocerebellar tracts (myelinating)
Intermediate gray
Central canal surrounded by ependyma
Lateral funiculus
Spinothalamic tracts? (dense reactive glia)
Ventral gray
Intraspinal tracts? (myelinating)
Ventral funiculus
Myelinated ventral rootlets
Ventral corticospinal tract (nonreactive glia)
Ventral commissure (myelinated)
Vestibulospinal tract?
Tectospinal tract?
Medial longitudinal fasciculus?
dense reactive glia

PLATE 29A

CR 310 mm
GW 37
Y117-61
Cervical Enlargement
Cell body stain

Areas (mm^2)

Central canal	.0079
Ependyma	.0230
Gray matter	6.6710
White matter	10.3340

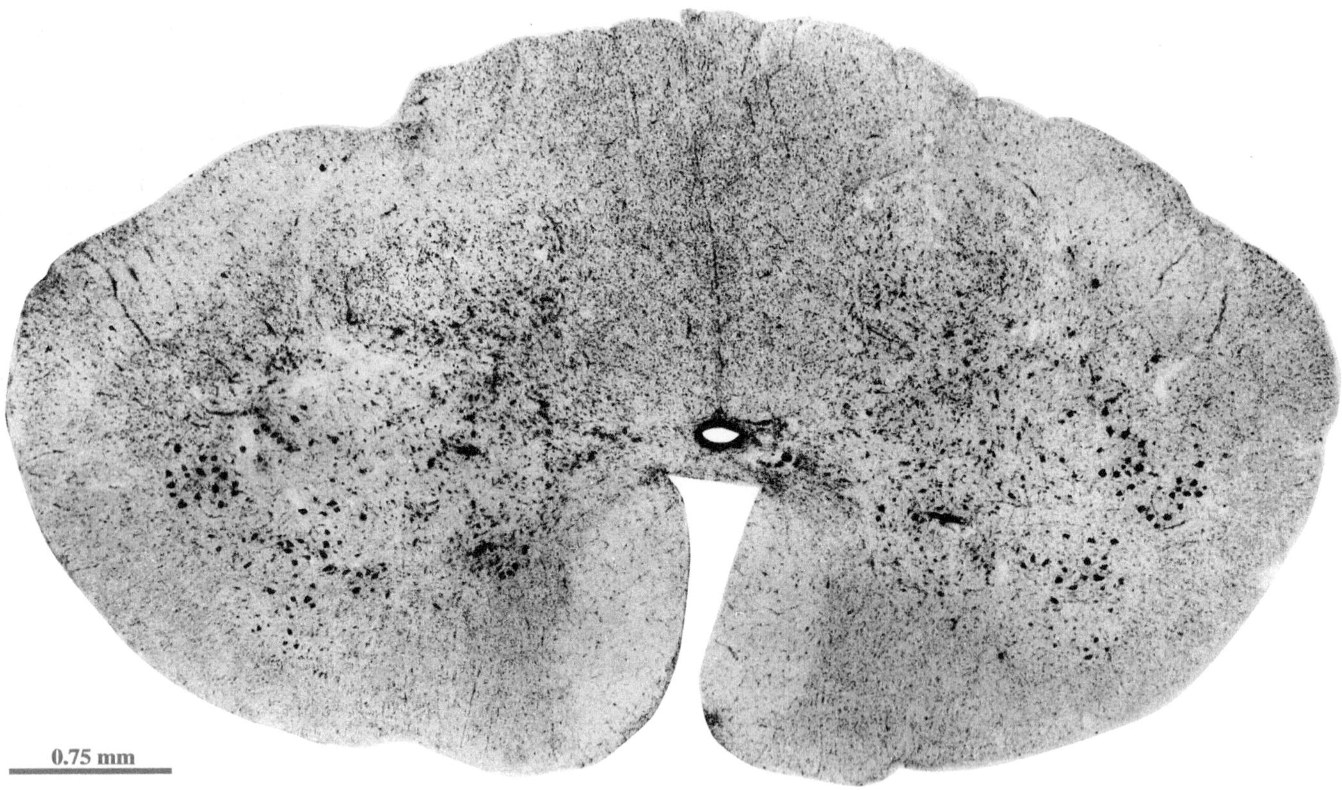

PLATE 29B

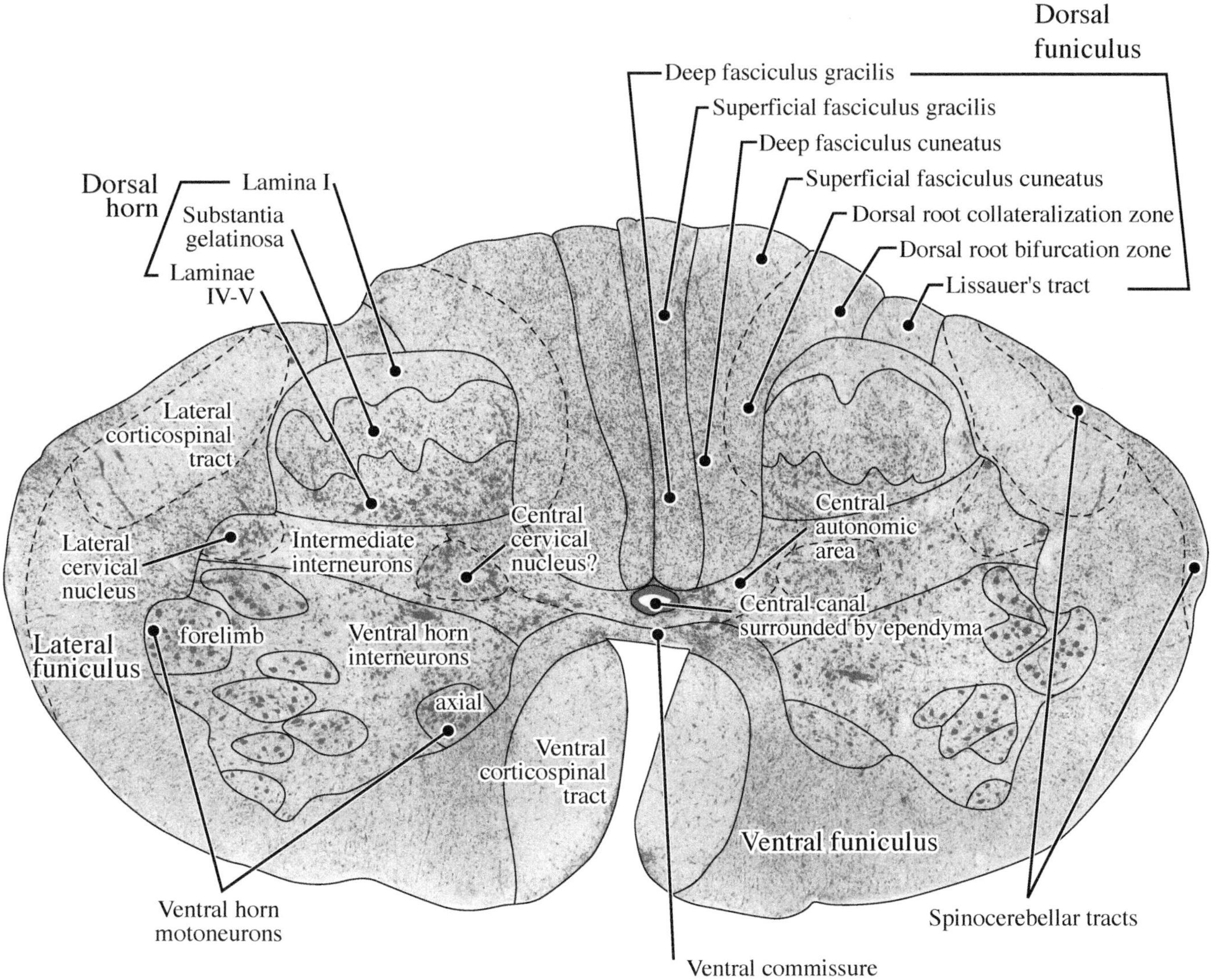
Dorsal funiculus
Deep fasciculus gracilis
Superficial fasciculus gracilis
Deep fasciculus cuneatus
Superficial fasciculus cuneatus
Dorsal root collateralization zone
Dorsal root bifurcation zone
Lissauer's tract
Dorsal horn
Lamina I
Substantia gelatinosa
Laminae IV-V
Lateral corticospinal tract
Lateral cervical nucleus
Intermediate interneurons
Central cervical nucleus?
Central autonomic area
Central canal surrounded by ependyma
forelimb
Ventral horn interneurons
Lateral funiculus
axial
Ventral corticospinal tract
Ventral funiculus
Ventral horn motoneurons
Spinocerebellar tracts
Ventral commissure

PLATE 30A

CR 310 mm
GW 37
Y117-61
Upper Thoracic
Myelin stain

Areas (mm²)

Central canal	.0037
Ependyma	.0086
Gray matter	2.1599
White matter	6.2545

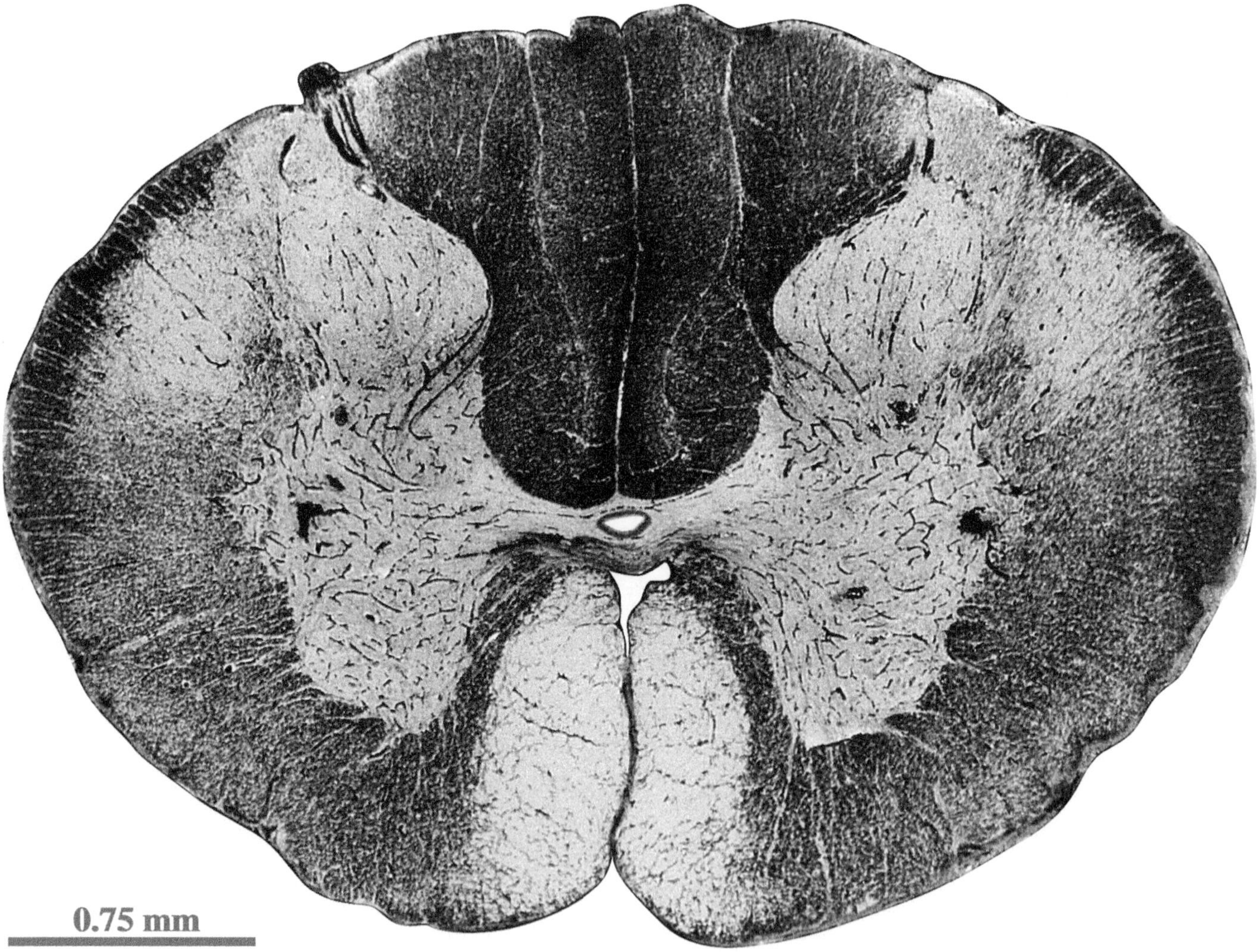

PLATE 30B

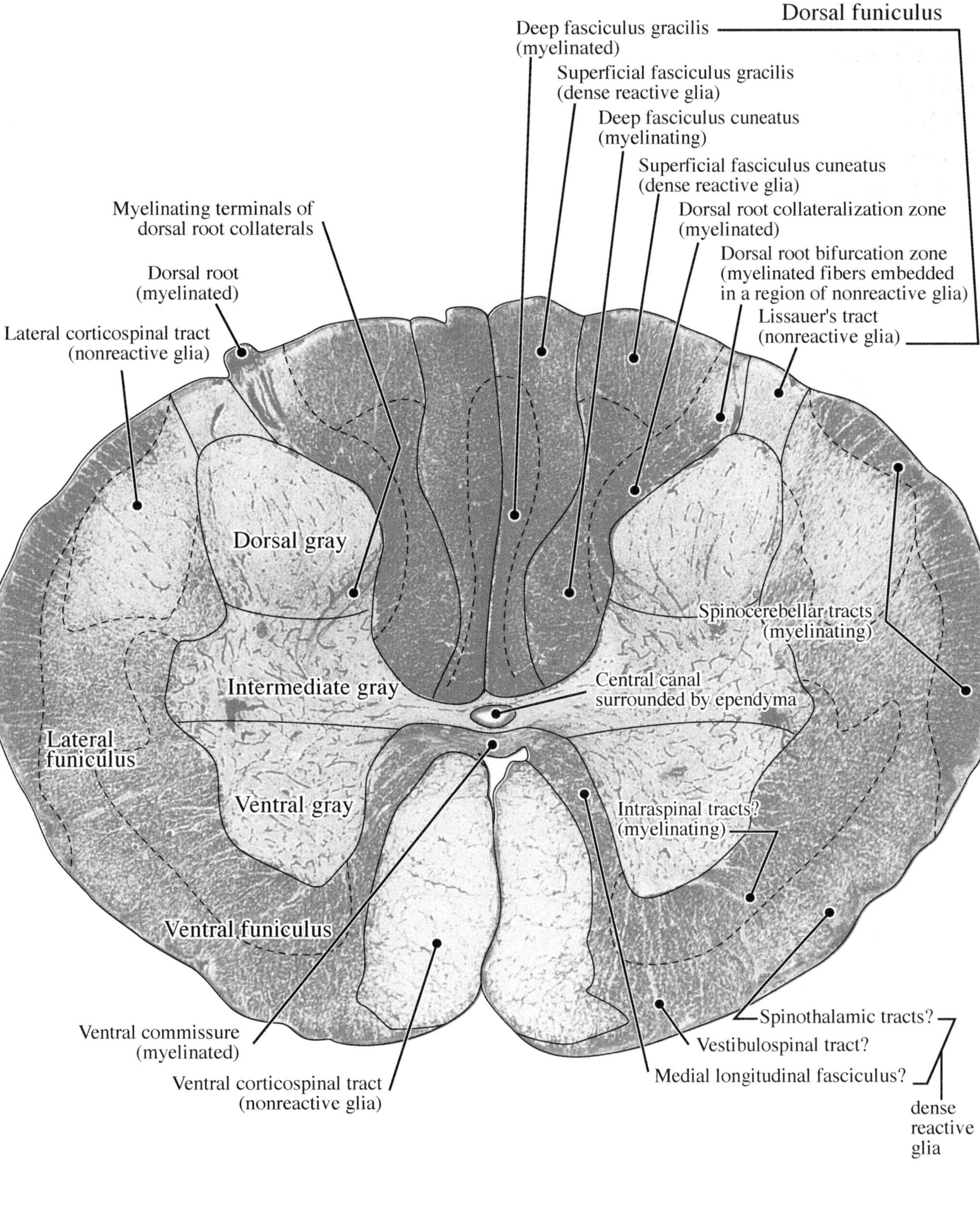
Dorsal funiculus
Deep fasciculus gracilis (myelinated)
Superficial fasciculus gracilis (dense reactive glia)
Deep fasciculus cuneatus (myelinating)
Superficial fasciculus cuneatus (dense reactive glia)
Dorsal root collateralization zone (myelinated)
Dorsal root bifurcation zone (myelinated fibers embedded in a region of nonreactive glia)
Lissauer's tract (nonreactive glia)
Myelinating terminals of dorsal root collaterals
Dorsal root (myelinated)
Lateral corticospinal tract (nonreactive glia)
Dorsal gray
Intermediate gray
Lateral funiculus
Ventral gray
Ventral funiculus
Spinocerebellar tracts (myelinating)
Central canal surrounded by ependyma
Intraspinal tracts? (myelinating)
Spinothalamic tracts?
Vestibulospinal tract?
Medial longitudinal fasciculus?
dense reactive glia
Ventral commissure (myelinated)
Ventral corticospinal tract (nonreactive glia)

PLATE 31A

CR 310 mm
GW 37
Y117-61
Upper Thoracic
Cell body stain

Areas (mm^2)

Central canal	.0030
Ependyma	.0152
Gray matter	2.3167
White matter	6.0916

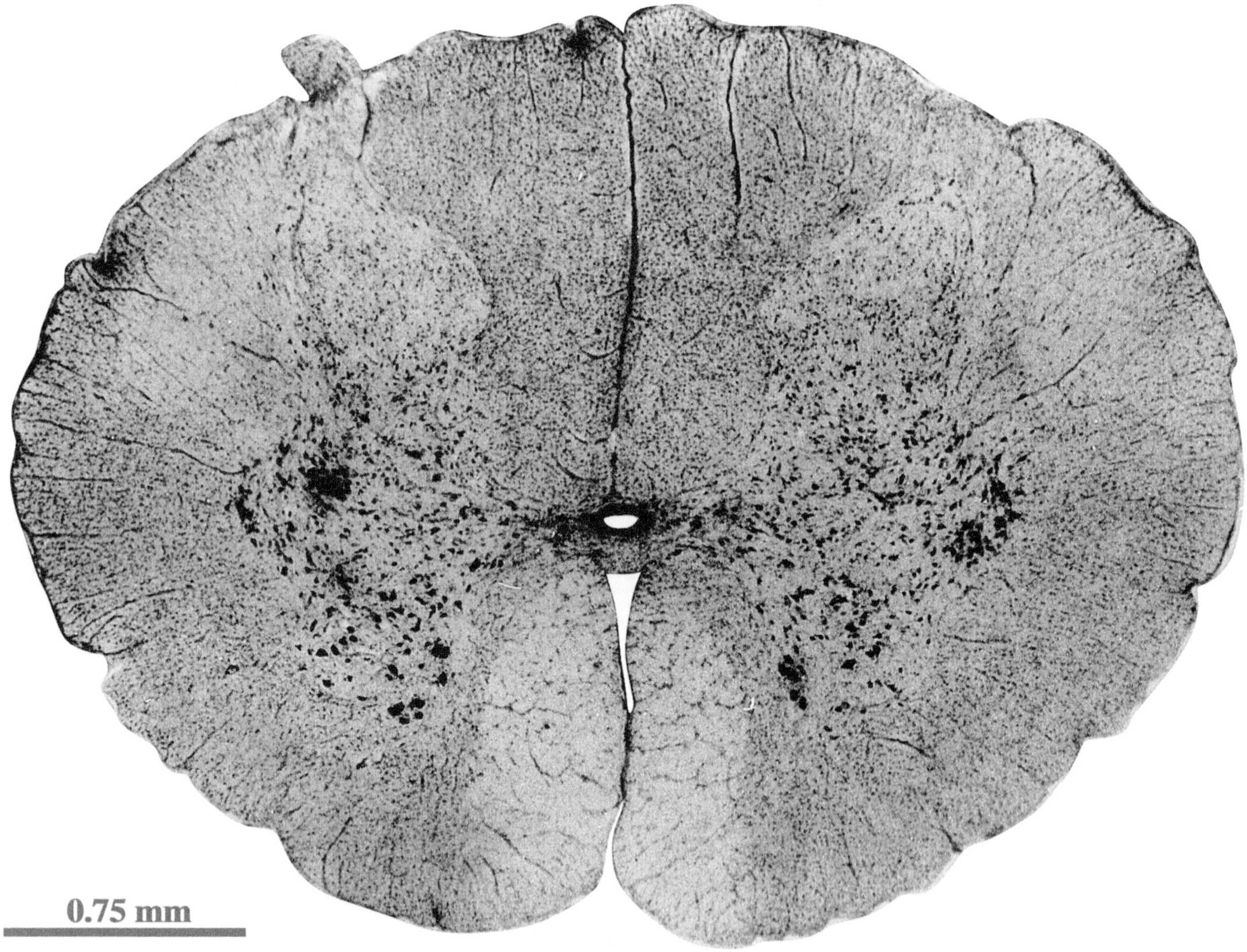

PLATE 31B

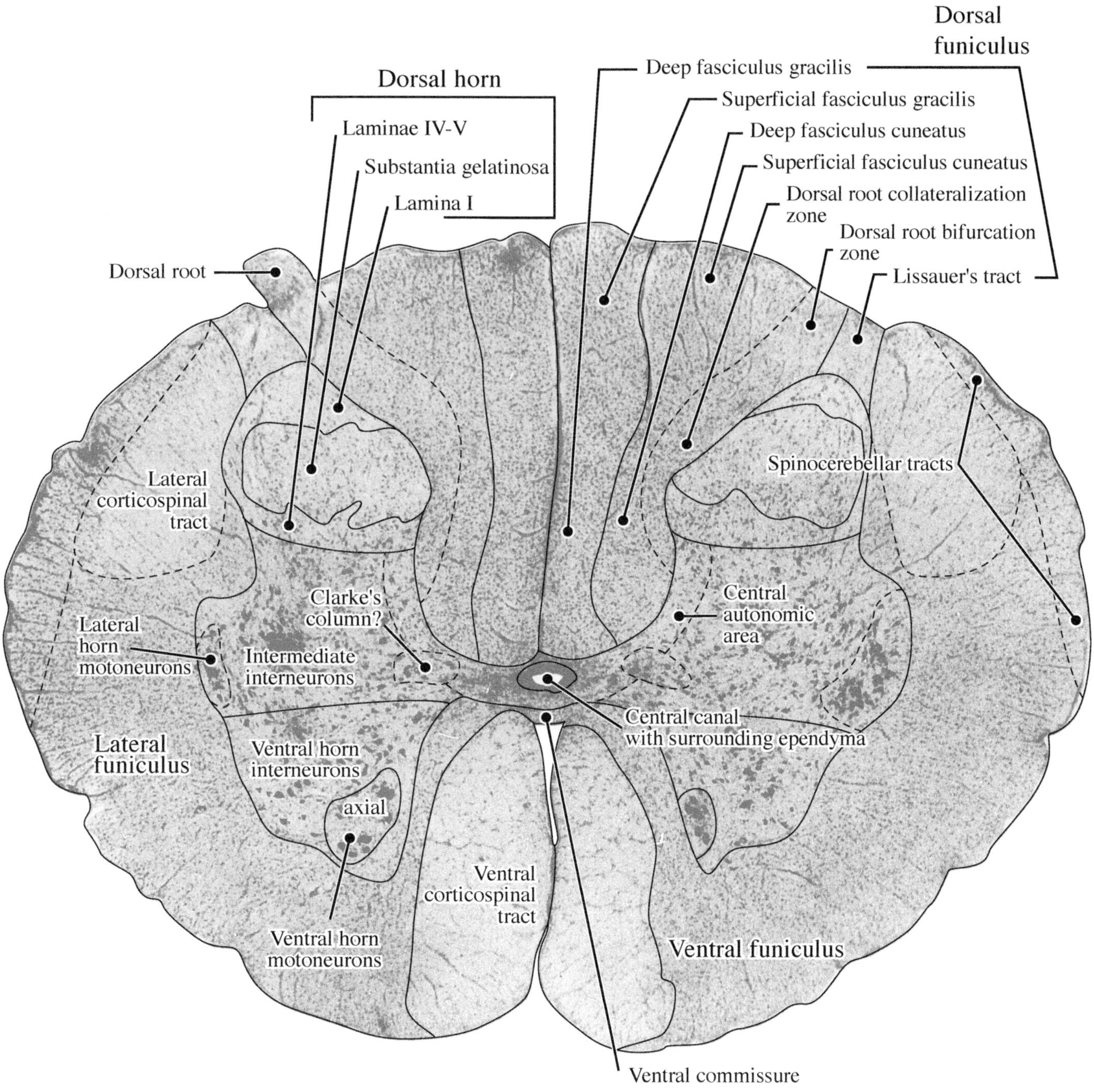
Dorsal horn
Laminae IV-V
Substantia gelatinosa
Lamina I
Deep fasciculus gracilis
Dorsal funiculus
Superficial fasciculus gracilis
Deep fasciculus cuneatus
Superficial fasciculus cuneatus
Dorsal root collateralization zone
Dorsal root bifurcation zone
Lissauer's tract
Dorsal root
Lateral corticospinal tract
Spinocerebellar tracts
Clarke's column?
Central autonomic area
Lateral horn motoneurons
Intermediate interneurons
Central canal with surrounding ependyma
Lateral funiculus
Ventral horn interneurons
axial
Ventral corticospinal tract
Ventral horn motoneurons
Ventral funiculus
Ventral commissure

PLATE 32A

CR 310 mm
GW 37
Y117-61
Middle Thoracic
Myelin stain

Areas (mm^2)

Central canal	.0029
Ependyma	.0069
Gray matter	1.5695
White matter	4.5747

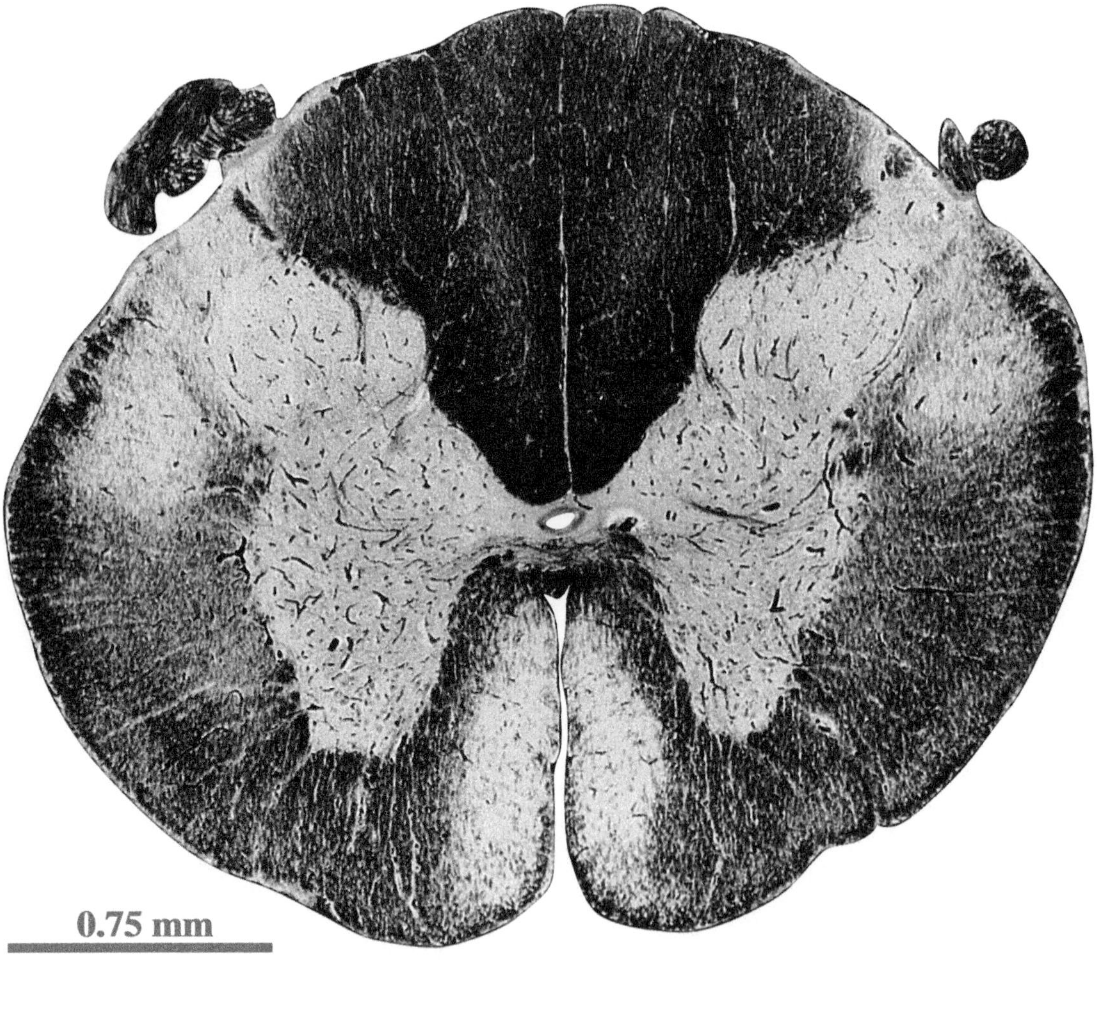

PLATE 32B

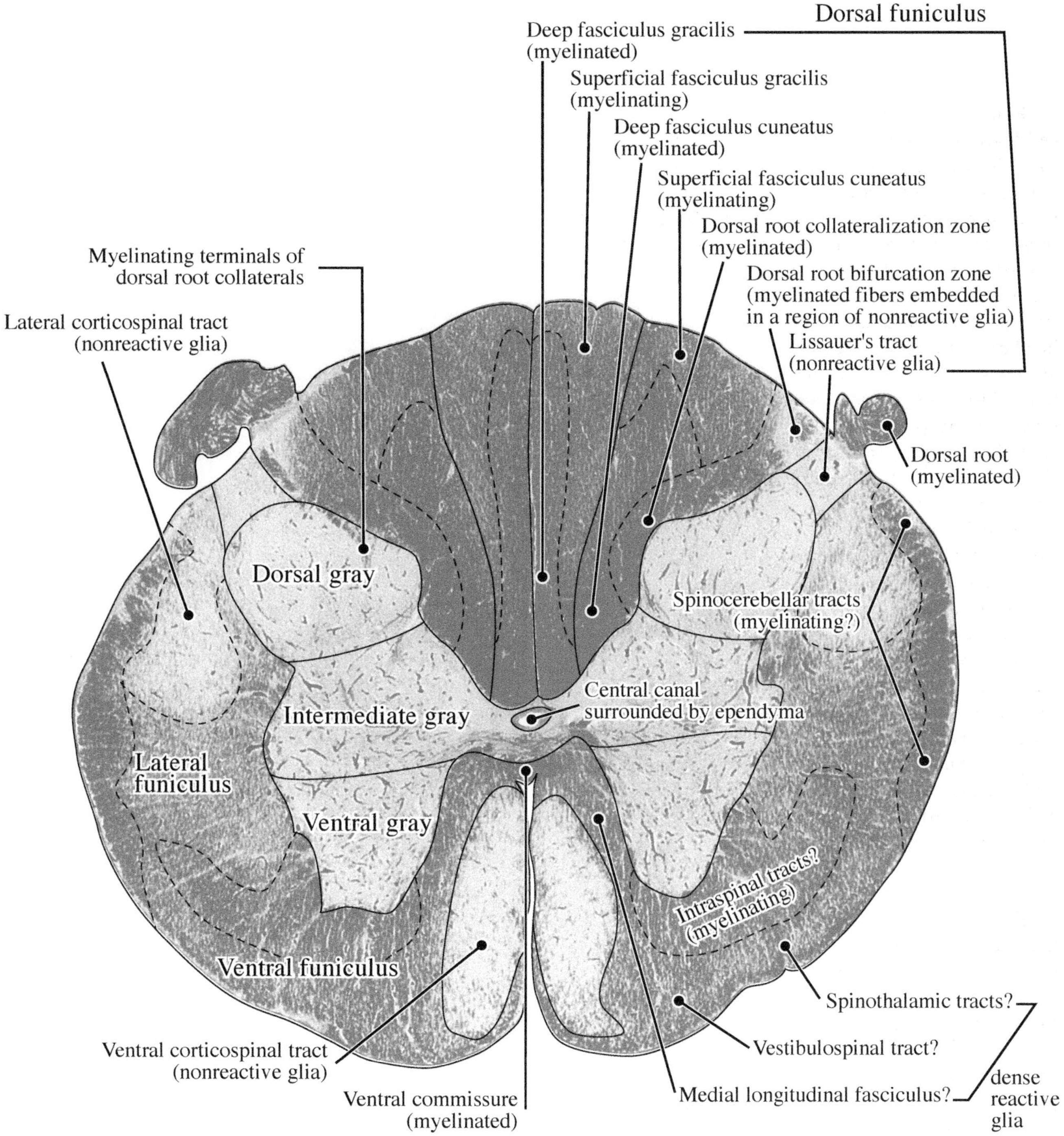
Dorsal funiculus
Deep fasciculus gracilis (myelinated)
Superficial fasciculus gracilis (myelinating)
Deep fasciculus cuneatus (myelinated)
Superficial fasciculus cuneatus (myelinating)
Dorsal root collateralization zone (myelinated)
Dorsal root bifurcation zone (myelinated fibers embedded in a region of nonreactive glia)
Lissauer's tract (nonreactive glia)
Dorsal root (myelinated)
Myelinating terminals of dorsal root collaterals
Lateral corticospinal tract (nonreactive glia)
Dorsal gray
Spinocerebellar tracts (myelinating?)
Central canal surrounded by ependyma
Intermediate gray
Lateral funiculus
Ventral gray
Intraspinal tracts? (myelinating)
Ventral funiculus
Spinothalamic tracts?
Vestibulospinal tract?
Medial longitudinal fasciculus?
dense reactive glia
Ventral corticospinal tract (nonreactive glia)
Ventral commissure (myelinated)

PLATE 33A

CR 310 mm
GW 37
Y117-61
Middle Thoracic
Cell body stain

Areas (mm^2)

Central canal	.0045
Ependyma	.0132
Gray matter	1.6976
White matter	4.9867

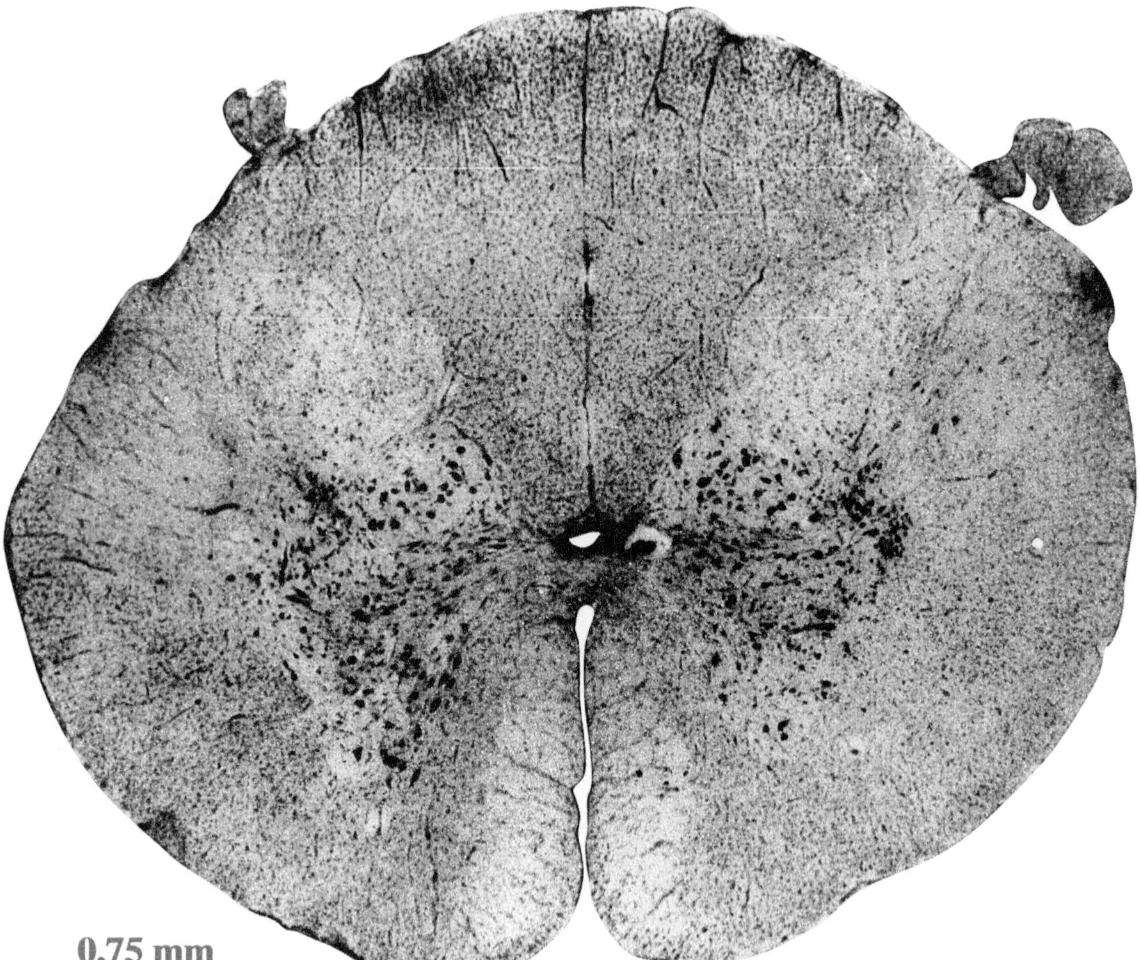

PLATE 33B

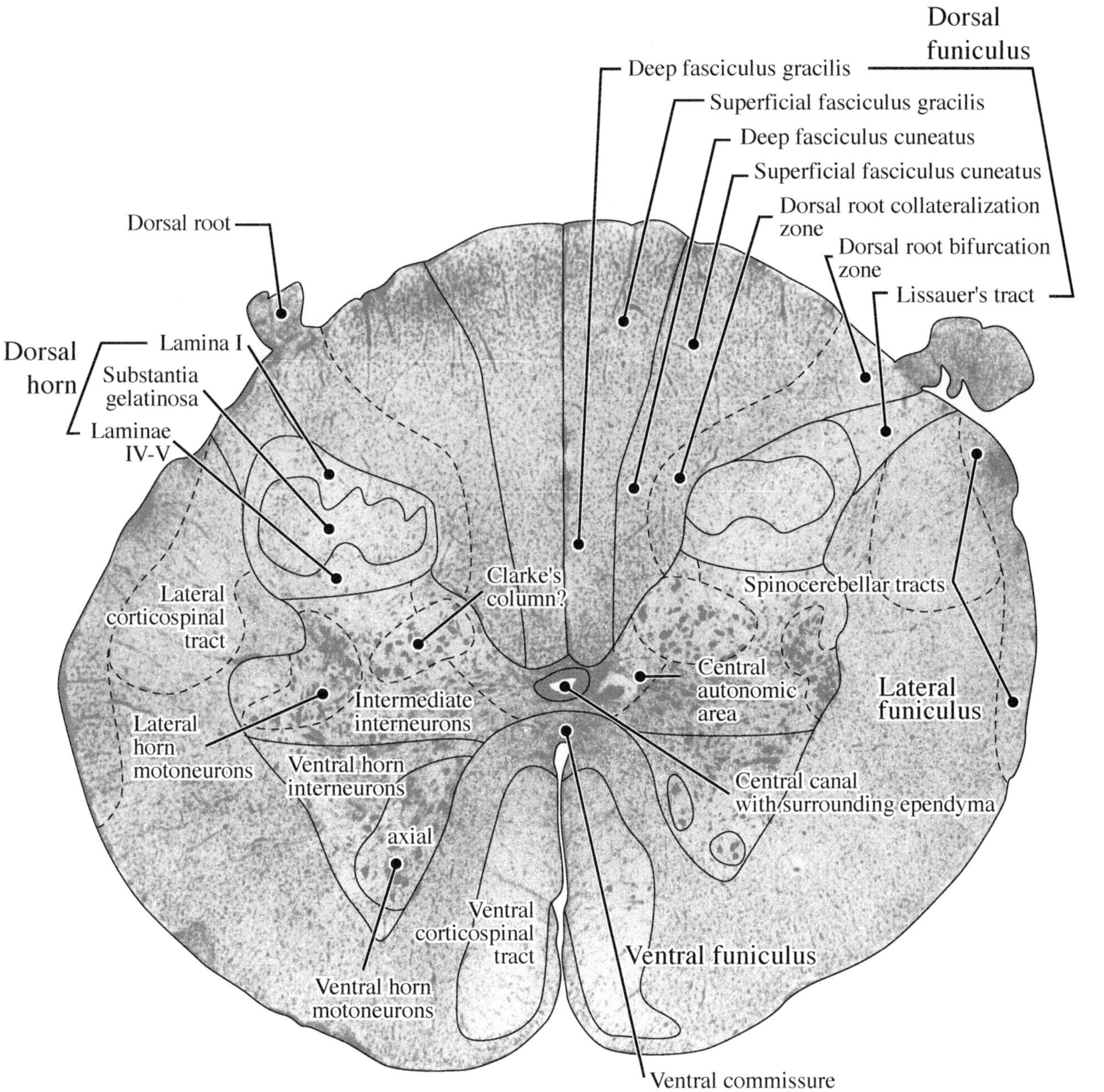
Dorsal funiculus
Deep fasciculus gracilis
Superficial fasciculus gracilis
Deep fasciculus cuneatus
Superficial fasciculus cuneatus
Dorsal root collateralization zone
Dorsal root bifurcation zone
Lissauer's tract
Dorsal root
Dorsal horn
Lamina I
Substantia gelatinosa
Laminae IV-V
Clarke's column?
Spinocerebellar tracts
Lateral corticospinal tract
Central autonomic area
Lateral funiculus
Intermediate interneurons
Lateral horn motoneurons
Ventral horn interneurons
Central canal with surrounding ependyma
axial
Ventral corticospinal tract
Ventral funiculus
Ventral horn motoneurons
Ventral commissure

PLATE 34A

CR 310 mm
GW 37
Y117-61
Lower Thoracic
Myelin stain

Areas (mm²)

Central canal	.0057
Ependyma	.0070
Gray matter	2.3538
White matter	4.8340

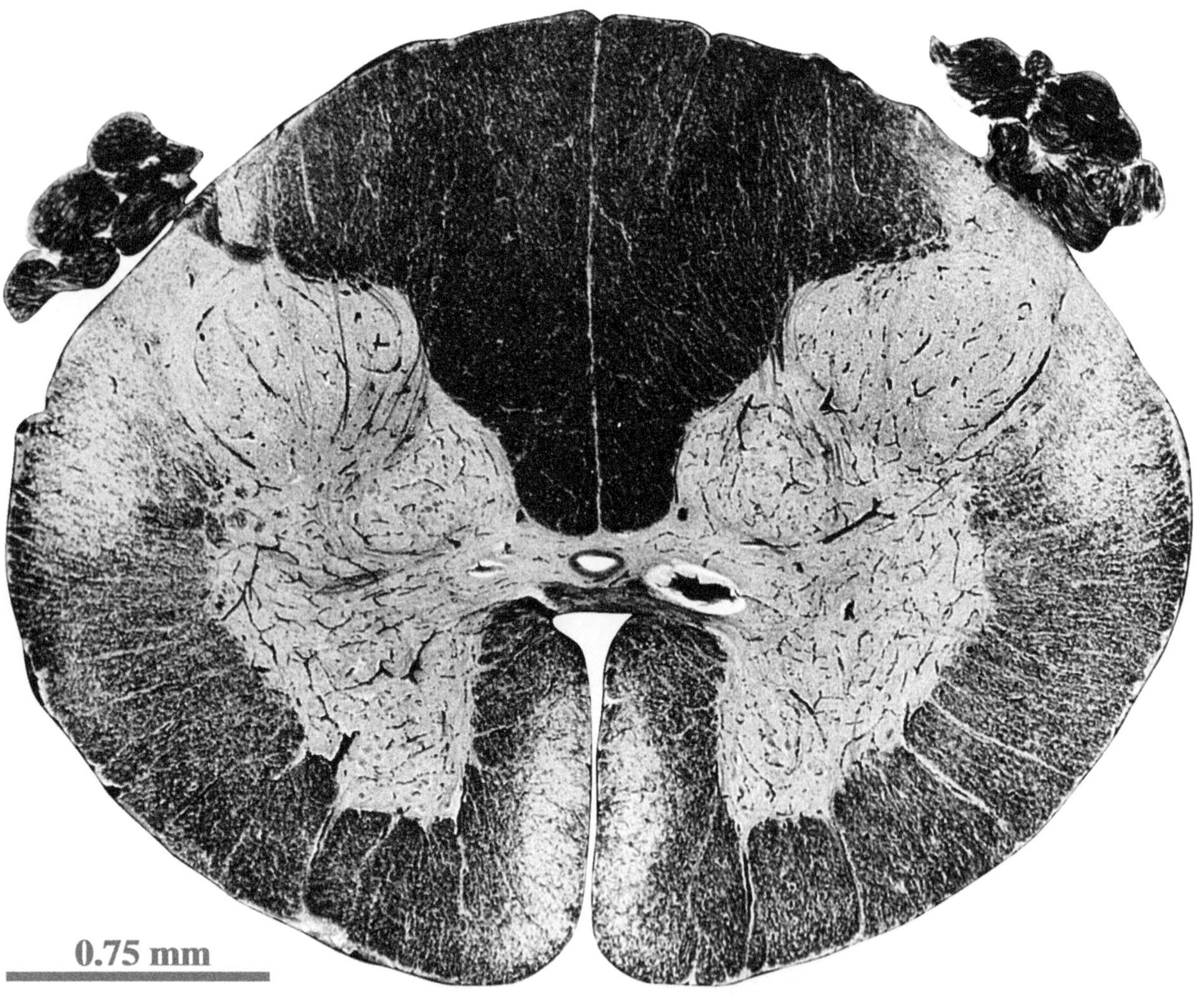

PLATE 34B

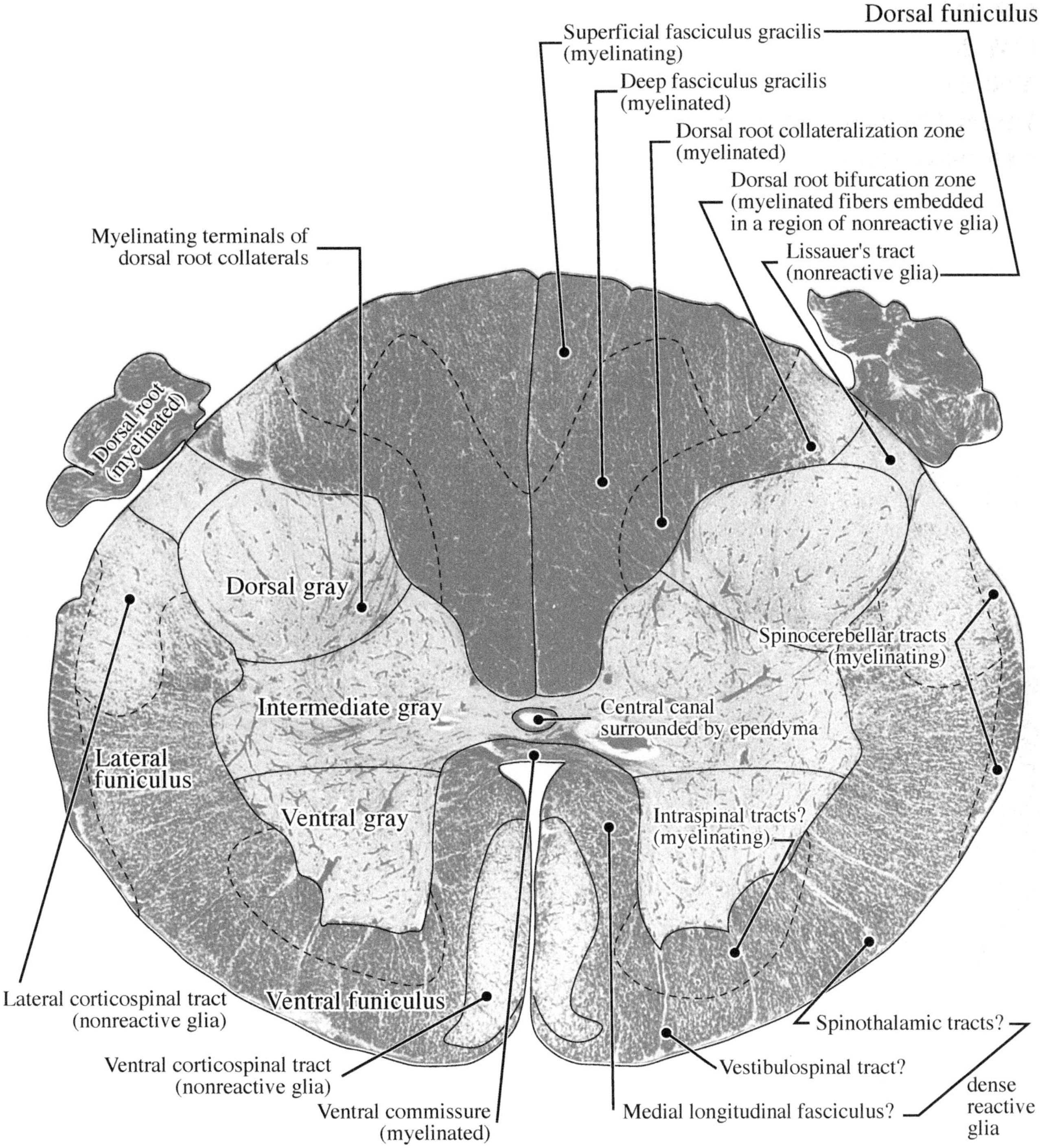

Dorsal funiculus
Superficial fasciculus gracilis (myelinating)
Deep fasciculus gracilis (myelinated)
Dorsal root collateralization zone (myelinated)
Dorsal root bifurcation zone (myelinated fibers embedded in a region of nonreactive glia)
Lissauer's tract (nonreactive glia)
Myelinating terminals of dorsal root collaterals
Dorsal root (myelinated)
Dorsal gray
Intermediate gray
Lateral funiculus
Ventral gray
Central canal surrounded by ependyma
Spinocerebellar tracts (myelinating)
Intraspinal tracts? (myelinating)
Ventral funiculus
Lateral corticospinal tract (nonreactive glia)
Ventral corticospinal tract (nonreactive glia)
Ventral commissure (myelinated)
Spinothalamic tracts?
Vestibulospinal tract?
Medial longitudinal fasciculus?
dense reactive glia

PLATE 35A

CR 310 mm
GW 37
Y117-61
Lower Thoracic
Cell body stain

Areas (mm^2)

Central canal	.0031
Ependyma	.0130
Gray matter	2.5442
White matter	5.3951

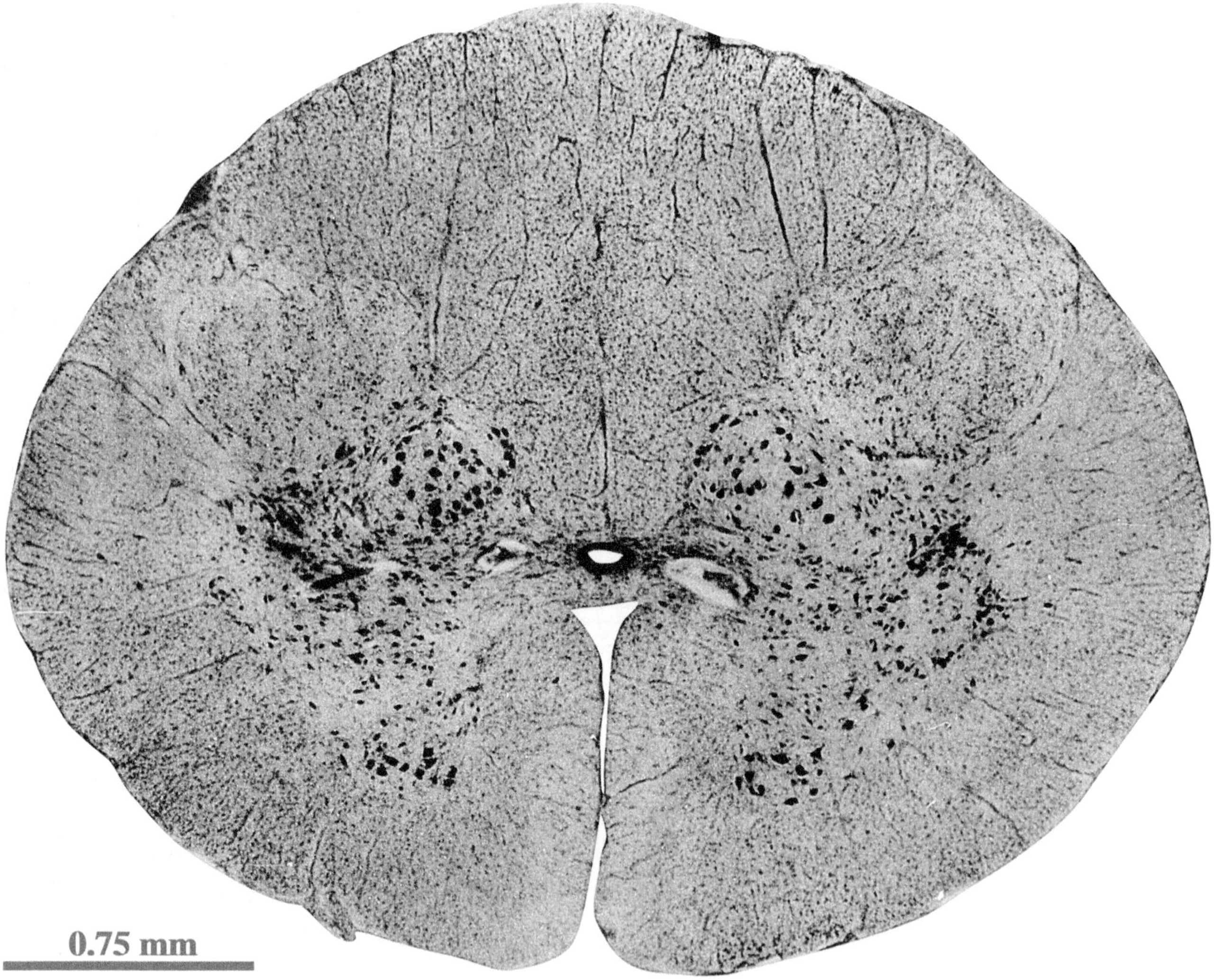

PLATE 35B

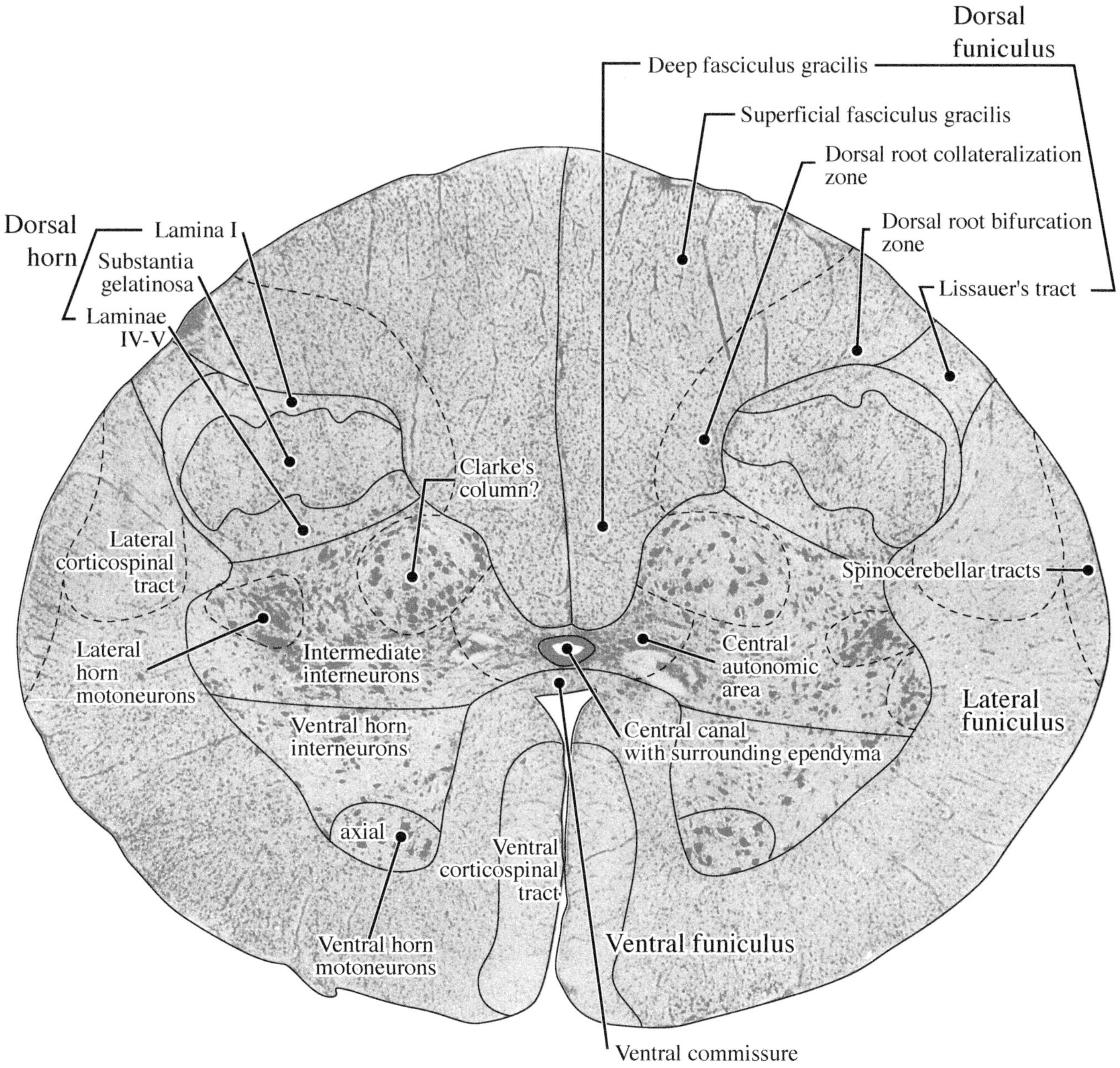
Dorsal funiculus
Deep fasciculus gracilis
Superficial fasciculus gracilis
Dorsal root collateralization zone
Dorsal root bifurcation zone
Lissauer's tract
Dorsal horn
Lamina I
Substantia gelatinosa
Laminae IV-V
Clarke's column?
Lateral corticospinal tract
Spinocerebellar tracts
Lateral horn motoneurons
Intermediate interneurons
Central autonomic area
Lateral funiculus
Ventral horn interneurons
Central canal with surrounding ependyma
axial
Ventral corticospinal tract
Ventral horn motoneurons
Ventral funiculus
Ventral commissure

PLATE 36A

CR 310 mm
GW 37
Y117-61
Lumbar
Myelin stain

Areas (mm²)

Central canal	.0109
Ependyma	.0113
Gray matter	5.4305
White matter	6.0778

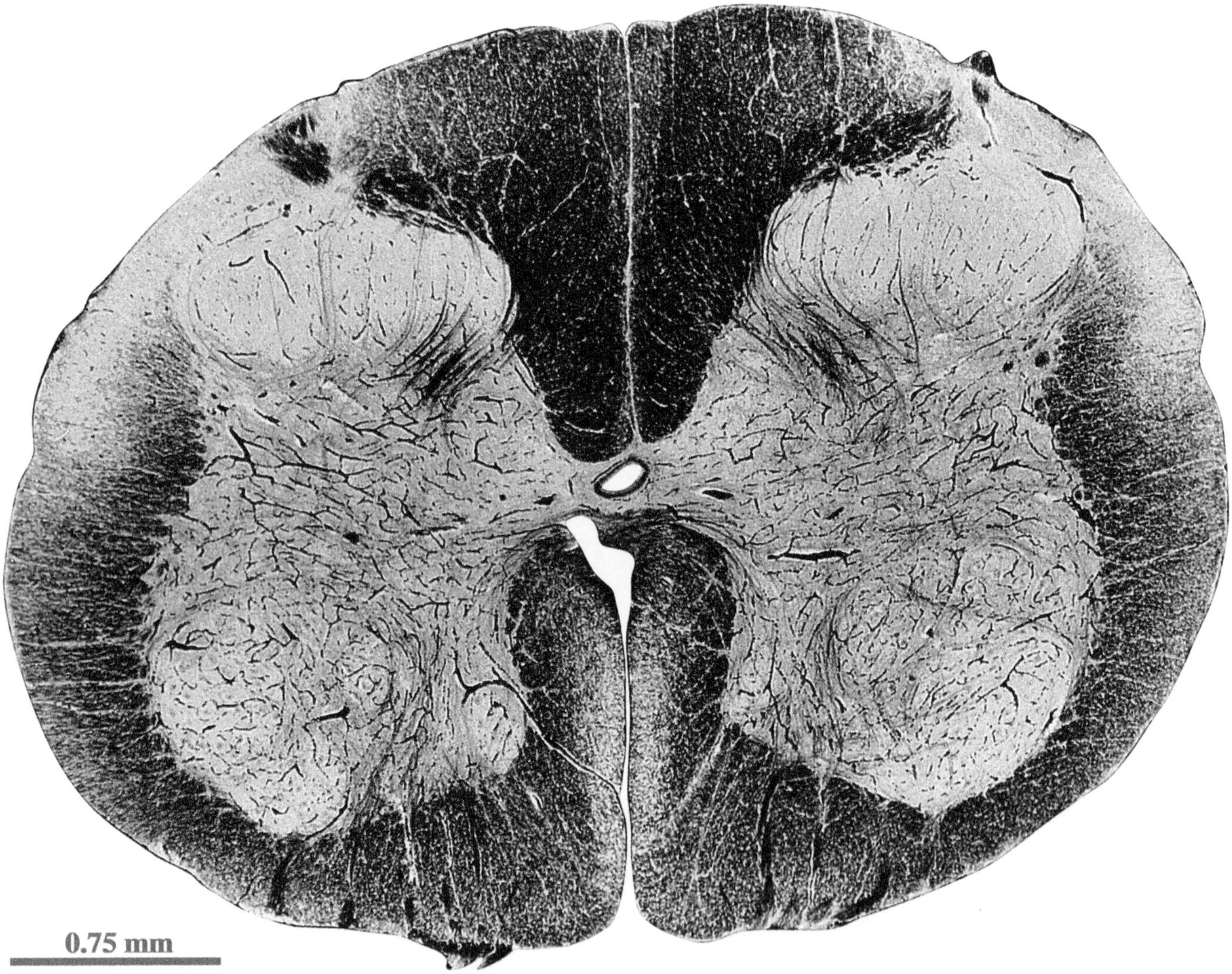

PLATE 36B

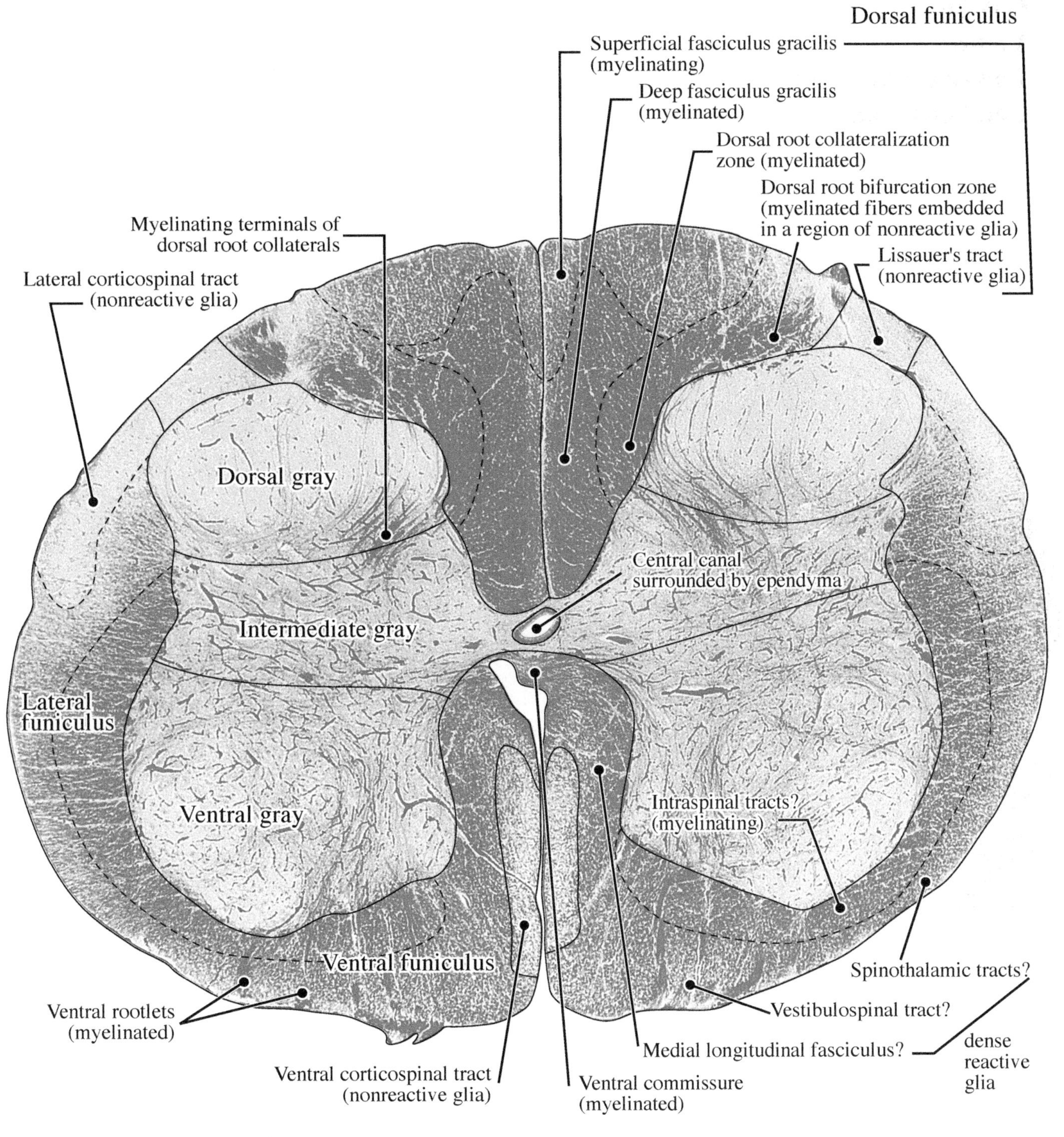
Dorsal funiculus
Superficial fasciculus gracilis (myelinating)
Deep fasciculus gracilis (myelinated)
Dorsal root collateralization zone (myelinated)
Dorsal root bifurcation zone (myelinated fibers embedded in a region of nonreactive glia)
Lissauer's tract (nonreactive glia)
Myelinating terminals of dorsal root collaterals
Lateral corticospinal tract (nonreactive glia)
Dorsal gray
Central canal surrounded by ependyma
Intermediate gray
Lateral funiculus
Ventral gray
Intraspinal tracts? (myelinating)
Ventral funiculus
Spinothalamic tracts?
Ventral rootlets (myelinated)
Vestibulospinal tract?
Medial longitudinal fasciculus?
dense reactive glia
Ventral corticospinal tract (nonreactive glia)
Ventral commissure (myelinated)

PLATE 37A

CR 310 mm
GW 37
Y117-61
Upper Lumbar
Cell body stain

Areas (mm^2)

Central canal	.0087
Ependyma	.0178
Gray matter	5.7810
White matter	6.4894

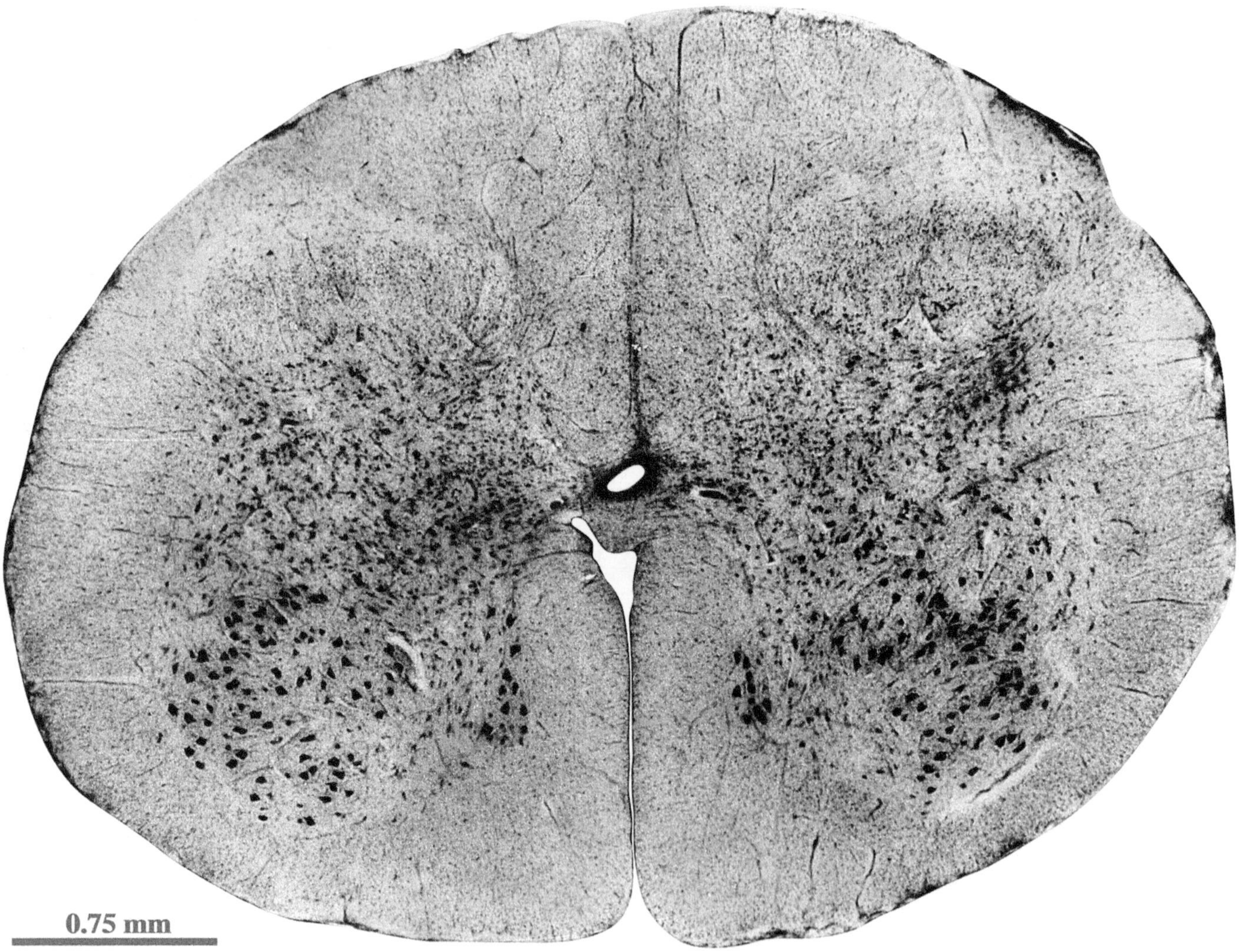

PLATE 37B

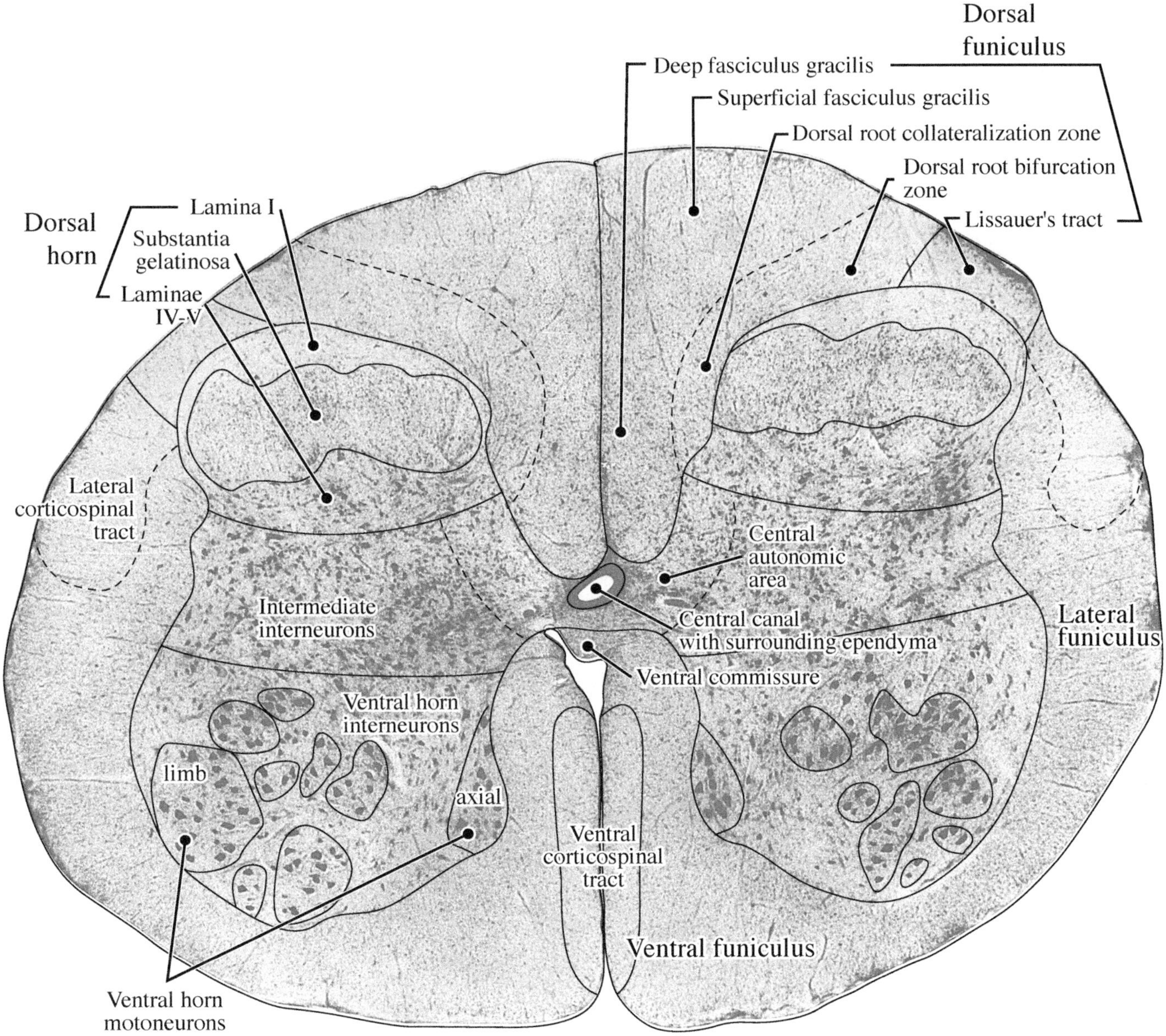
Dorsal funiculus
Deep fasciculus gracilis
Superficial fasciculus gracilis
Dorsal root collateralization zone
Dorsal root bifurcation zone
Lissauer's tract
Dorsal horn
Lamina I
Substantia gelatinosa
Laminae IV-V
Lateral corticospinal tract
Central autonomic area
Intermediate interneurons
Central canal with surrounding ependyma
Lateral funiculus
Ventral commissure
Ventral horn interneurons
limb
axial
Ventral corticospinal tract
Ventral funiculus
Ventral horn motoneurons

PLATE 38A

CR 310 mm
GW 37
Y117-61
Lumbar Enlargement
Myelin stain

Areas (mm^2)

Central canal	.0043
Ependyma	.0087
Gray matter	6.6891
White matter	5.5948

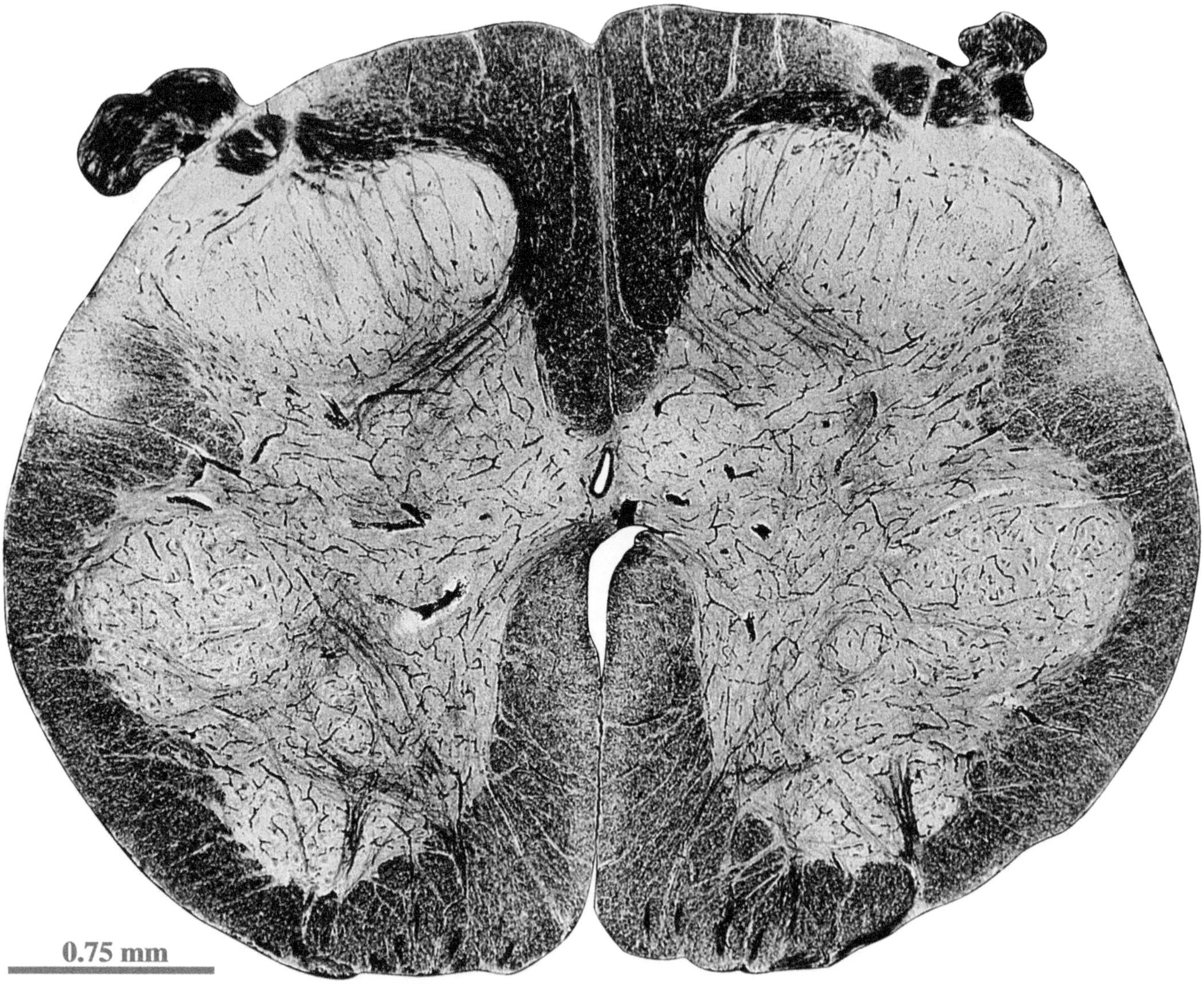

PLATE 38B

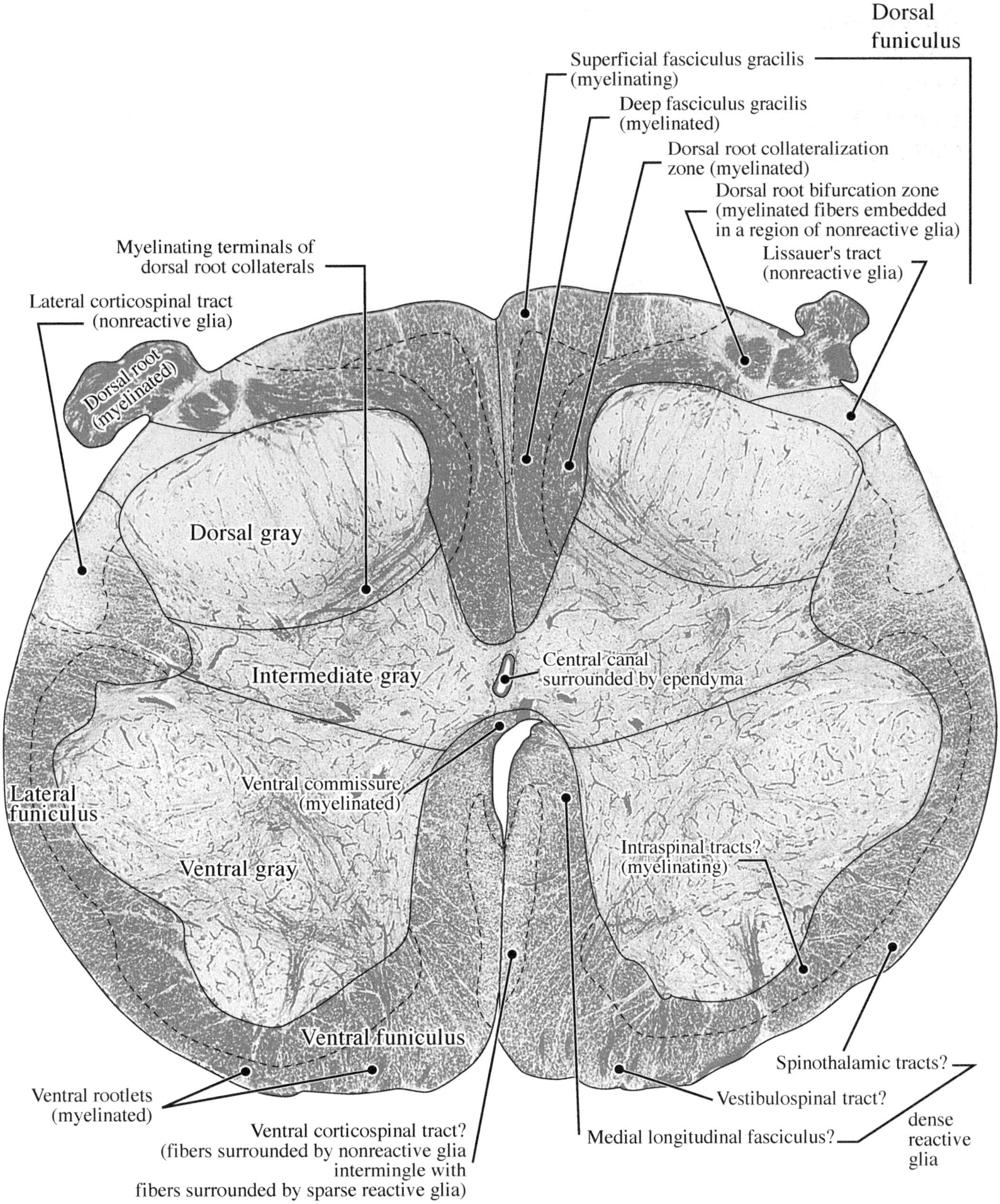
Dorsal funiculus
Superficial fasciculus gracilis (myelinating)
Deep fasciculus gracilis (myelinated)
Dorsal root collateralization zone (myelinated)
Dorsal root bifurcation zone (myelinated fibers embedded in a region of nonreactive glia)
Lissauer's tract (nonreactive glia)
Myelinating terminals of dorsal root collaterals
Lateral corticospinal tract (nonreactive glia)
Dorsal root (myelinated)
Dorsal gray
Central canal surrounded by ependyma
Intermediate gray
Ventral commissure (myelinated)
Lateral funiculus
Intraspinal tracts? (myelinating)
Ventral gray
Ventral funiculus
Spinothalamic tracts?
Vestibulospinal tract?
Ventral rootlets (myelinated)
Ventral corticospinal tract? (fibers surrounded by nonreactive glia intermingle with fibers surrounded by sparse reactive glia)
Medial longitudinal fasciculus?
dense reactive glia

PLATE 39A

CR 310 mm
GW 37
Y117-61
Lumbar Enlargement
Cell body stain

Areas (mm²)

Central canal	.0036
Ependyma	.0140
Gray matter	6.7877
White matter	6.0663

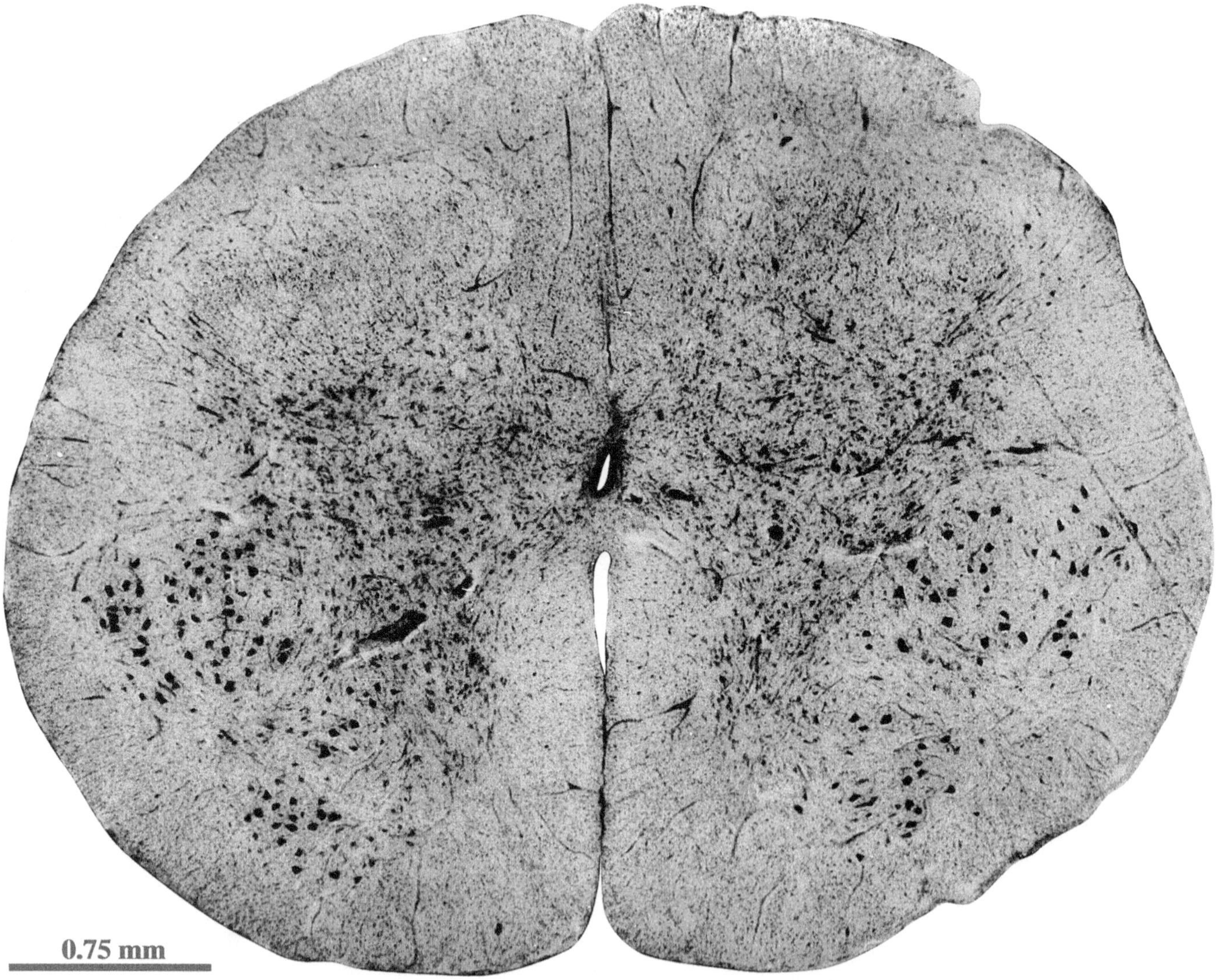

PLATE 39B

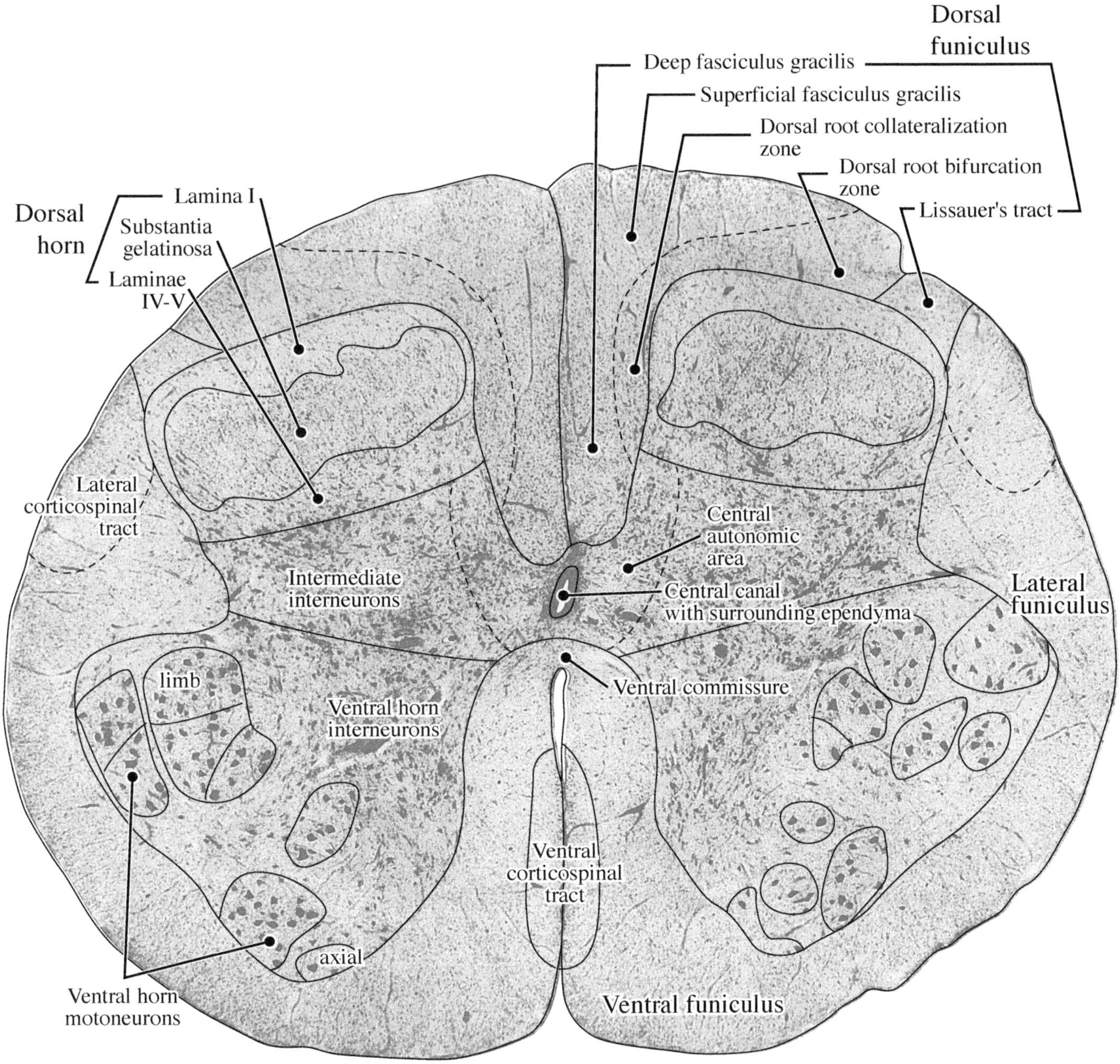
Dorsal funiculus
Deep fasciculus gracilis
Superficial fasciculus gracilis
Dorsal root collateralization zone
Dorsal root bifurcation zone
Lissauer's tract
Dorsal horn
Lamina I
Substantia gelatinosa
Laminae IV-V
Lateral corticospinal tract
Intermediate interneurons
Central autonomic area
Central canal with surrounding ependyma
Lateral funiculus
Ventral commissure
limb
Ventral horn interneurons
Ventral corticospinal tract
axial
Ventral horn motoneurons
Ventral funiculus

PLATE 40A

CR 310 mm
GW 37
Y117-61
Sacral
Myelin stain

Areas (mm^2)

Central canal	.0050
Ependyma	.0160
Gray matter	4.3017
White matter	2.1841

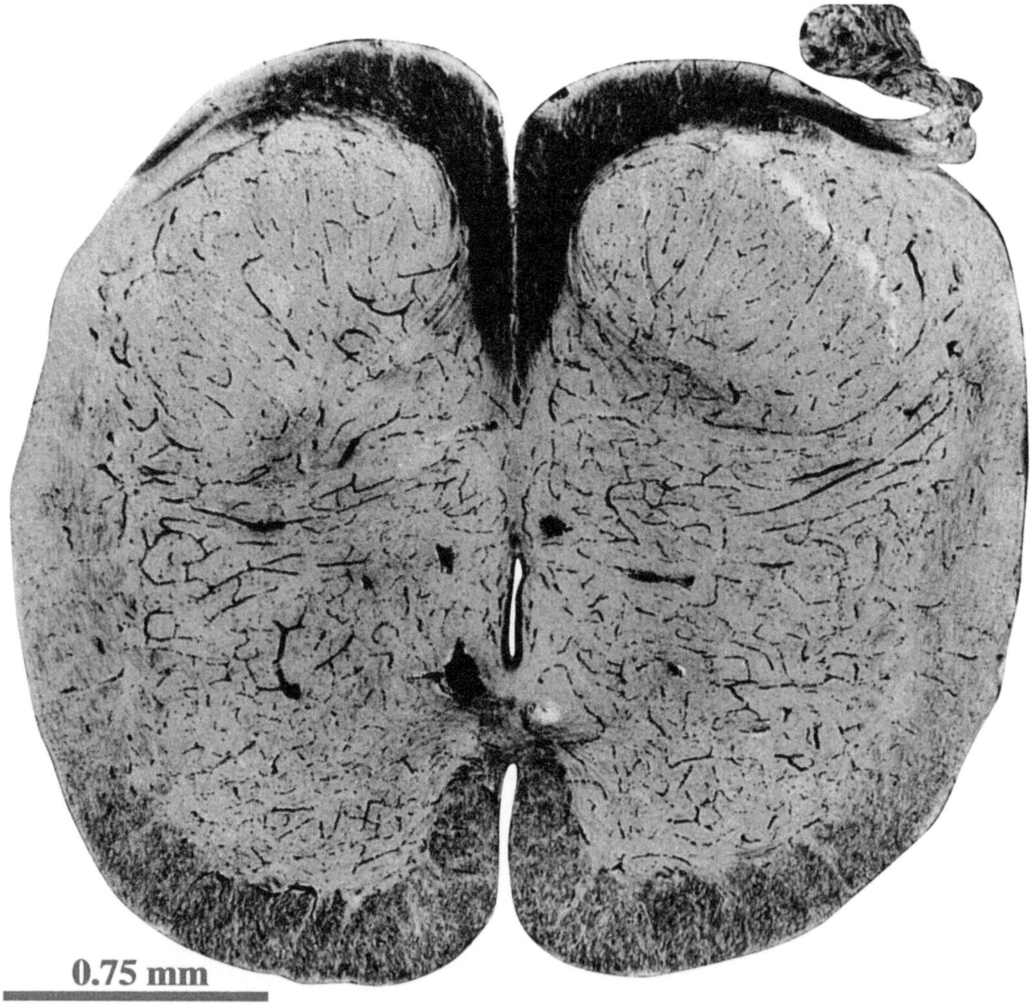

PLATE 40B

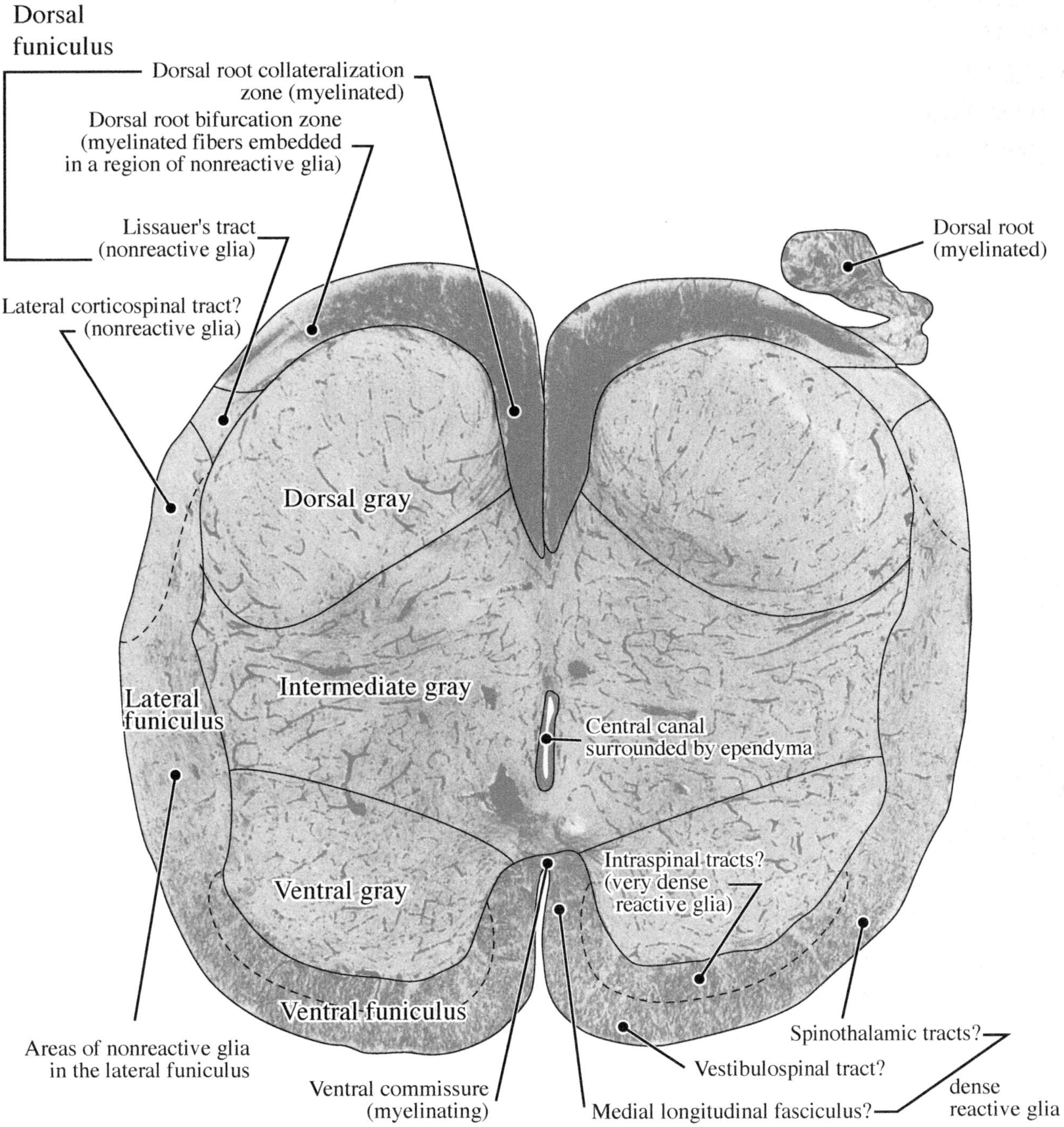
Dorsal
funiculus
Dorsal root collateralization
zone (myelinated)
Dorsal root bifurcation zone
(myelinated fibers embedded
in a region of nonreactive glia)
Lissauer's tract
(nonreactive glia)
Lateral corticospinal tract?
(nonreactive glia)
Dorsal root
(myelinated)
Dorsal gray
Intermediate gray
Lateral
funiculus
Central canal
surrounded by ependyma
Intraspinal tracts?
(very dense
reactive glia)
Ventral gray
Ventral funiculus
Areas of nonreactive glia
in the lateral funiculus
Ventral commissure
(myelinating)
Medial longitudinal fasciculus?
Vestibulospinal tract?
Spinothalamic tracts?
dense
reactive glia

PLATE 41A

CR 310 mm
GW 37
Y117-61
Sacral
Cell body stain

Areas (mm²)

Central canal	.0044
Ependyma	.0271
Gray matter	4.4850
White matter	2.2665

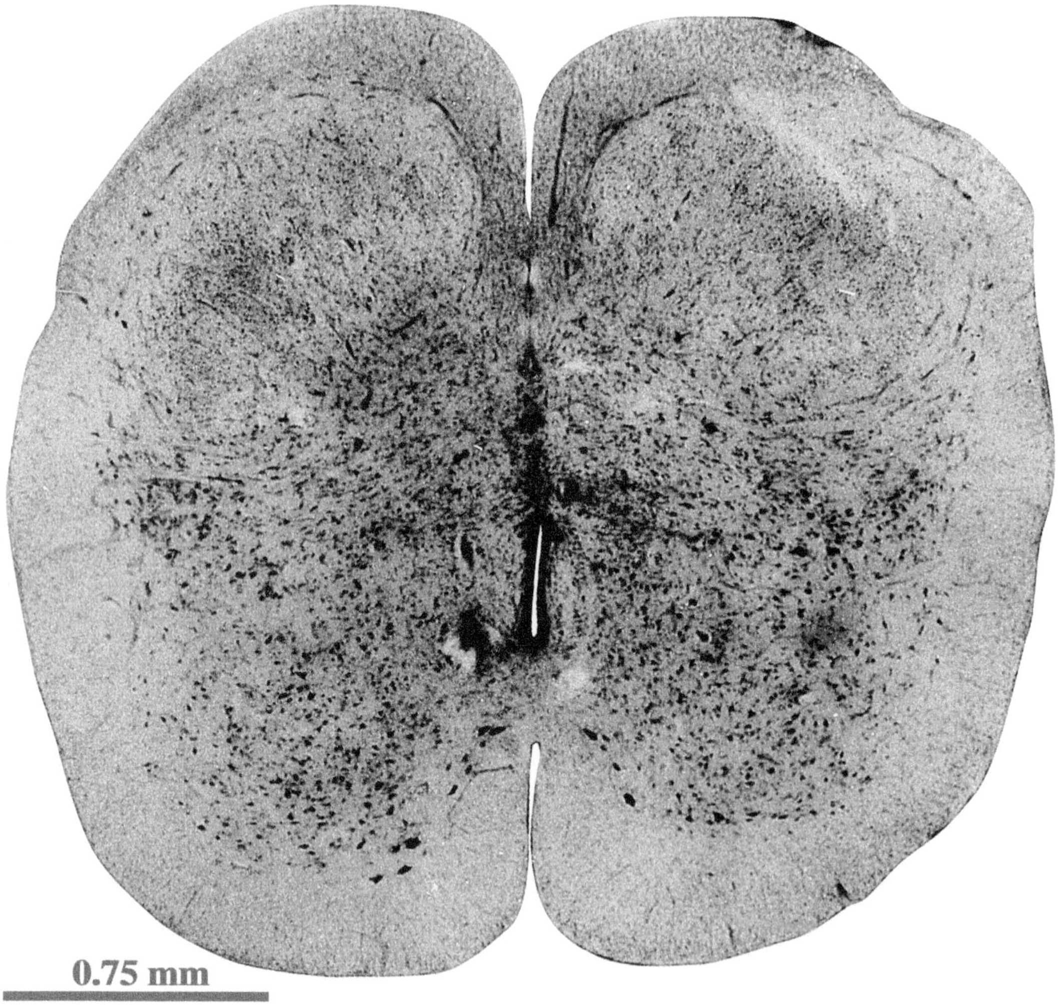

PLATE 41B

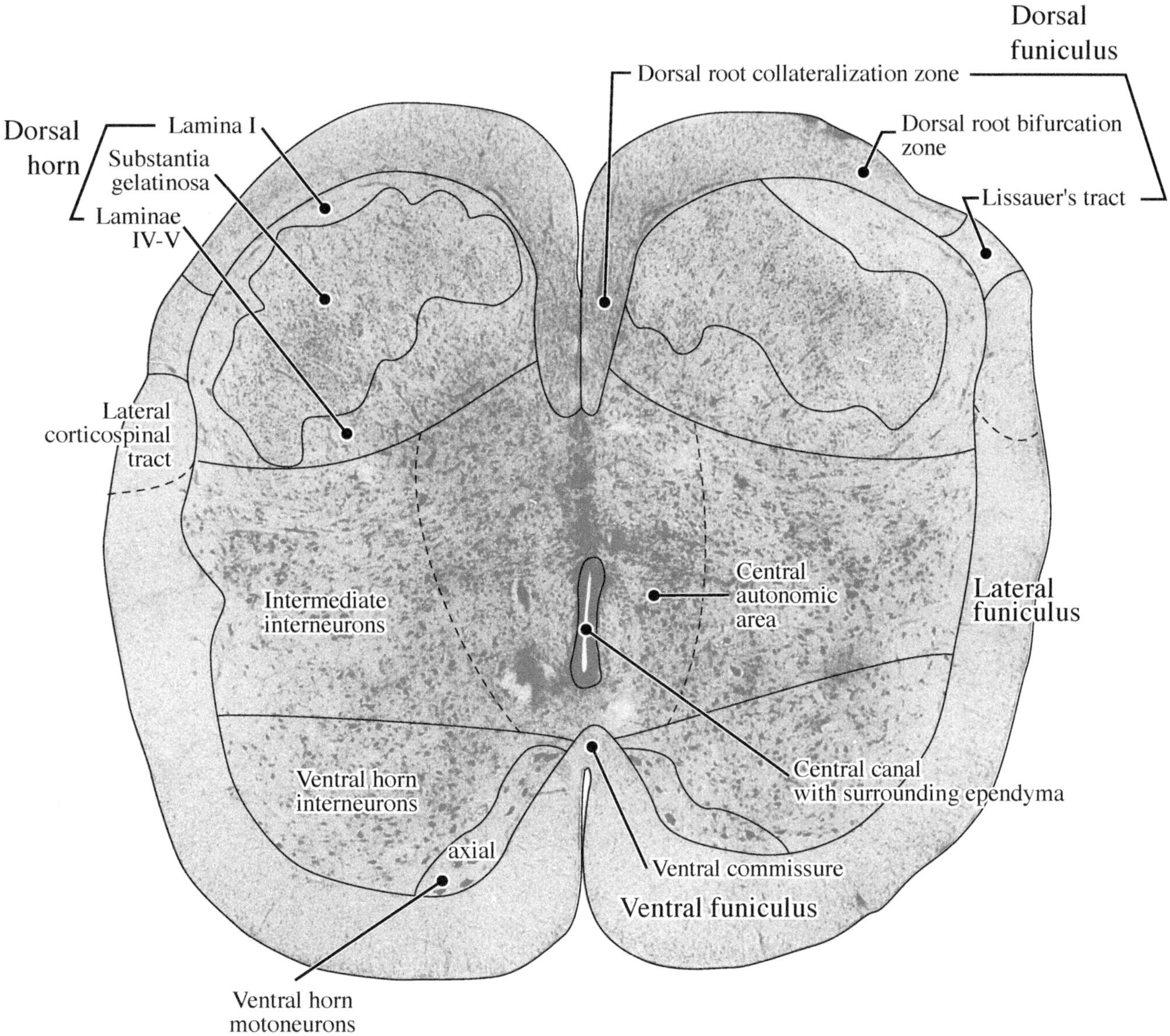
Dorsal funiculus
Dorsal root collateralization zone
Dorsal horn
Lamina I
Substantia gelatinosa
Laminae IV-V
Dorsal root bifurcation zone
Lissauer's tract
Lateral corticospinal tract
Central autonomic area
Intermediate interneurons
Lateral funiculus
Central canal with surrounding ependyma
Ventral horn interneurons
axial
Ventral commissure
Ventral funiculus
Ventral horn motoneurons

PART VI: Y299-62 CR 350 mm (4 Days)

Plate 42 is a survey of matched myelin-stained and cell-body-stained sections from Y299-62, a specimen in the Yakovlev Collection. All sections are shown at the same scale. The boxes enclosing each section list the approximate level and the total area of the section in square millimeters (mm^2). Full-page normal-contrast photographs of each section are in **Plates 43A–54A.** Low-contrast photographs with superimposed labels and outlines of structural details are in **Plates 43B–54B.** Twenty-two myelin-stained sections and 22 cell-body-stained sections were photographed ranging from the cervical enlargement to sacral/coccygeal levels. In this specimen, the myelin-stained and cell-body-stained sections were preserved on separate large glass plates without any section numbers. The 44 photographic prints were intuitively arranged in order from upper cervical to sacral/coccygeal levels, using internal features such as the size of the corticospinal tracts, and the width of the ventral horn. Then, myelin- and cell-body-stained sections were matched and 6 different levels were analyzed. Since many sections in the middle thoracic region are damaged, that level is not illustrated.

As in the previous specimens, the cross-sectional area of a myelin-stained section is smaller than the matching cell-body-stained section. The myelin staining procedure consistently produces greater tissue shrinkage than the cell-body staining procedure. Using the total areas of the myelin-stained sections, the overall size differences between levels indicate the following comparisons. The cervical enlargement has the largest cross-sectional area, 65% larger than the lumbar enlargement. The sacral/coccygeal level has the smallest cross-sectional area, 37% smaller than the lower thoracic level.

The structural labels in this specimen delineate the degree of myelination using the density of staining as a guide. The most densely stained areas are considered myelinated; the next most densely stained areas contain reactive glia ranging from dense to sparse. It is interesting to note that the white matter in the cell-body-stained sections show a uniform density of glia, and various fiber tracts are difficult to delineate. But again, the lateral and ventral corticospinal tracts generally stand out as having slightly lower concentrations of glia. The corticospinal tracts are the last to myelinate because their axons arrive late in the spinal cord, and they originate from younger neurons in the primary motor cortex. Other fiber tracts are either of local origin in the spinal cord or come from early-generated areas in the brainstem. With the exception of the corticospinal tracts, the spinocerebellar tracts, and components in the dorsal funiculus, other fiber tracts are only tentatively identified in the cell-body-stained sections.

Myelination in this specimen is advanceing because there is no longer a gradient between superficial and deep parts of the gracile and cuneate fasciculi, and more myelinated axons are in the subgelatinosal plexus (*Columns 2 and 3*, **Table 5**). Myelination follows a medial (early) to lateral (late) gradient of reactive and proliferating glia in the lateral corticospinal tract. In the rest of the ventral and lateral funiculi, deep fibers appear to myelinate ahead of more superficial fibers. That gradient cuts across the fiber tracts in these funiculi.

Table 5: Glia types and concentration in the white matter in a 4-day infant

Name	Myelination	Reactive glia	Proliferating glia
DORSAL ROOT	myelinated	---	**sparse
VENTRAL ROOT	myelinated	---	---
DORSAL FUNICULUS:			
dorsal root bif. zone	*many fibers	---	sparse
dorsal root col. zone	myelinated	---	sparse
fas. gracilis	myelinated	---	sparse
fas. cuneatus	myelinated	---	sparse
Lissauer's tract	---	none	sparse
LATERAL and VENTRAL FUNICULI:			
lat. corticospinal tract	---	very sparse	sparse
ven. corticospinal tract	---	very sparse	sparse
spinocerebellar tracts	myelinated	---	sparse
ven. commissure	myelinated	---	dense
intraspinal tracts	many fibers	---	dense
lat. reticulospinal tract	some fibers	very sparse	sparse
spinocephalic tracts	†some fibers	•very sparse	dense
med. long. fasciculus	some fibers	---	sparse
vestibulospinal tract	some fibers	---	sparse

** dense band in a sparse field (part of the boundary cap?)
* intermingled in a bed of nonreactive glia
† few fibers at the cervical level
• peripheral fibers at the cervical level

PLATE 42

CR 350 mm, INFANT, 4 DAYS, Y299-62

MYELIN STAIN	CELL BODY STAIN
Plates 43A, 43B **Cervical Enlargement** Total area: 20.063 mm^2 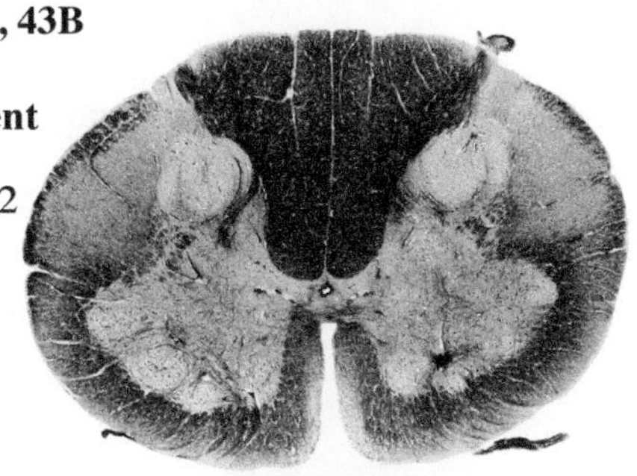	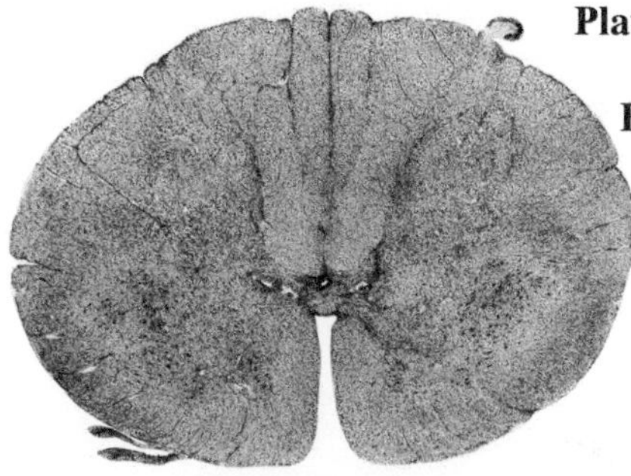 **Plates 44A, 44B** **Cervical Enlargement** Total area: 21.370 mm^2
Plates 45A, 45B **Upper Thoracic** Total area: 10.177 mm^2 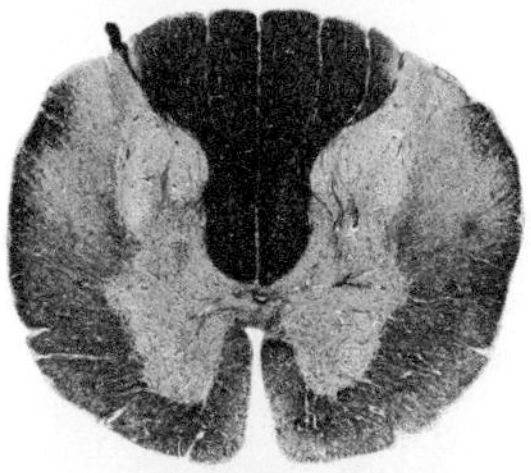	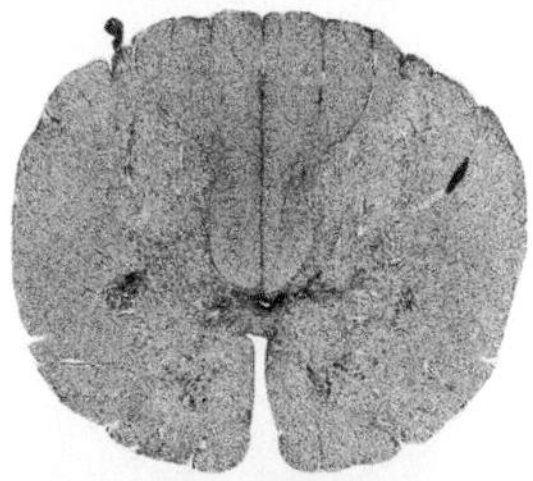 **Plates 46A, 46B** **Upper Thoracic** Total area: 10.514 mm^2
Plates 47A, 47B **Lower Thoracic** Total area: 7.7961 mm^2 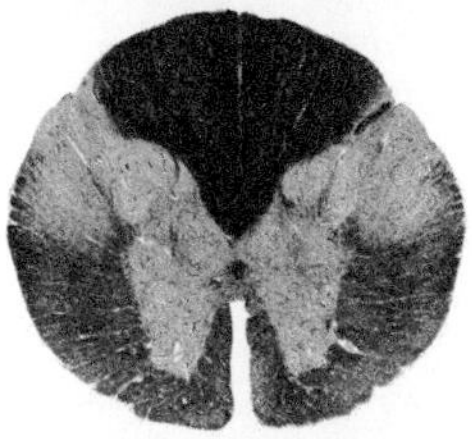	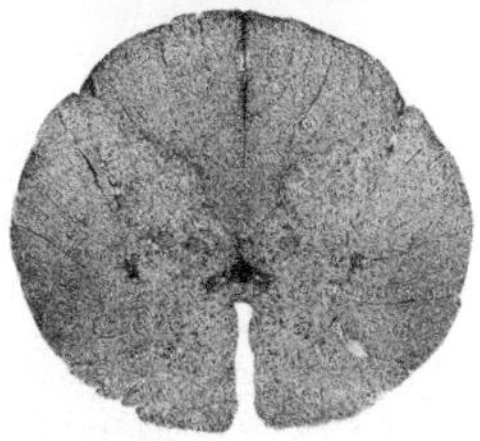 **Plates 48A, 48B** **Lower Thoracic** Total area: 8.2168 mm^2
Plates 49A, 49B **Upper Lumbar** Total area: 11.735 mm^2 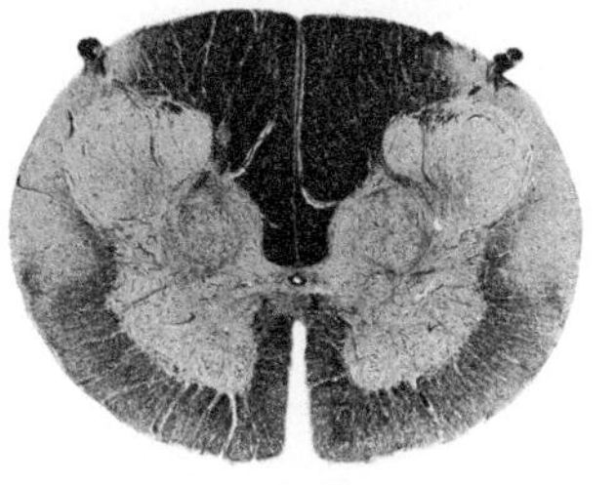	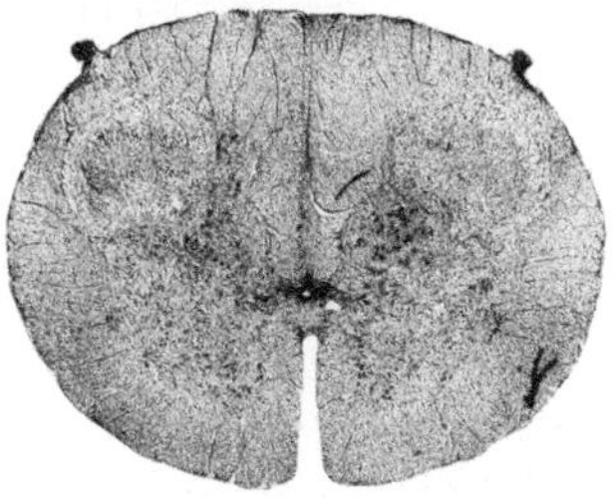 **Plates 50A, 50B** **Upper Lumbar** Total area: 12.730 mm^2
Plates 51A, 51 **Lumbar Enlargement** Total area: 12.152 mm^2 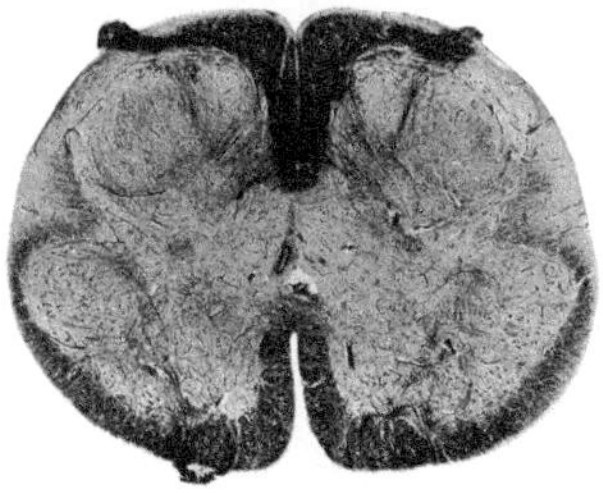	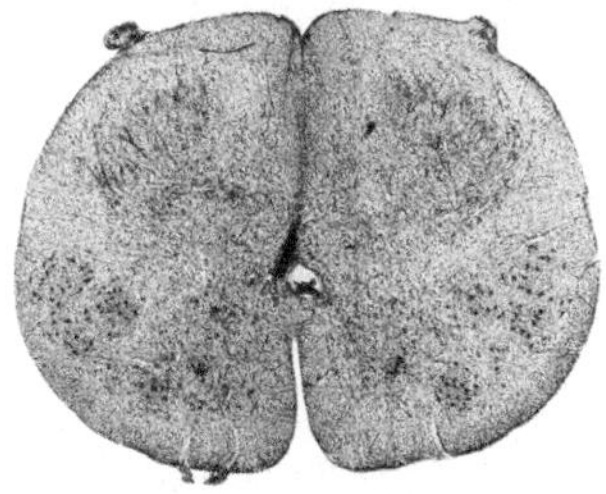 **Plates 52A, 52B** **Lumbar Enlargement** Total area: 12.162 mm^2
Plates 53A, 53 **Sacral/Coccygeal** Total area: 5.6792 mm^2 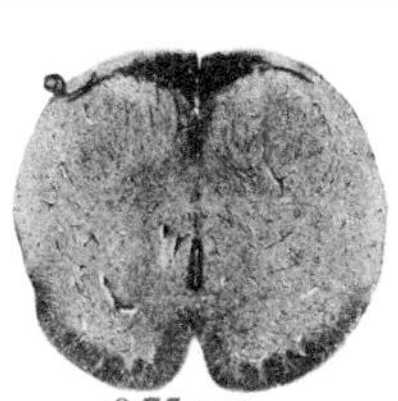 0.75 mm	 **Plates 54A, 54B** **Sacral/Coccygeal** Total area: 5.9052 mm^2 0.75 mm

PLATE 43A

CR 350 mm
Infant, 4 Days
Y299-62
Cervical Enlargement
Myelin stain

Areas (mm^2)

Central canal	.0029
Ependyma	.0119
Gray matter	6.9875
White matter	13.0610

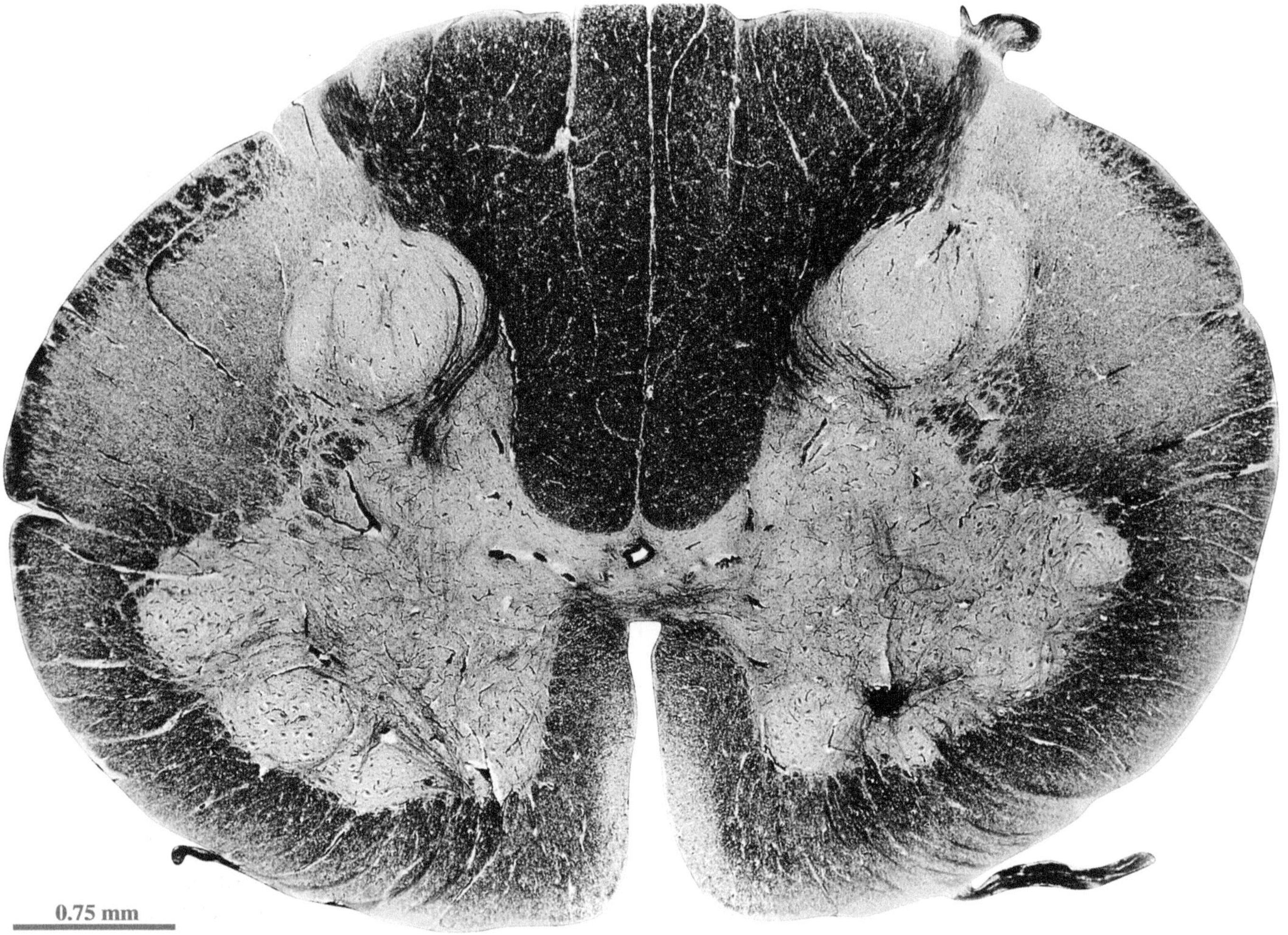

PLATE 43B

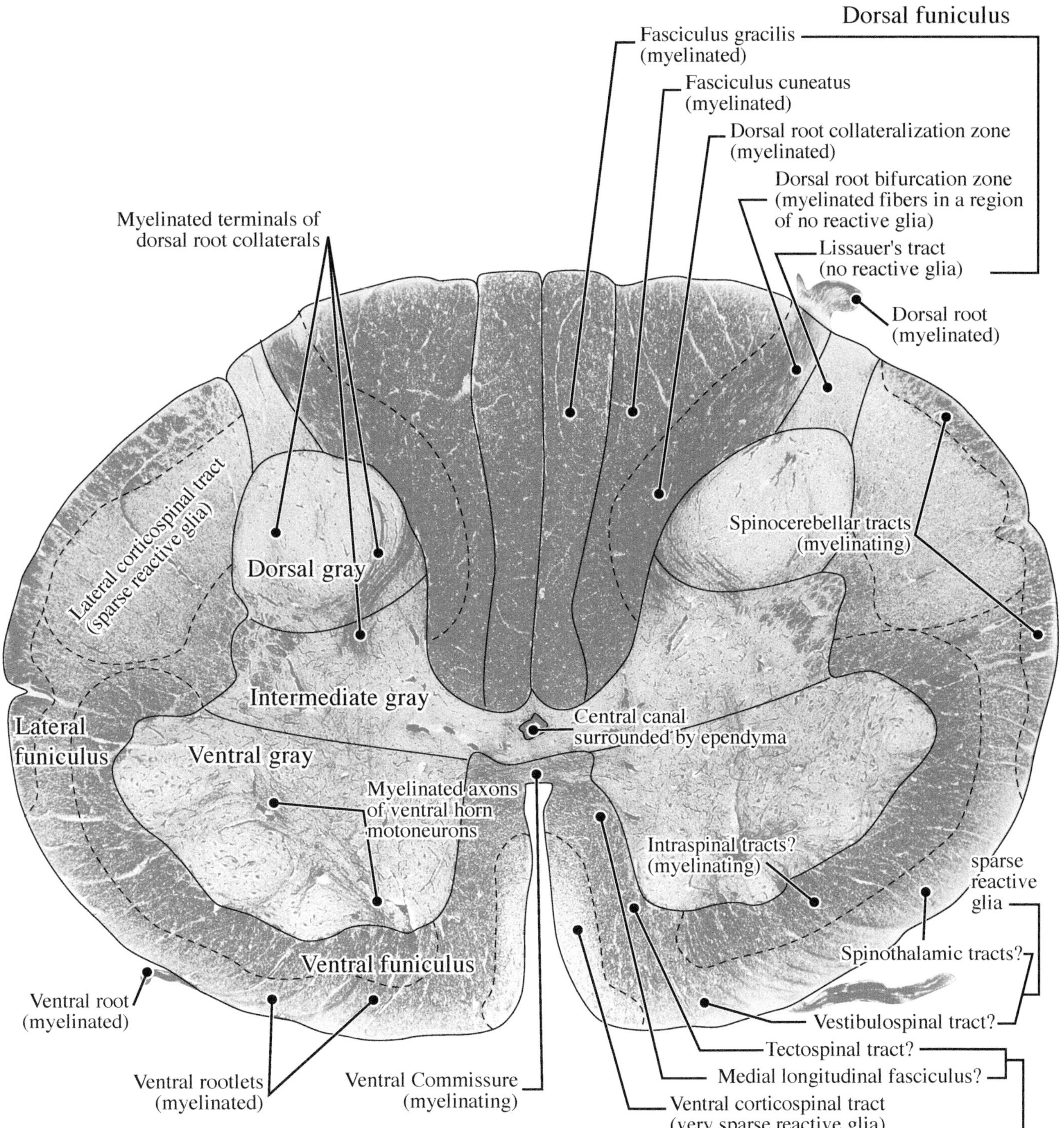
Dorsal funiculus
Fasciculus gracilis (myelinated)
Fasciculus cuneatus (myelinated)
Dorsal root collateralization zone (myelinated)
Dorsal root bifurcation zone (myelinated fibers in a region of no reactive glia)
Lissauer's tract (no reactive glia)
Dorsal root (myelinated)
Myelinated terminals of dorsal root collaterals
Lateral corticospinal tract (sparse reactive glia)
Dorsal gray
Spinocerebellar tracts (myelinating)
Intermediate gray
Central canal surrounded by ependyma
Lateral funiculus
Ventral gray
Myelinated axons of ventral horn motoneurons
Intraspinal tracts? (myelinating)
sparse reactive glia
Spinothalamic tracts?
Ventral funiculus
Ventral root (myelinated)
Vestibulospinal tract?
Tectospinal tract?
Medial longitudinal fasciculus?
Ventral rootlets (myelinated)
Ventral Commissure (myelinating)
Ventral corticospinal tract (very sparse reactive glia)
dense reactive glia

PLATE 44A

CR 350 mm
Infant, 4 Days
Y299-62
Cervical Enlargement
Cell body stain

Areas (mm²)

Central canal	.0019
Ependyma	.0167
Gray matter	7.8633
White matter	13.4880

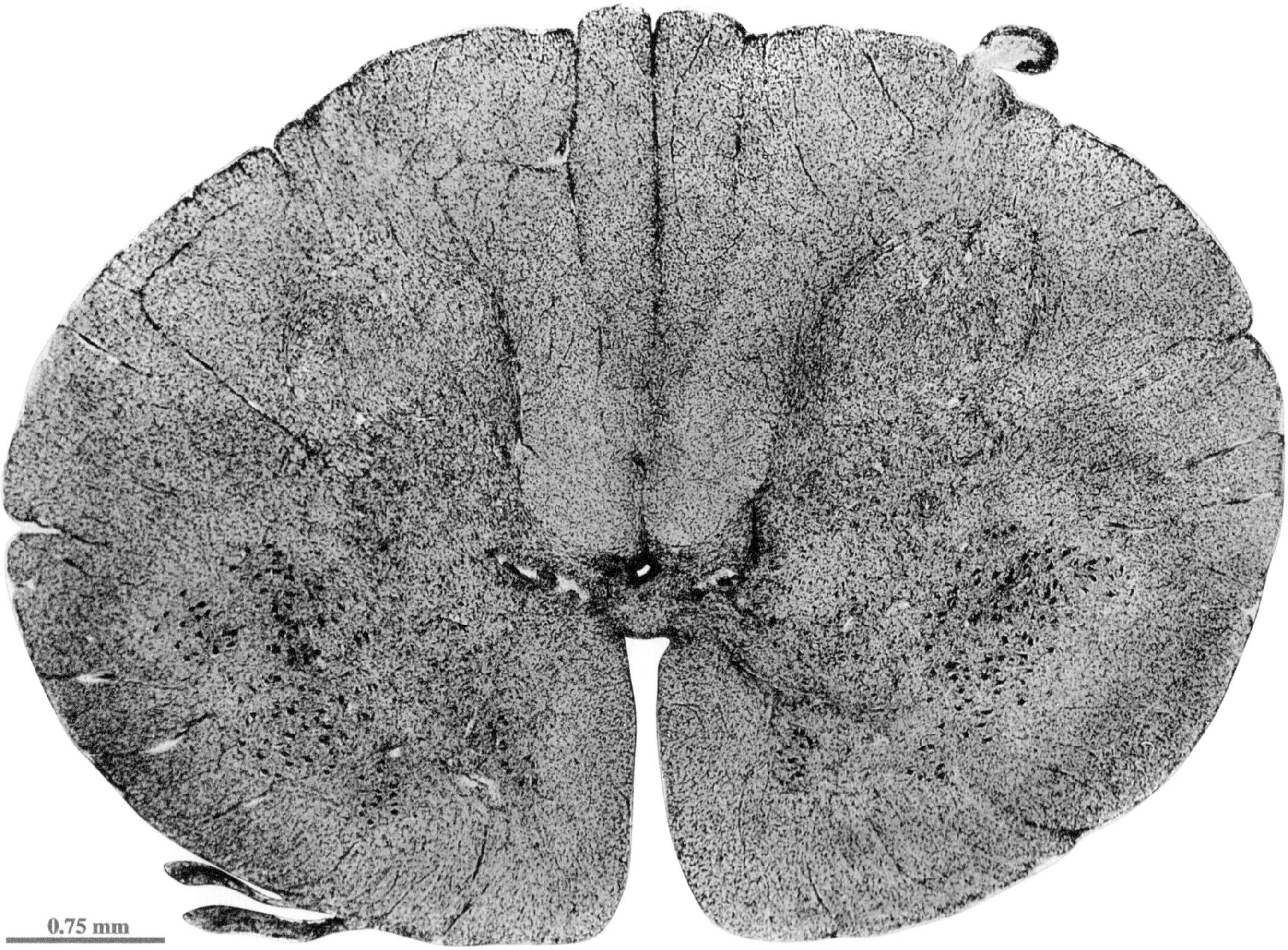

PLATE 44B

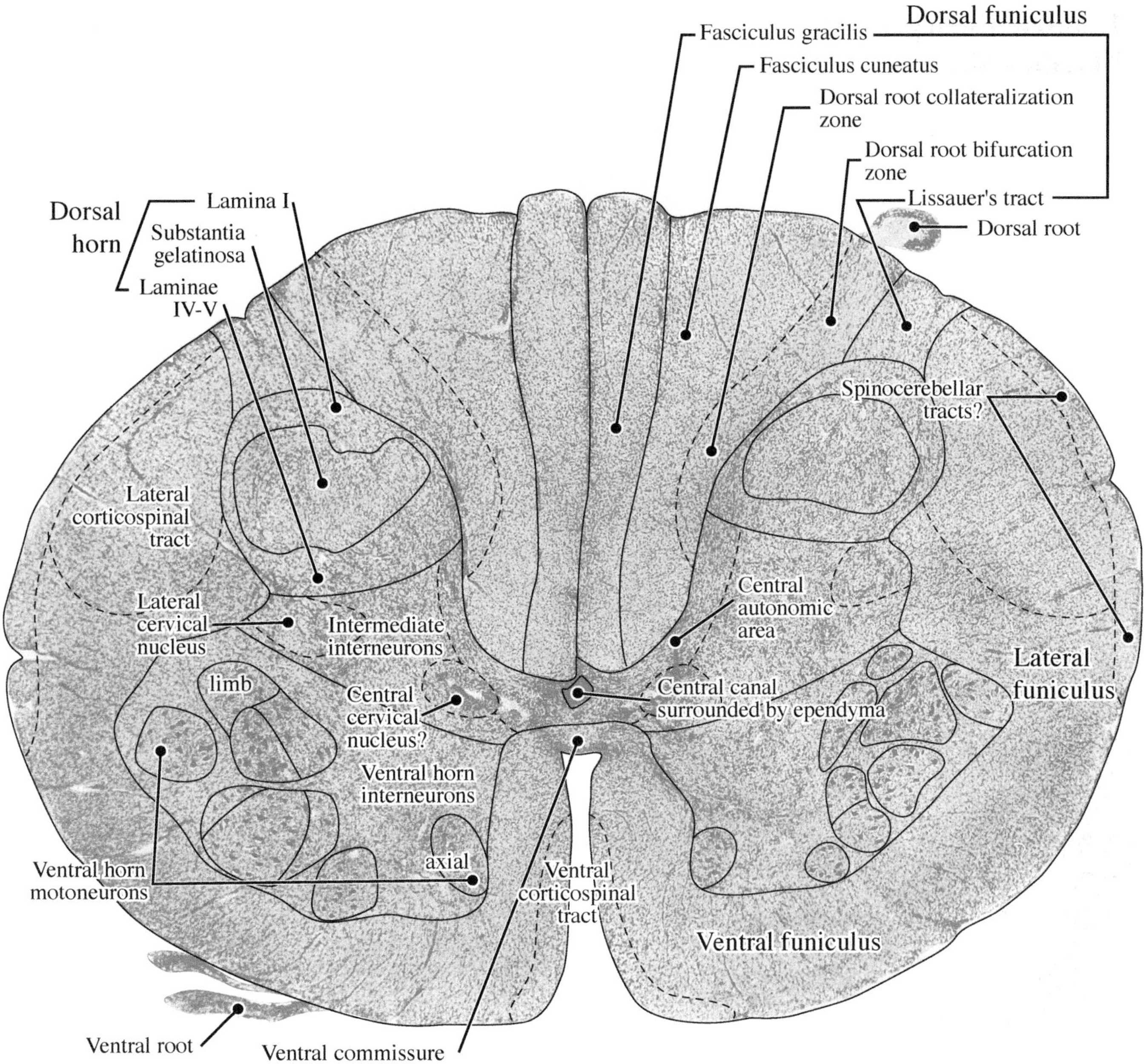
Dorsal funiculus
Fasciculus gracilis
Fasciculus cuneatus
Dorsal root collateralization zone
Dorsal root bifurcation zone
Lissauer's tract
Dorsal root
Dorsal horn
Lamina I
Substantia gelatinosa
Laminae IV-V
Spinocerebellar tracts?
Lateral corticospinal tract
Lateral cervical nucleus
Intermediate interneurons
Central autonomic area
Lateral funiculus
limb
Central cervical nucleus?
Central canal surrounded by ependyma
Ventral horn interneurons
axial
Ventral horn motoneurons
Ventral corticospinal tract
Ventral funiculus
Ventral root
Ventral commissure

PLATE 45A

CR 350 mm
Infant, 4 Days
Y299-62
Upper Thoracic
Myelin stain

Areas (mm²)

Central canal	.0016
Ependyma	.0081
Gray matter	2.8495
White matter	7.3181

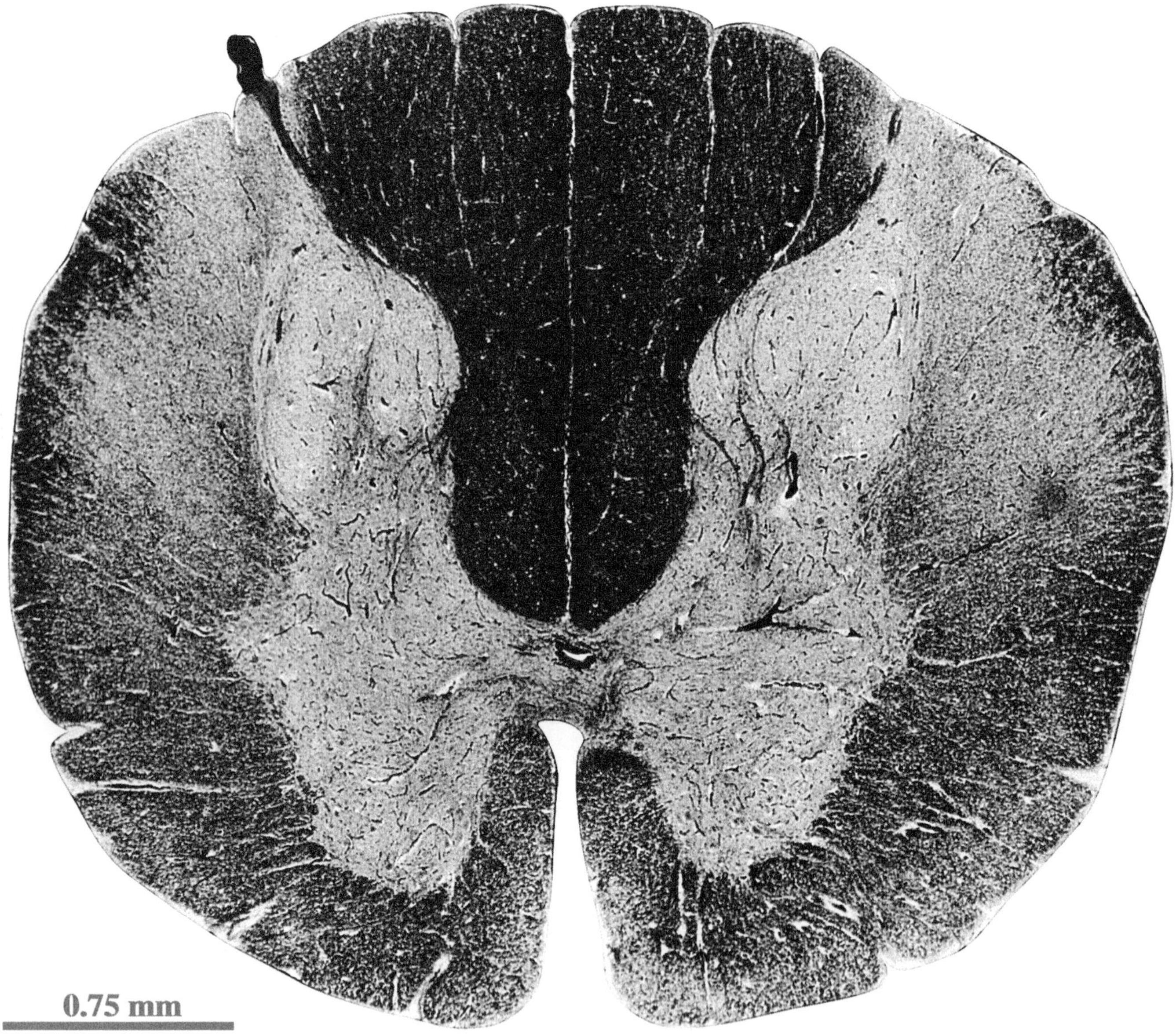

PLATE 45B

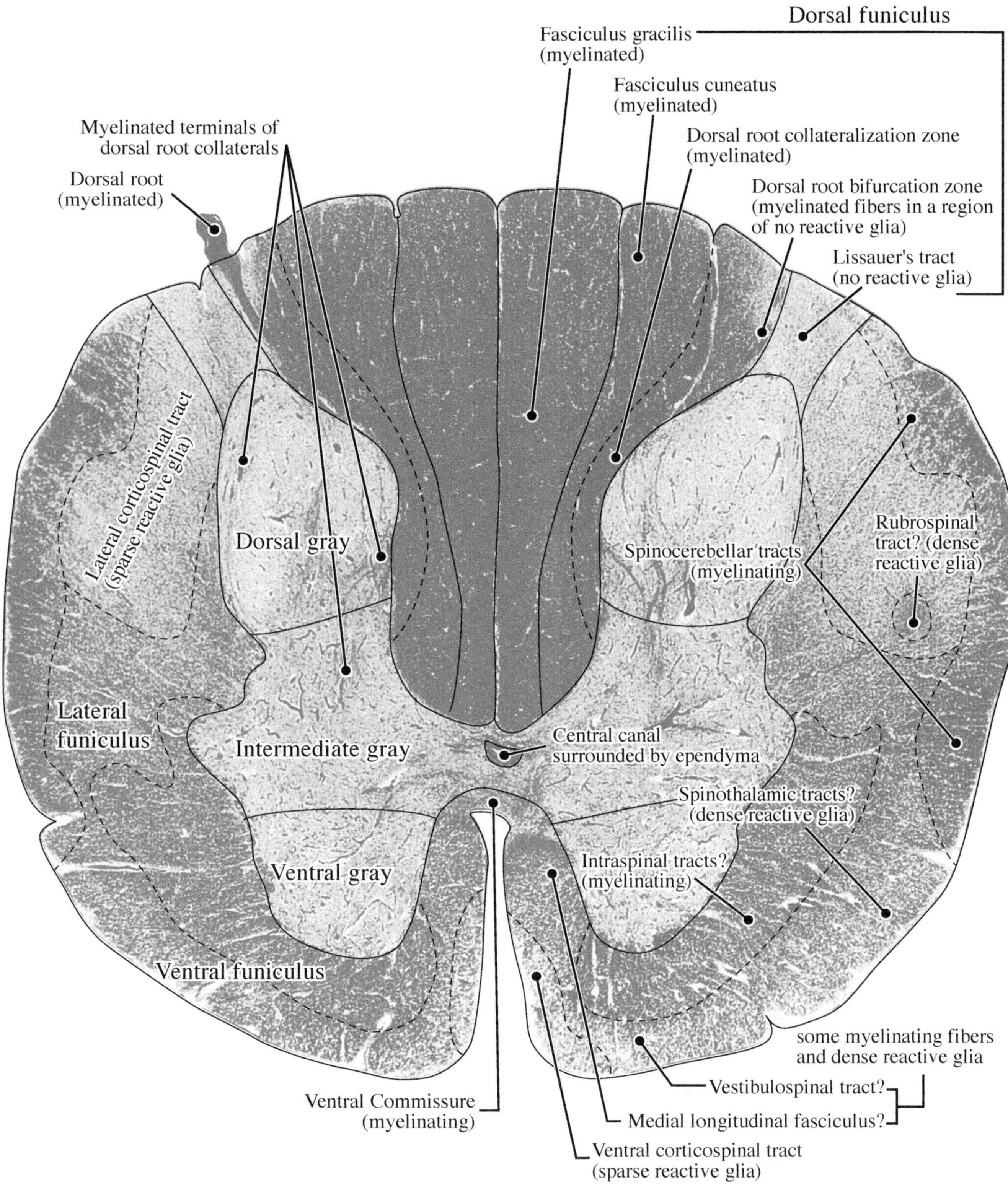
Dorsal funiculus
Fasciculus gracilis (myelinated)
Fasciculus cuneatus (myelinated)
Dorsal root collateralization zone (myelinated)
Dorsal root bifurcation zone (myelinated fibers in a region of no reactive glia)
Lissauer's tract (no reactive glia)
Myelinated terminals of dorsal root collaterals
Dorsal root (myelinated)
Lateral corticospinal tract (sparse reactive glia)
Dorsal gray
Spinocerebellar tracts (myelinating)
Rubrospinal tract? (dense reactive glia)
Lateral funiculus
Intermediate gray
Central canal surrounded by ependyma
Spinothalamic tracts? (dense reactive glia)
Ventral gray
Intraspinal tracts? (myelinating)
Ventral funiculus
some myelinating fibers and dense reactive glia
Vestibulospinal tract?
Medial longitudinal fasciculus?
Ventral Commissure (myelinating)
Ventral corticospinal tract (sparse reactive glia)

PLATE 46A

CR 350 mm
Infant, 4 Days
Y299-62
Upper Thoracic
Cell body stain

Areas (mm^2)

Central canal	.0019
Ependyma	.0141
Gray matter	2.9713
White matter	7.5265

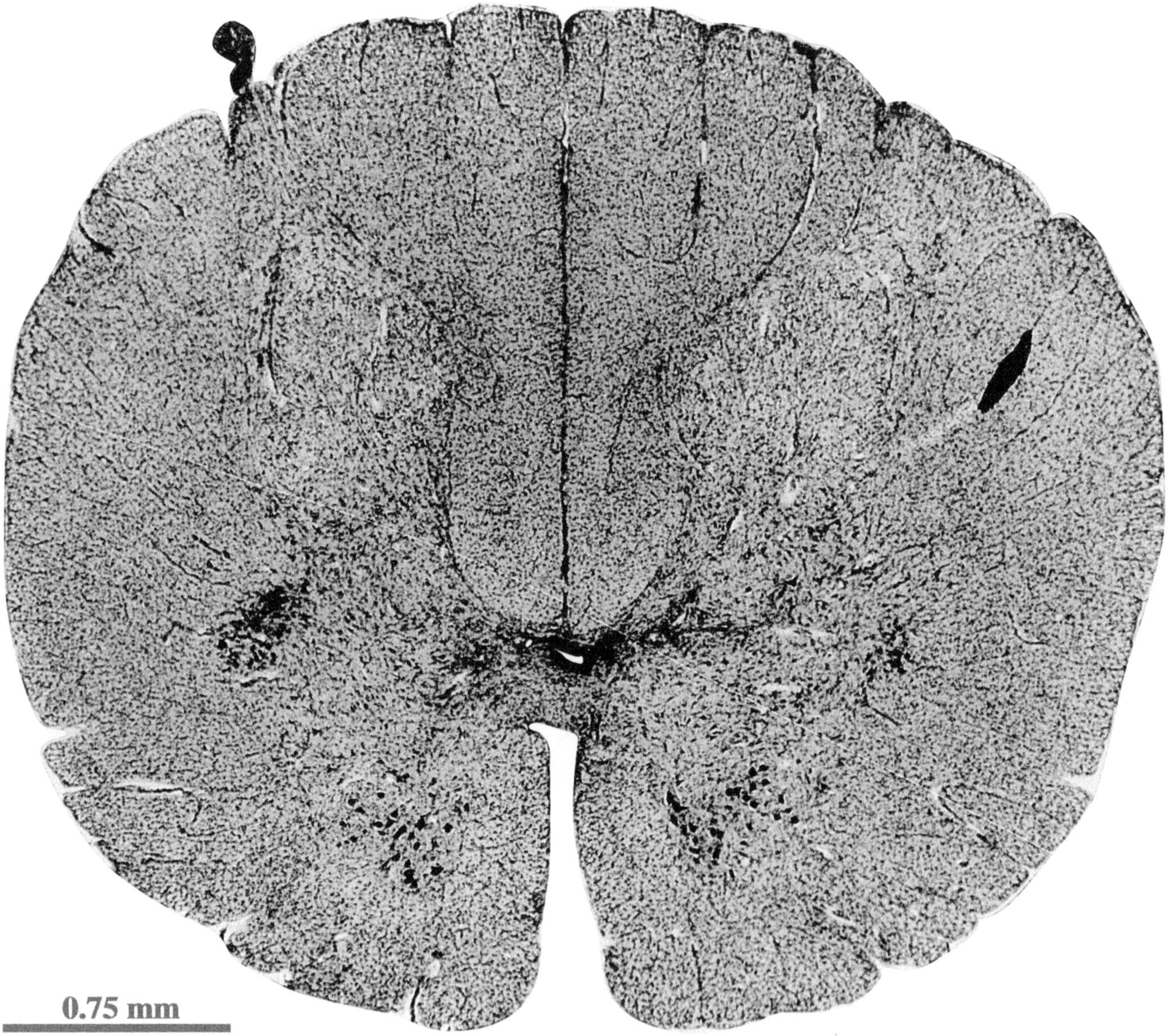

PLATE 46B

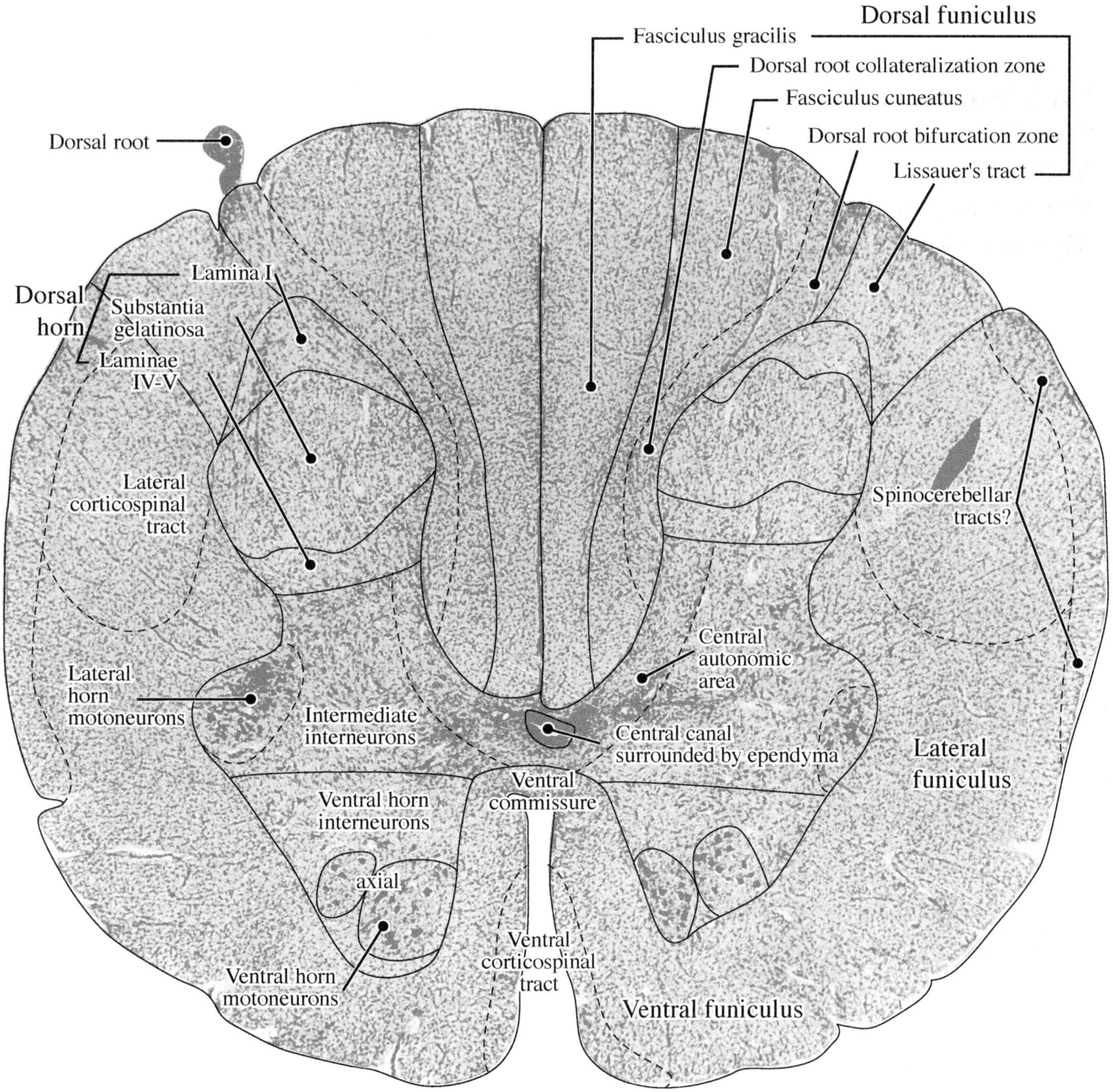
Dorsal funiculus
Fasciculus gracilis
Dorsal root collateralization zone
Fasciculus cuneatus
Dorsal root bifurcation zone
Lissauer's tract
Dorsal root
Dorsal horn
Lamina I
Substantia gelatinosa
Laminae IV-V
Lateral corticospinal tract
Spinocerebellar tracts?
Central autonomic area
Lateral horn motoneurons
Intermediate interneurons
Central canal surrounded by ependyma
Lateral funiculus
Ventral commissure
Ventral horn interneurons
axial
Ventral horn motoneurons
Ventral corticospinal tract
Ventral funiculus

PLATE 47A

CR 350 mm
Infant, 4 Days
Y299-62
Lower Thoracic
Myelin stain

Areas (mm²)

Central canal	.0005
Ependyma	.0047
Gray matter	1.9740
White matter	5.8169

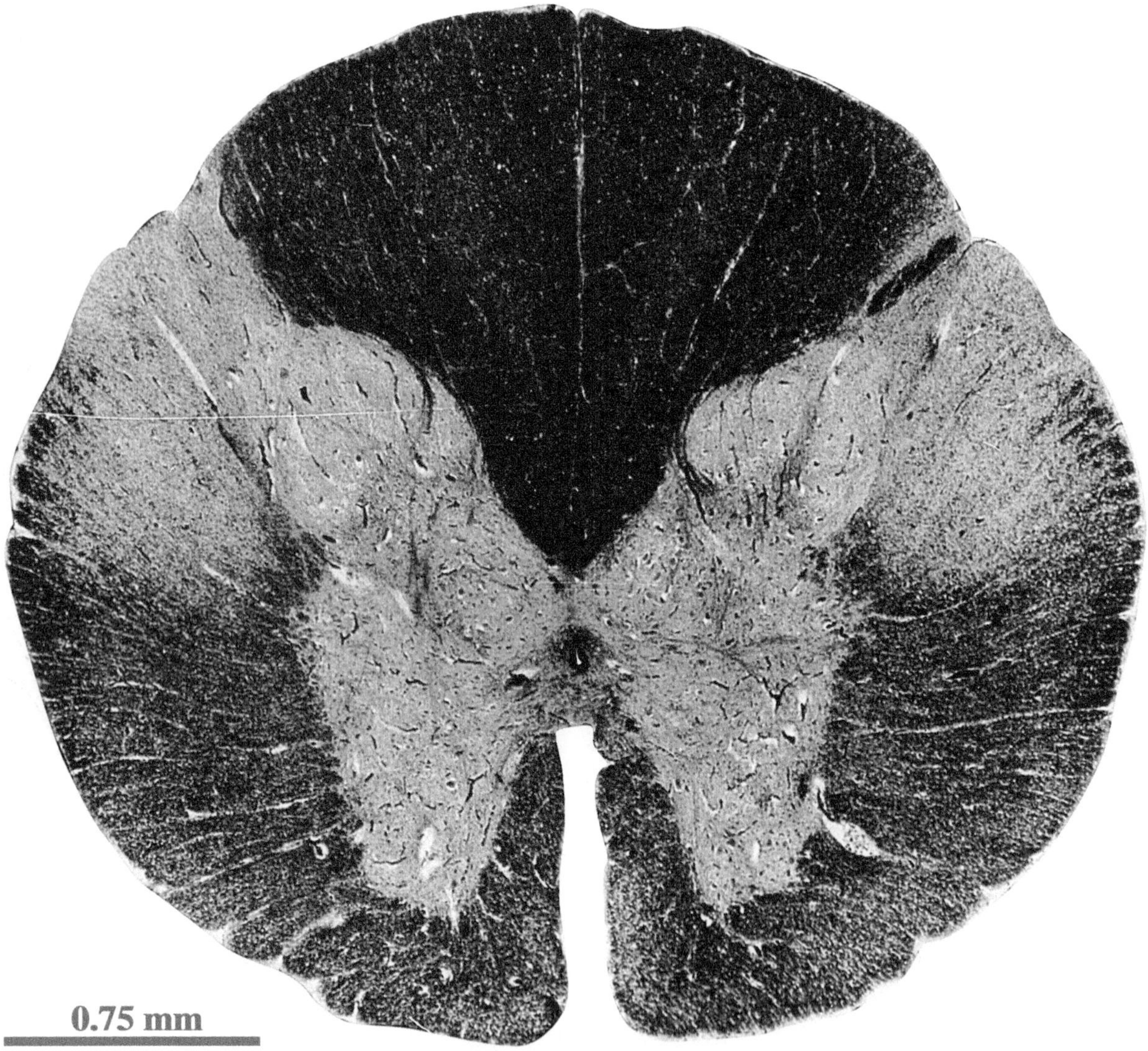

0.75 mm

PLATE 47B

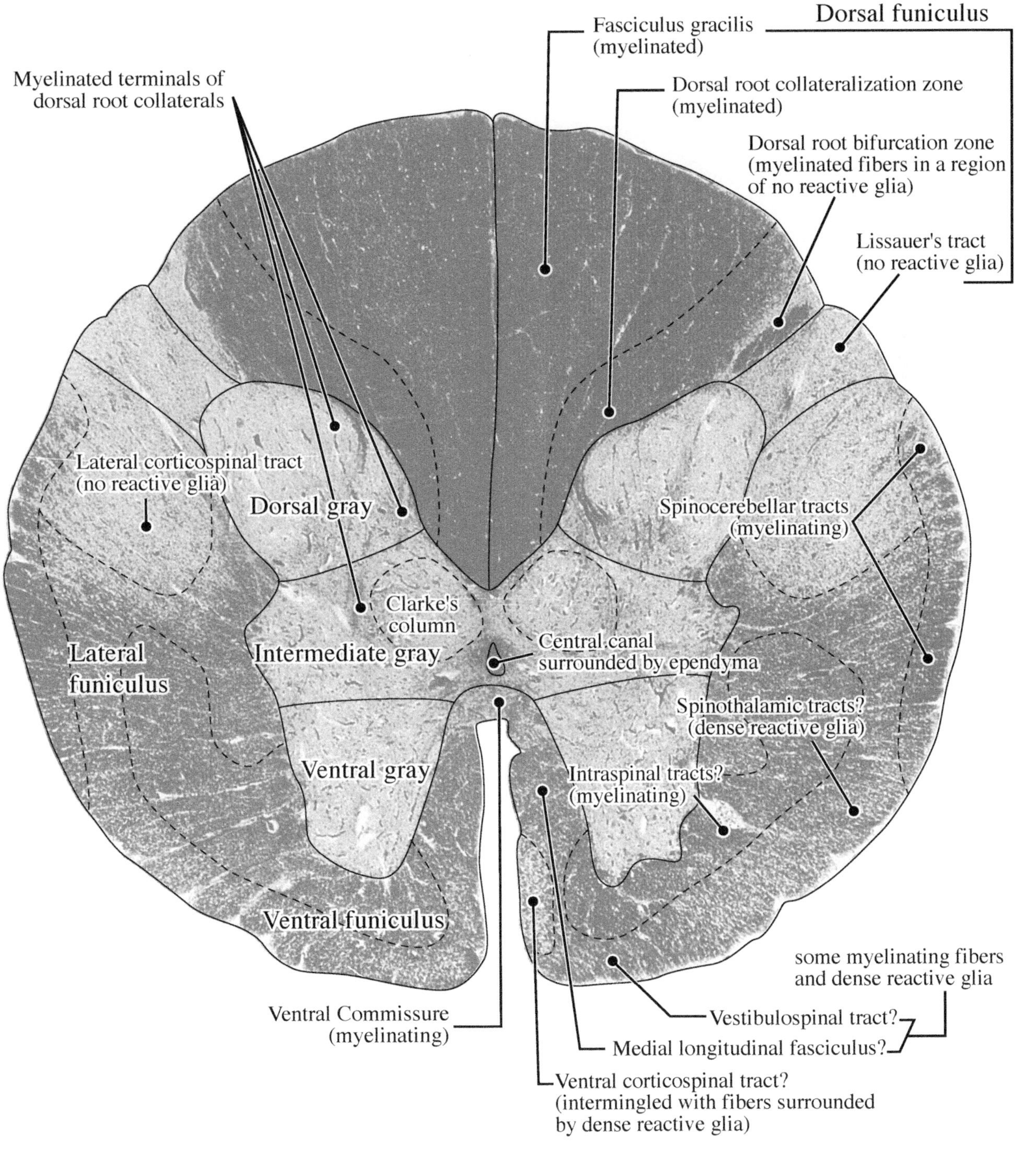
Dorsal funiculus
Fasciculus gracilis (myelinated)
Dorsal root collateralization zone (myelinated)
Dorsal root bifurcation zone (myelinated fibers in a region of no reactive glia)
Lissauer's tract (no reactive glia)
Myelinated terminals of dorsal root collaterals
Lateral corticospinal tract (no reactive glia)
Dorsal gray
Spinocerebellar tracts (myelinating)
Clarke's column
Lateral funiculus
Intermediate gray
Central canal surrounded by ependyma
Spinothalamic tracts? (dense reactive glia)
Ventral gray
Intraspinal tracts? (myelinating)
Ventral funiculus
some myelinating fibers and dense reactive glia
Ventral Commissure (myelinating)
Vestibulospinal tract?
Medial longitudinal fasciculus?
Ventral corticospinal tract? (intermingled with fibers surrounded by dense reactive glia)

PLATE 48A

CR 350 mm
Infant, 4 Days
Y299-62
Lower Thoracic
Cell body stain

Areas (mm^2)

Central canal	.0015
Ependyma	.0148
Gray matter	2.1221
White matter	6.0785

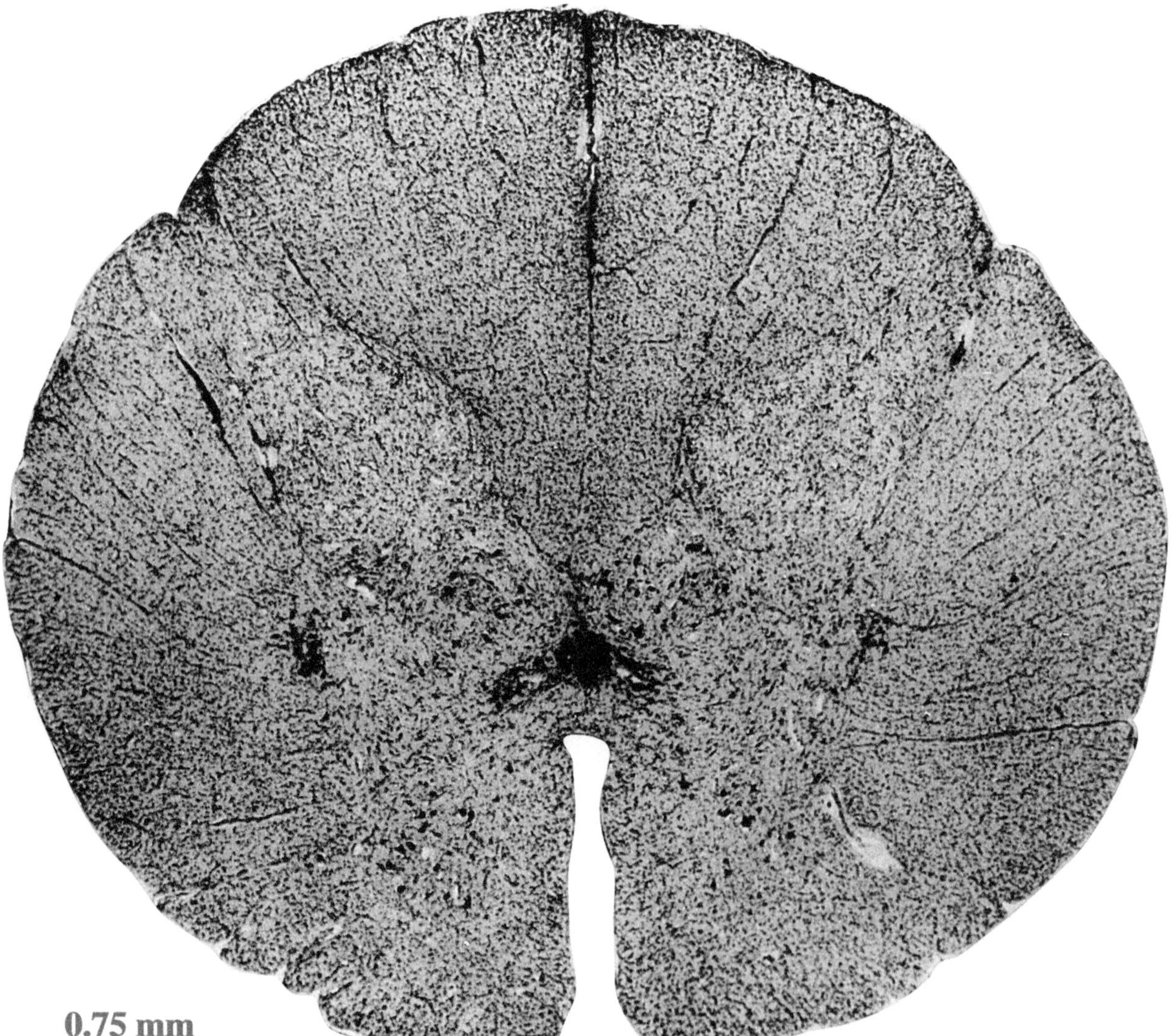

0.75 mm

PLATE 48B

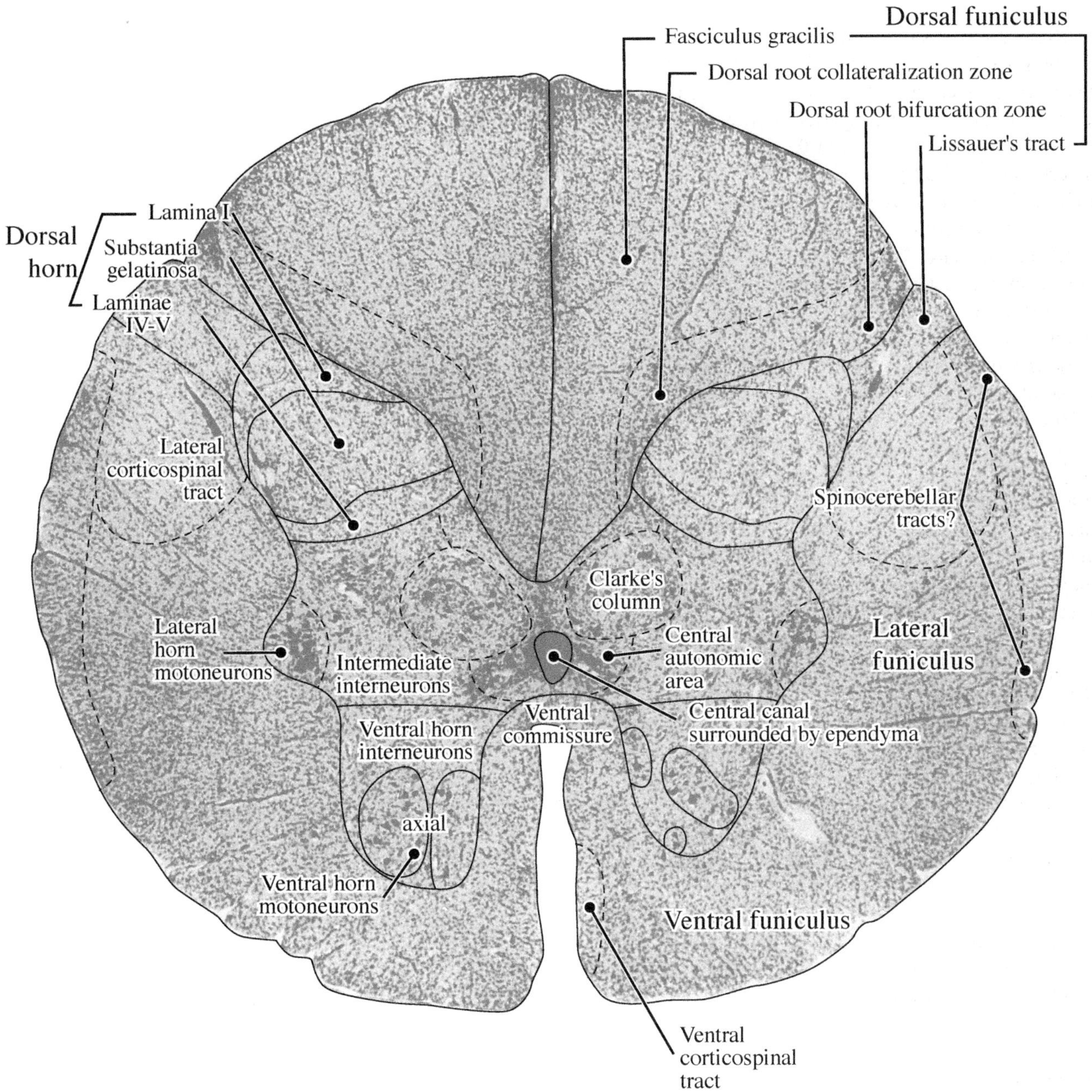
Dorsal funiculus
Fasciculus gracilis
Dorsal root collateralization zone
Dorsal root bifurcation zone
Lissauer's tract
Dorsal horn
Lamina I
Substantia gelatinosa
Laminae IV-V
Lateral corticospinal tract
Spinocerebellar tracts?
Clarke's column
Lateral horn motoneurons
Intermediate interneurons
Central autonomic area
Lateral funiculus
Central canal surrounded by ependyma
Ventral commissure
Ventral horn interneurons
axial
Ventral horn motoneurons
Ventral funiculus
Ventral corticospinal tract

PLATE 49A

CR 350 mm
Infant, 4 Days
Y299-62
Upper Lumbar
Myelin stain

Areas (mm²)

Central canal	.0019
Ependyma	.0087
Gray matter	4.7867
White matter	6.9382

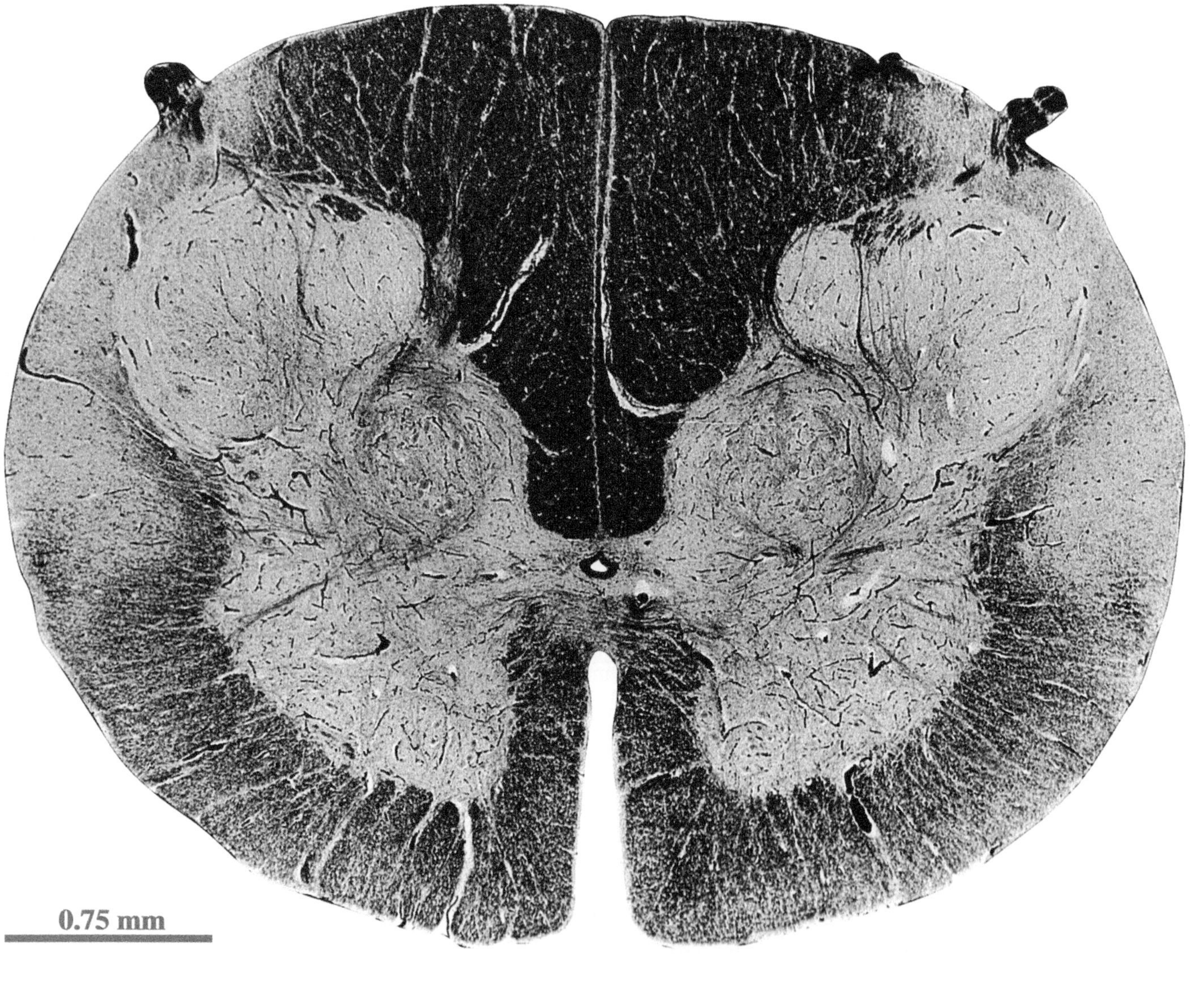

PLATE 49B

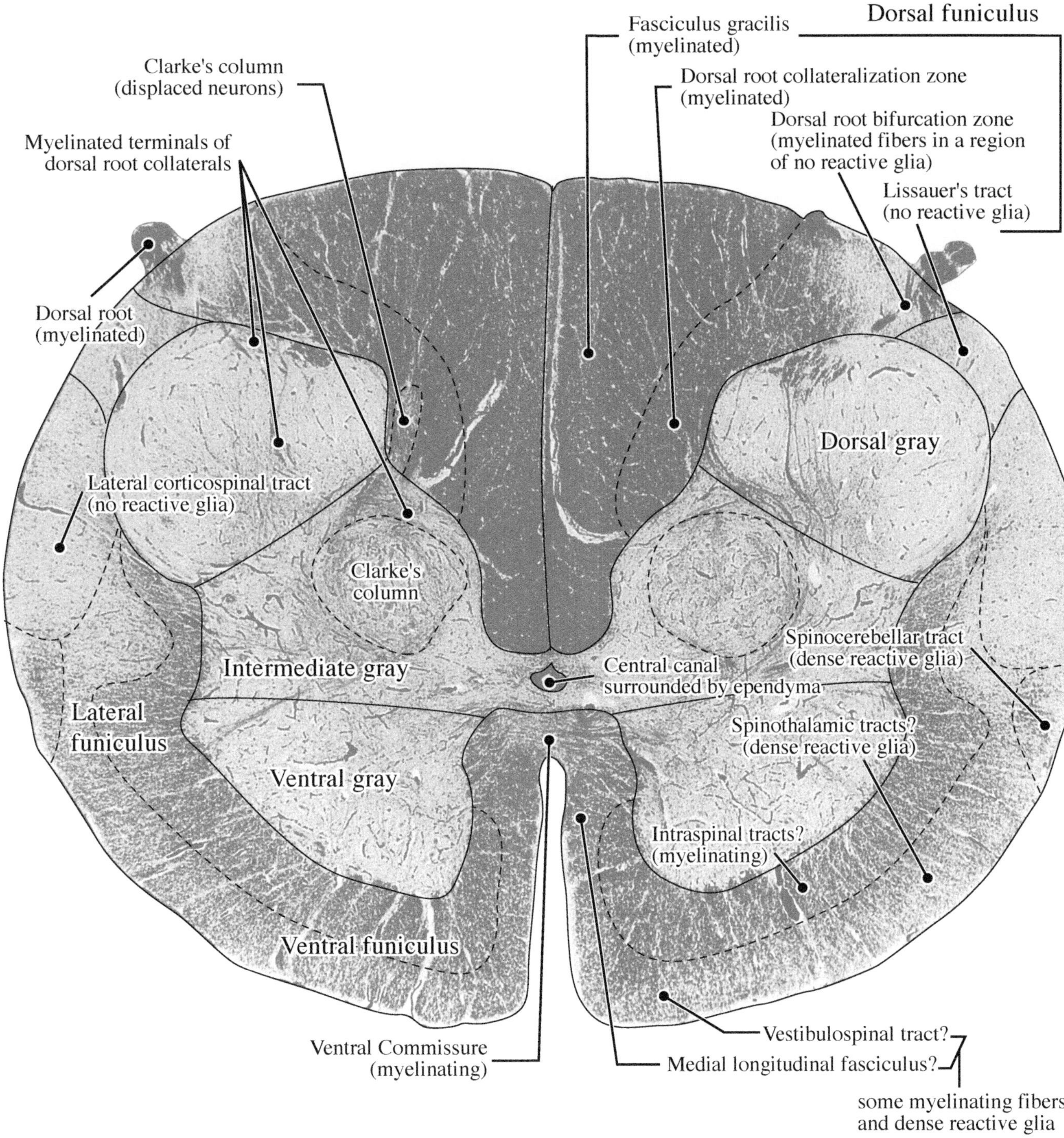
Dorsal funiculus
Fasciculus gracilis
(myelinated)
Clarke's column
(displaced neurons)
Dorsal root collateralization zone
(myelinated)
Dorsal root bifurcation zone
(myelinated fibers in a region
of no reactive glia)
Myelinated terminals of
dorsal root collaterals
Lissauer's tract
(no reactive glia)
Dorsal root
(myelinated)
Dorsal gray
Lateral corticospinal tract
(no reactive glia)
Clarke's
column
Spinocerebellar tract
(dense reactive glia)
Intermediate gray
Central canal
surrounded by ependyma
Lateral
funiculus
Spinothalamic tracts?
(dense reactive glia)
Ventral gray
Intraspinal tracts?
(myelinating)
Ventral funiculus
Vestibulospinal tract?
Ventral Commissure
(myelinating)
Medial longitudinal fasciculus?
some myelinating fibers
and dense reactive glia

PLATE 50A

CR 350 mm
Infant, 4 Days
Y299-62
Upper Lumbar
Cell body stain

Areas (mm^2)

Central canal	.0014
Ependyma	.0147
Gray matter	5.2589
White matter	7.4545

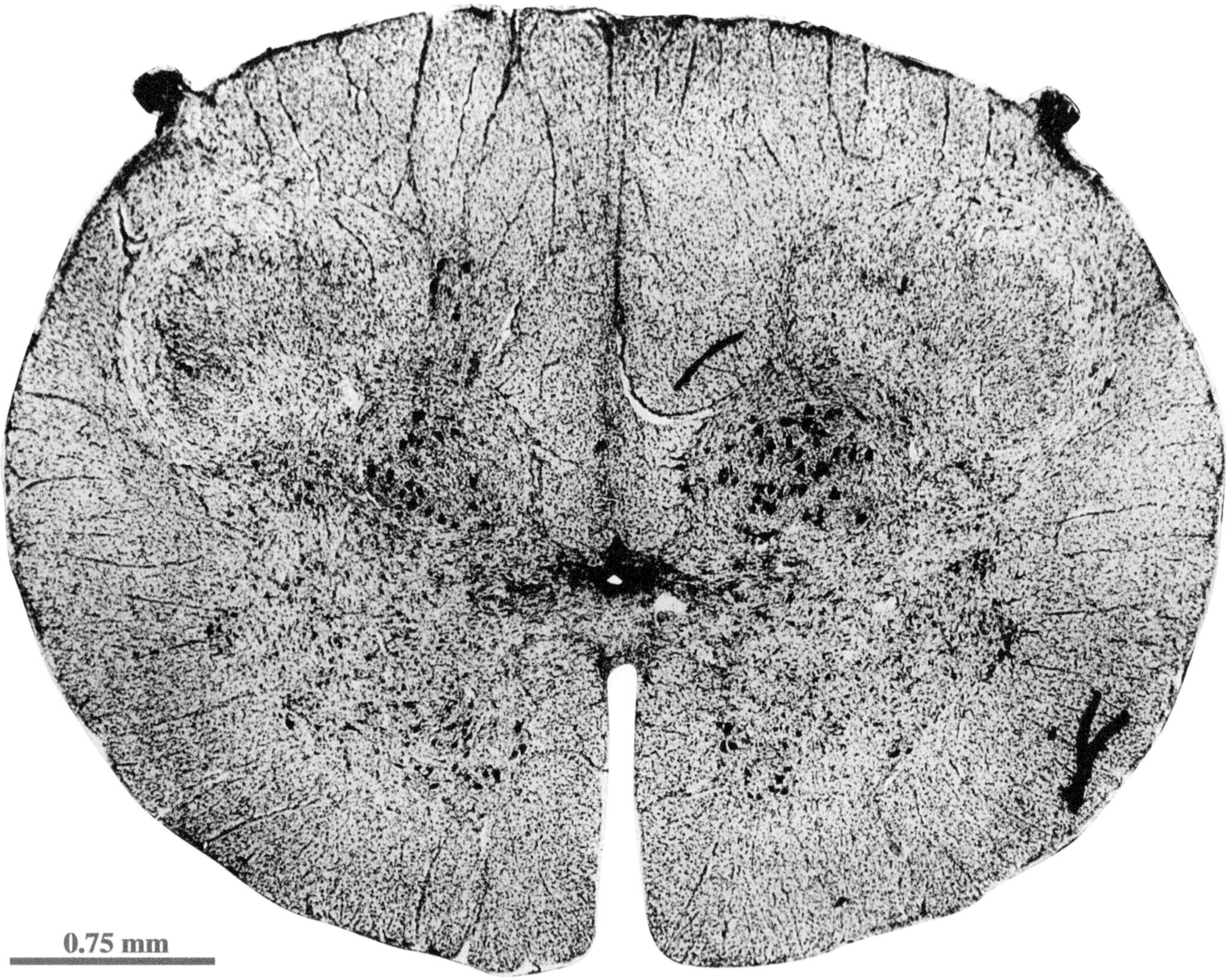

PLATE 50B

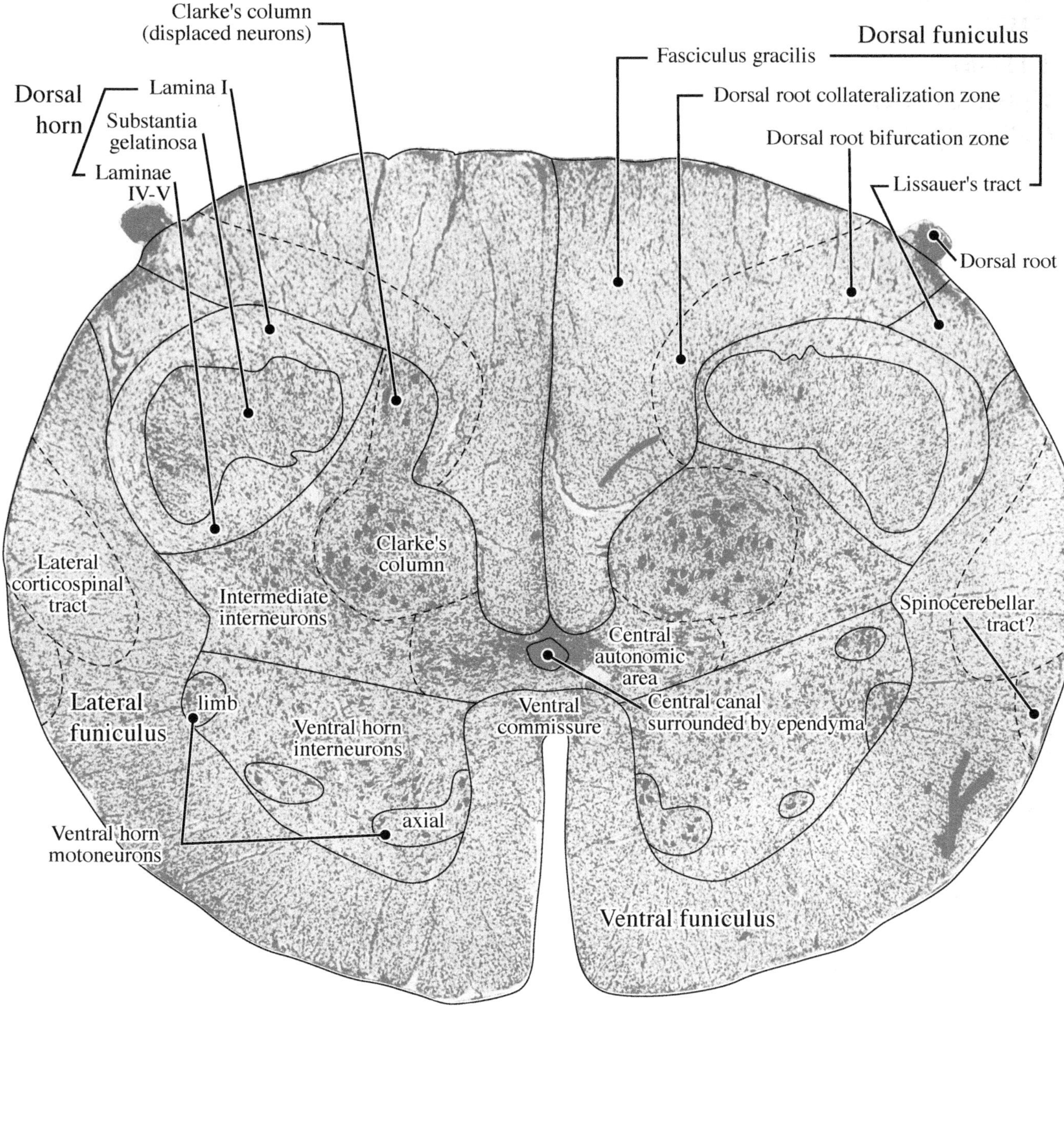
Clarke's column
(displaced neurons)
Dorsal horn
Lamina I
Substantia gelatinosa
Laminae IV-V
Dorsal funiculus
Fasciculus gracilis
Dorsal root collateralization zone
Dorsal root bifurcation zone
Lissauer's tract
Dorsal root
Clarke's column
Lateral corticospinal tract
Intermediate interneurons
Spinocerebellar tract?
Central autonomic area
Central canal surrounded by ependyma
Lateral funiculus
limb
Ventral commissure
Ventral horn interneurons
axial
Ventral horn motoneurons
Ventral funiculus

PLATE 51A

CR 350 mm
Infant, 4 Days
Y299-62
Lumbar Enlargement
Myelin stain

Areas (mm^2)

Central canal	.0020
Ependyma	.0104
Gray matter	8.0120
White matter	4.1273

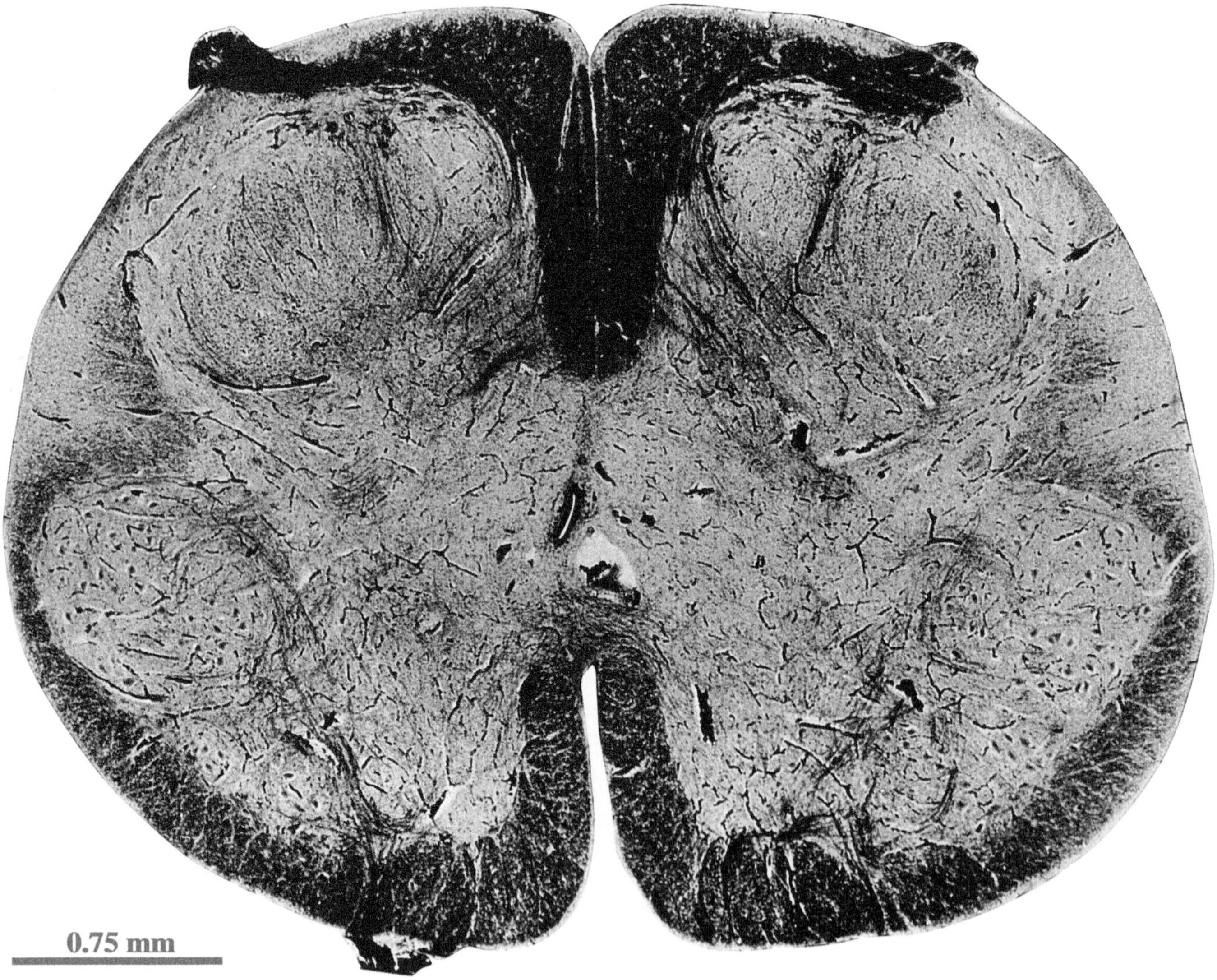

PLATE 51B

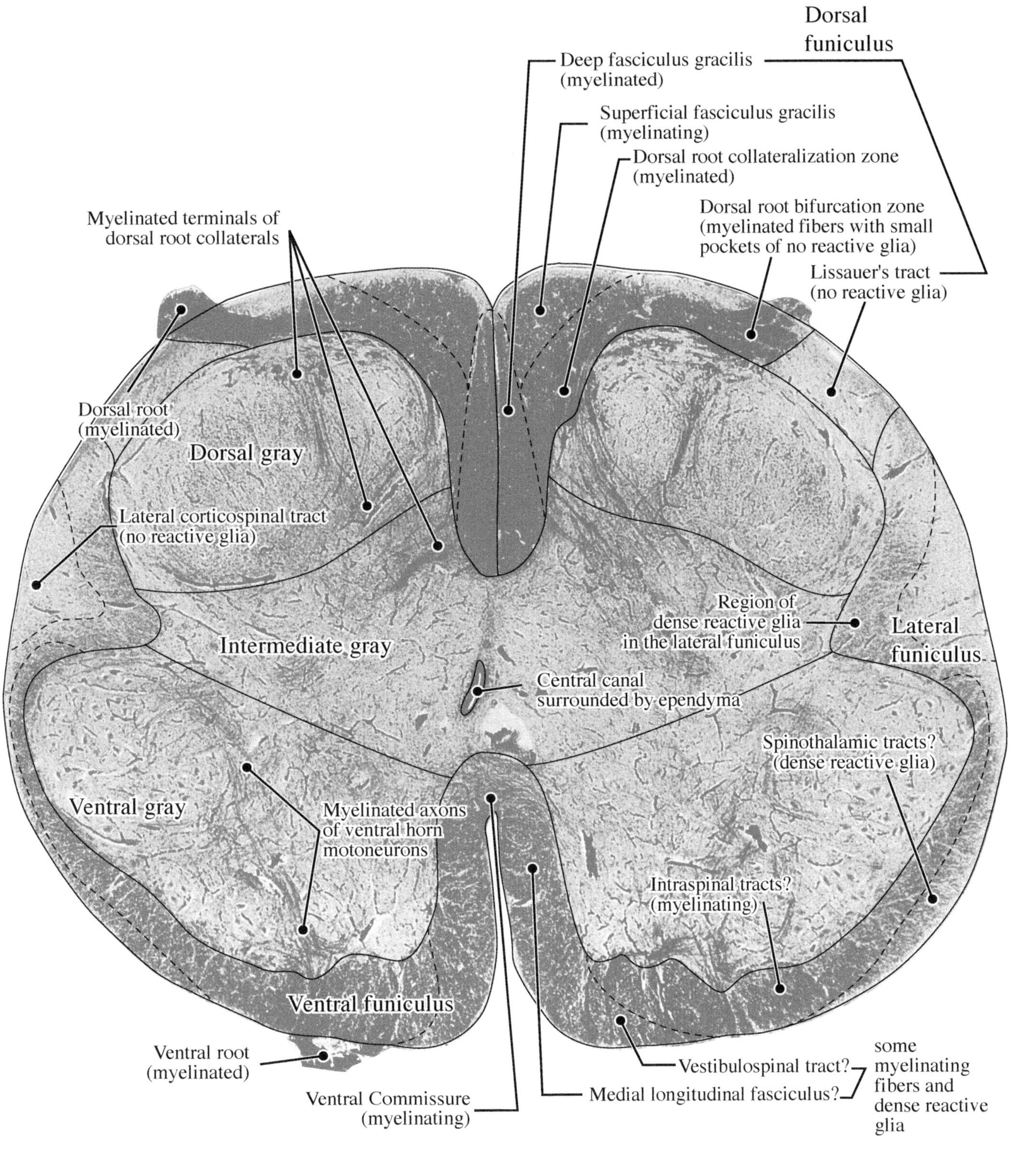
Dorsal
funiculus
Deep fasciculus gracilis
(myelinated)
Superficial fasciculus gracilis
(myelinating)
Dorsal root collateralization zone
(myelinated)
Dorsal root bifurcation zone
(myelinated fibers with small
pockets of no reactive glia)
Lissauer's tract
(no reactive glia)
Myelinated terminals of
dorsal root collaterals
Dorsal root
(myelinated)
Dorsal gray
Lateral corticospinal tract
(no reactive glia)
Region of
dense reactive glia
in the lateral funiculus
Lateral
funiculus
Intermediate gray
Central canal
surrounded by ependyma
Spinothalamic tracts?
(dense reactive glia)
Ventral gray
Myelinated axons
of ventral horn
motoneurons
Intraspinal tracts?
(myelinating)
Ventral funiculus
Ventral root
(myelinated)
Vestibulospinal tract?
Medial longitudinal fasciculus?
some
myelinating
fibers and
dense reactive
glia
Ventral Commissure
(myelinating)

PLATE 52A

CR 350 mm
Infant, 4 Days
Y299-62
Lumbar Enlargement
Cell body stain

Areas (mm^2)

Central canal	.0046
Ependyma	.0302
Gray matter	7.9475
White matter	4.1796

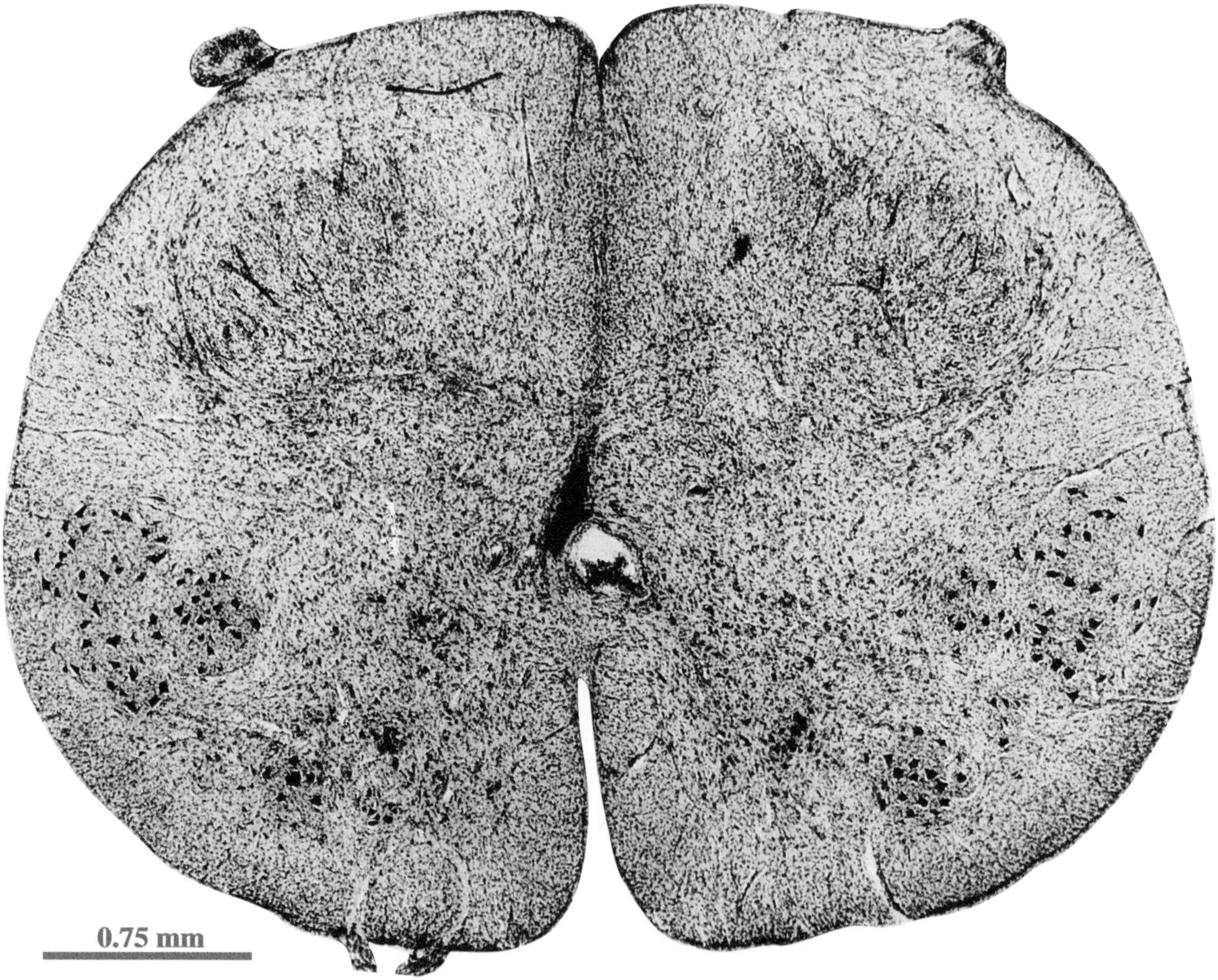

PLATE 52B

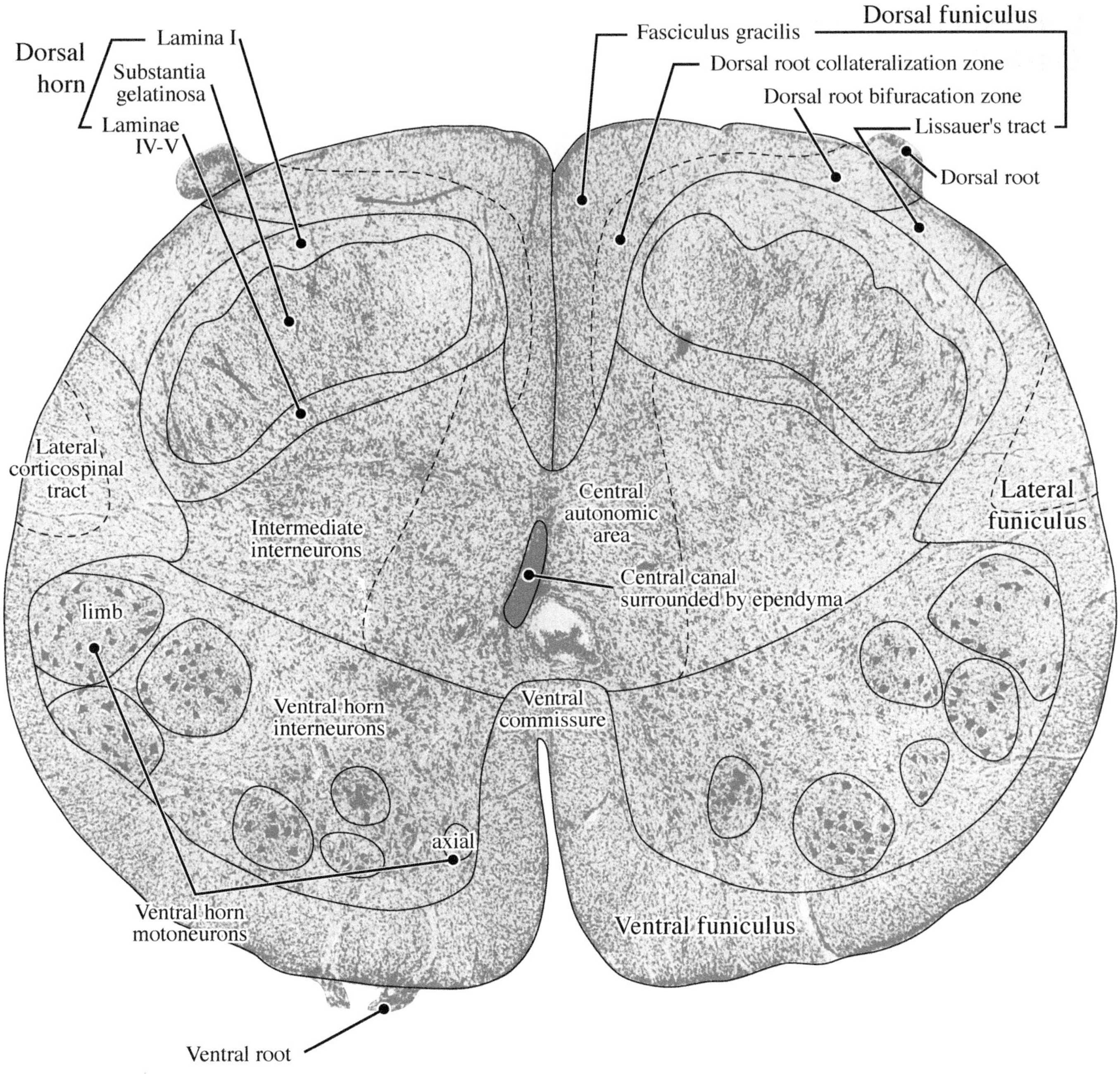
Dorsal horn
Lamina I
Substantia gelatinosa
Laminae IV-V
Fasciculus gracilis
Dorsal funiculus
Dorsal root collateralization zone
Dorsal root bifuracation zone
Lissauer's tract
Dorsal root
Lateral corticospinal tract
Central autonomic area
Lateral funiculus
Intermediate interneurons
Central canal surrounded by ependyma
limb
Ventral horn interneurons
Ventral commissure
axial
Ventral horn motoneurons
Ventral funiculus
Ventral root

PLATE 53A

CR 350 mm
Infant, 4 Days
Y299-62
Sacral/Coccygeal
Myelin stain

Areas (mm^2)

Central canal	.0036
Ependyma	.0228
Gray matter	3.8397
White matter	1.8131

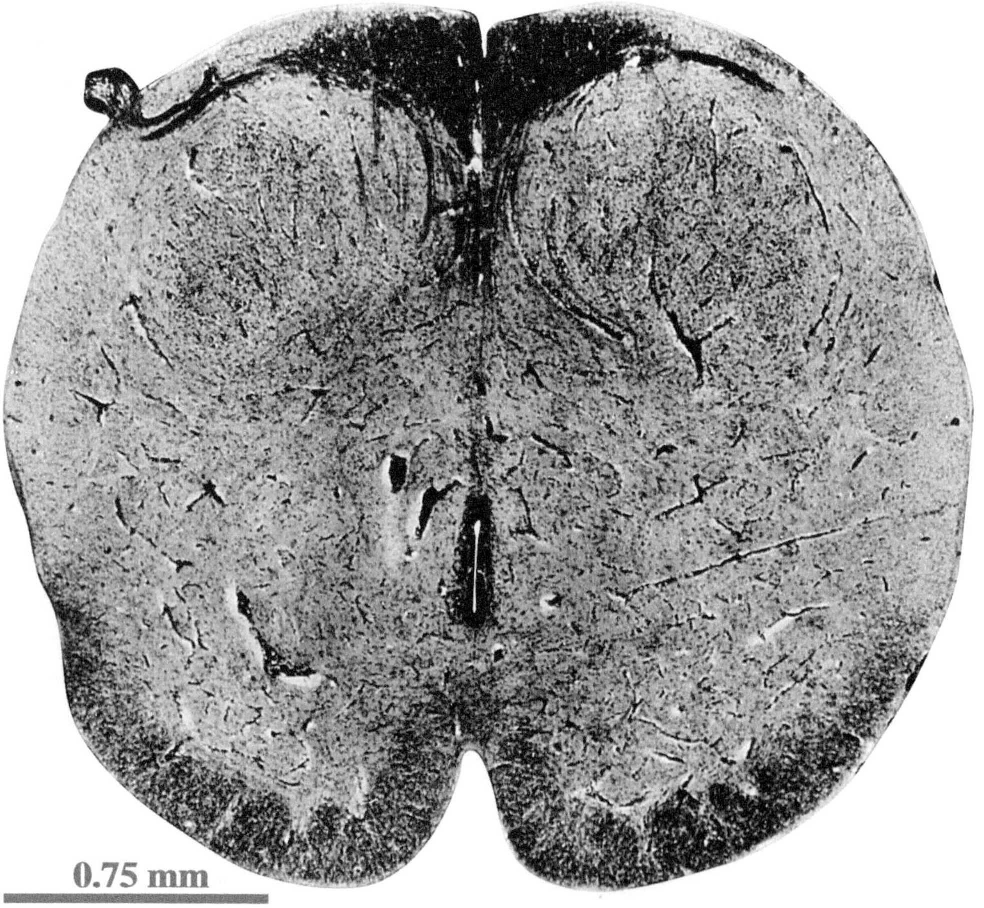

PLATE 53B

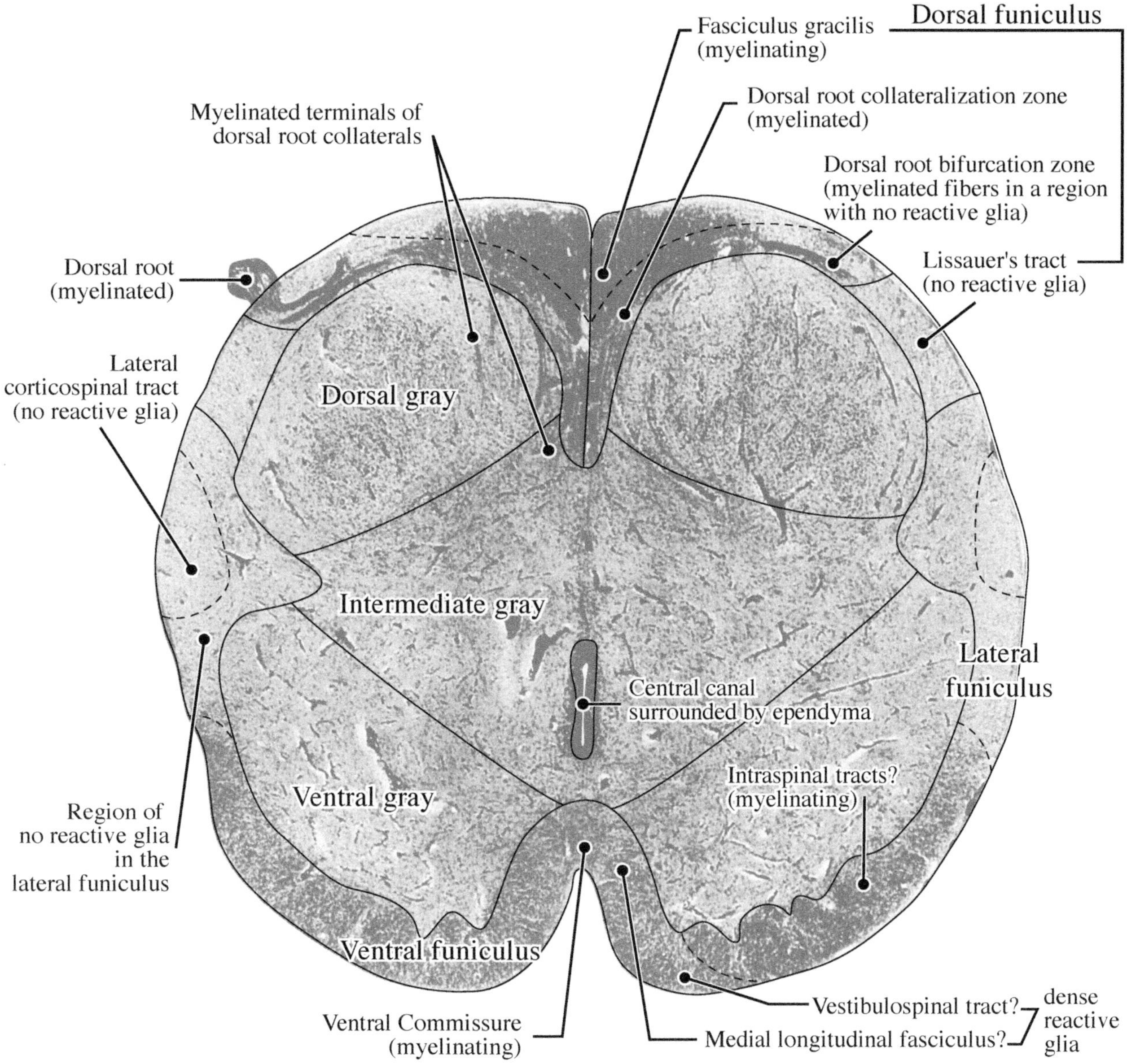

Dorsal funiculus
Fasciculus gracilis (myelinating)
Dorsal root collateralization zone (myelinated)
Myelinated terminals of dorsal root collaterals
Dorsal root bifurcation zone (myelinated fibers in a region with no reactive glia)
Lissauer's tract (no reactive glia)
Dorsal root (myelinated)
Lateral corticospinal tract (no reactive glia)
Dorsal gray
Intermediate gray
Central canal surrounded by ependyma
Lateral funiculus
Intraspinal tracts? (myelinating)
Ventral gray
Region of no reactive glia in the lateral funiculus
Ventral funiculus
Ventral Commissure (myelinating)
Vestibulospinal tract?
Medial longitudinal fasciculus?
dense reactive glia

PLATE 54A

CR 350 mm
Infant, 4 Days
Y299-62
Sacral/Coccygeal
Cell body stain

Areas (mm^2)

Central canal	.0037
Ependyma	.0281
Gray matter	4.0451
White matter	1.8283

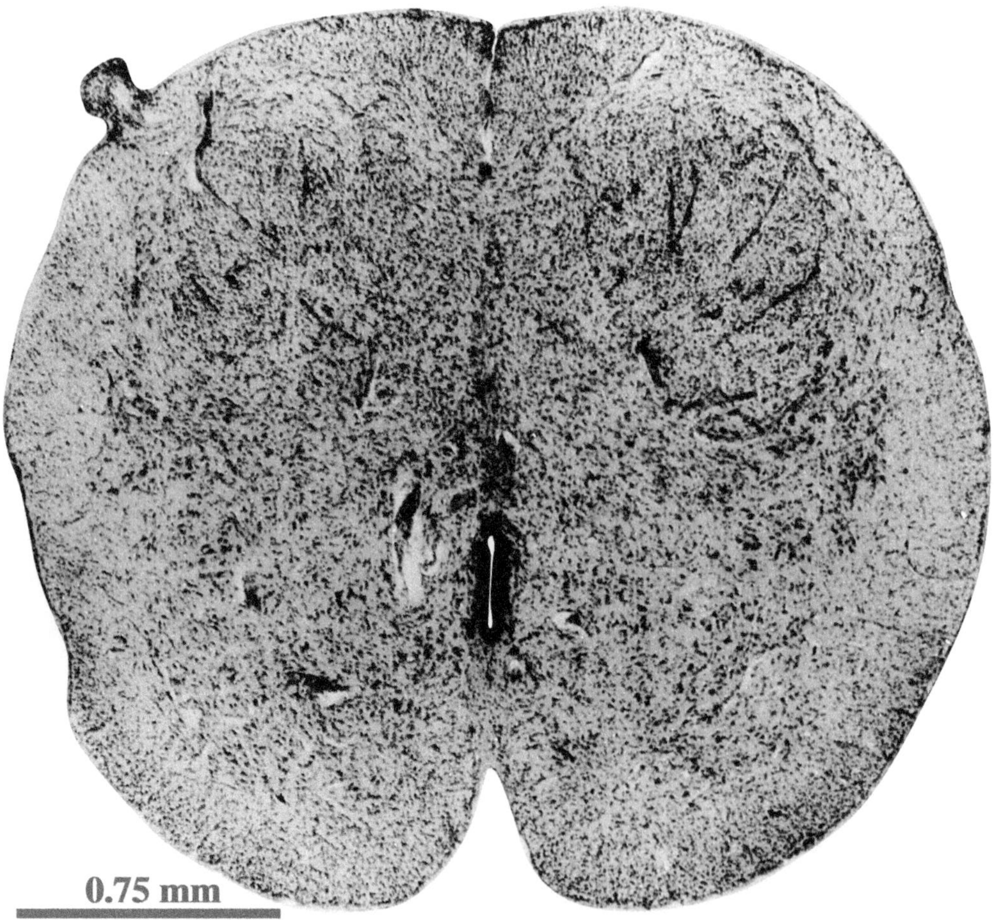

PLATE 54B

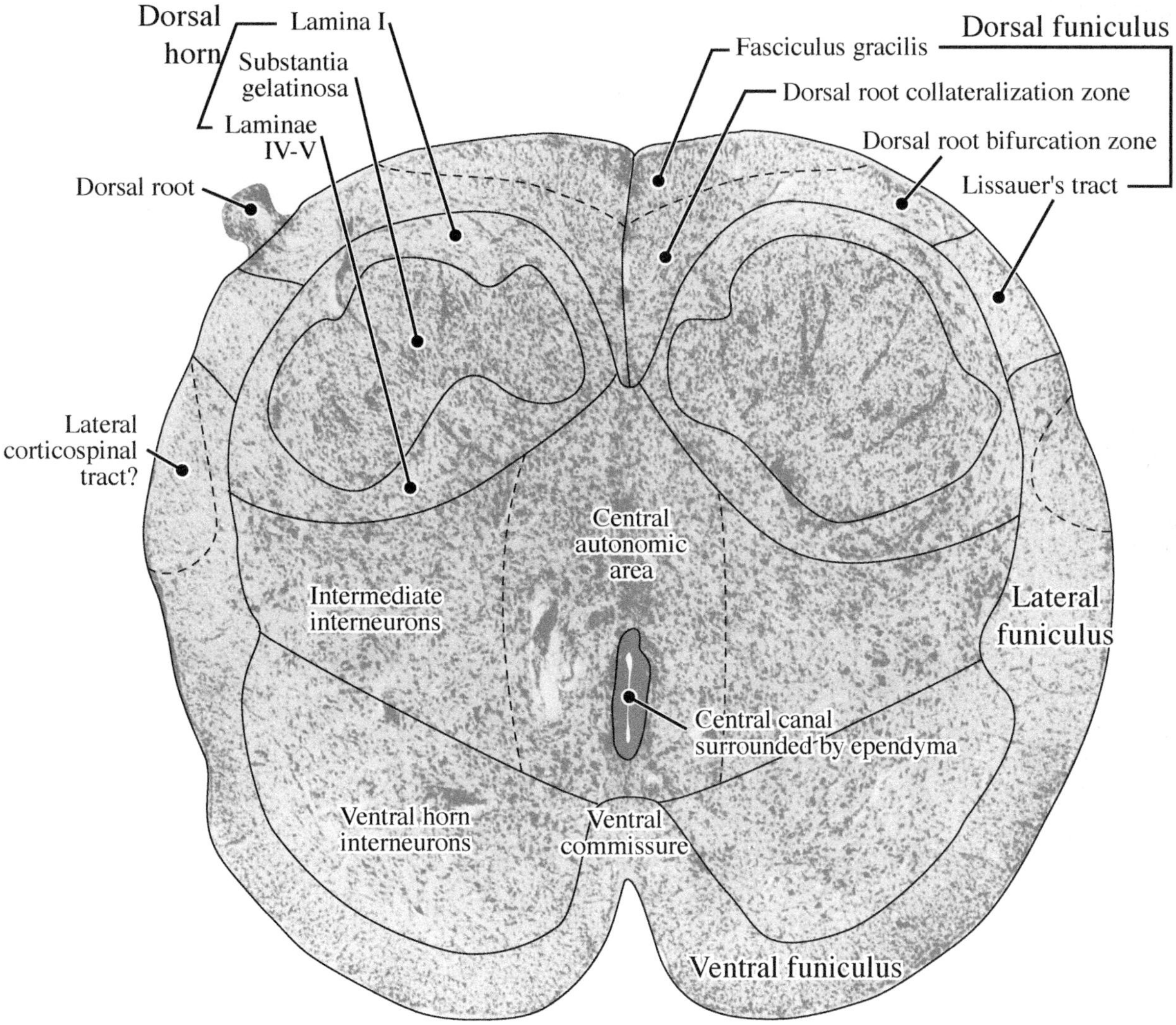
Dorsal horn
Lamina I
Substantia gelatinosa
Laminae IV-V
Dorsal root
Lateral corticospinal tract?
Fasciculus gracilis
Dorsal funiculus
Dorsal root collateralization zone
Dorsal root bifurcation zone
Lissauer's tract
Central autonomic area
Intermediate interneurons
Lateral funiculus
Central canal surrounded by ependyma
Ventral horn interneurons
Ventral commissure
Ventral funiculus

PART VII: Y23-60
CR 410 mm (4 Weeks)

Plate 55 is a survey of matched myelin-stained and cell-body-stained sections from Y23-60, a specimen in the Yakovlev Collection. All sections are shown at the same scale. The boxes enclosing each section list the approximate level and the total area of the section in square millimeters (mm^2). Full-page normal-contrast photographs of each section are in **Plates 56A–65A.** Low-contrast photographs with superimposed labels and outlines of structural details are in **Plates 56B–65B.** In this specimen, the myelin-stained and cell-body-stained sections were preserved on separate large glass plates without any section numbers. In addition, some sections were flipped over on the slides, and it appeared that the section order was mixed-up before they were placed on the plates. Sixteen myelin-stained sections and 16 cell-body-stained sections were photographed ranging from upper cervical to lumbar enlargement levels. The 32 photographic prints were intuitively arranged in order from upper cervical to sacral/coccygeal levels, using internal features such as the size of the corticospinal tracts, and the width of the ventral horn. Then, myelin and cell-body stained sections were matched and 6 different levels were analyzed. We could not find matching cell-body-stained sections at the cervical enlargement and upper lumbar levels. Therefore, only the myelin-stained sections are illustrated. As in the previous specimens, the cross section area of a myelin-stained section is smaller than the matching cell-body-stained section with the exception of the upper thoracic level. At that level, the cell-body and myelin-stained sections are perfectly matched; probably the cell body stained section is 3-4 sections below the myelin stained section. Using the total areas of the myelin-stained sections, the overall size differences between levels indicate the following comparisons. The cervical enlargement has the largest cross-sectional area, 20% larger than the lumbar enlargement. The two thoracic levels have smaller cross sectional areas, 74% smaller than the cervical enlargement, and 45% smaller than the lumbar enlargement.

Myelination gradients are in the lateral corticospinal tract (internal first, external last) and the spinocephalic tract (internal first, external last). The anterior corticospinal tract is quite small in this specimen. It cannot be detected at low thoracic levels and shows bilateral asymmetry (the right tract is larger). The ventral commissure is either myelinating or is myelinated at GW 37; however, this 4-day infant, appears to have "regressed," especially at cervical levels where the anterior corticospinal tract is small but present. That is possibly due to the crossing of unmyelinated axons from the ventral corticospinal tract. Prior to this time, axons in the ventral corticospinal tract are present, but they have not yet grown decussating collaterals. Since the entire ventral corticospinal tract contains unmyelinated axons at this stage, no doubt the unmyelinated fibers in the ventral commissure represent these axons (Ranson and Clark, 1959). The gray matter in the myelin-stained sections shows heavy fascicles of myelinated fibers penetrating the dorsal horn from above and medially where it joins the intermediate gray. Myelinated fibers in fine fascicles are also scattered throughout the ventral horn. These various elements are only labeled in the lumbar enlargement where there is a high ratio of gray matter to white matter. Many of these myelinated axons are from the large neurons in the dorsal root ganglion penetrating the dorsal horn from above and below, arborizing in the intermediate gray (especially around Clarke's column), and penetrating the ventral horn to terminate on the alpha motoneurons. Some of the myelinated axons are from the motoneurons in the ventral horn that later organize into fascicles to form the ventral rootlets of spinal nerves. Finally, some of the myelinated axons may be preterminal segments penetrating the gray matter from the intraspinal tract (propriospinal).

Table 6: Glia types and concentration in the white matter in a 4-week infant

Name	Myelination	Reactive glia	Proliferating glia
DORSAL ROOT	myelinated	---	•sparse
VENTRAL ROOT	myelinated	---	---
DORSAL FUNICULUS:			
dorsal root bif. zone	*many fibers	---	sparse
dorsal root col. zone	myelinated	---	sparse
fasciculus gracilis	myelinated	---	sparse
fasciculus cuneatus	myelinated	---	sparse
Lissauer's tract	---	none	sparse
LATERAL and VENTRAL FUNICULI:			
lat. corticospinal tract	---	†gradient	sparse
ven. corticospinal tract	---	sparse	sparse
spinocerebellar tracts	myelinated	---	sparse
ventral commissure	**myelinated	---	sparse
intraspinal tract	myelinated	---	sparse
spinocephalic tract	††gradient	††gradient	sparse
med. long. fasciculus	many fibers	---	sparse
vestibulospinal tract	some fibers	---	sparse

- • dense band in a sparse field in **Plate 64**
- * intermingled in a bed of nonreactive glia
- † dense to sparse at cervical and thoracic levels, sparse at remaining levels
- ** some unmyelinated fibers at cervical levels
- †† at cervical levels: myelinating internally, reactive glia externally; at other levels: myelinated or myelinating

PLATE 55

CR 410 mm, INFANT, 4 WEEKS, Y23-60

MYELIN STAIN | **CELL BODY STAIN**

Plates 56A, 56B
Upper Cervical
Total area:
15.186 mm^2

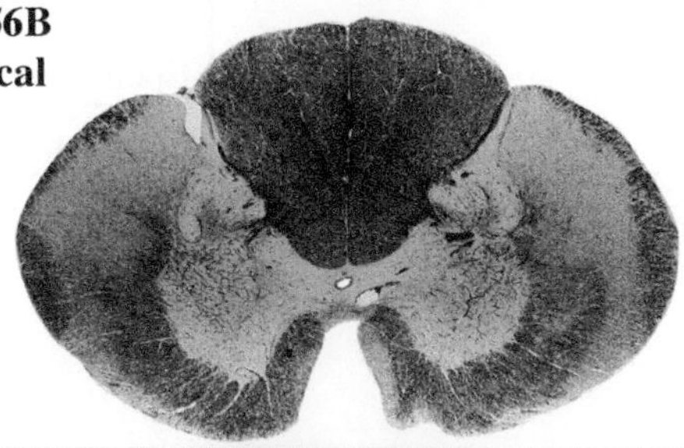

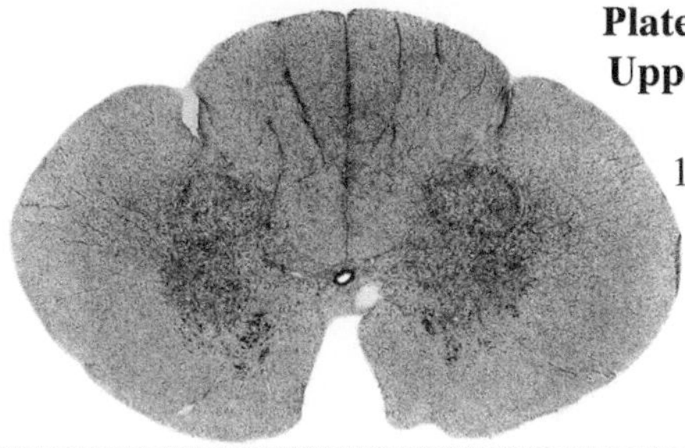

Plates 57A, 57B
Upper Cervical
Total area:
15.947 mm^2

Plates 58A, 58B
Cervical Enlargement
Total area:
17.136 mm^2

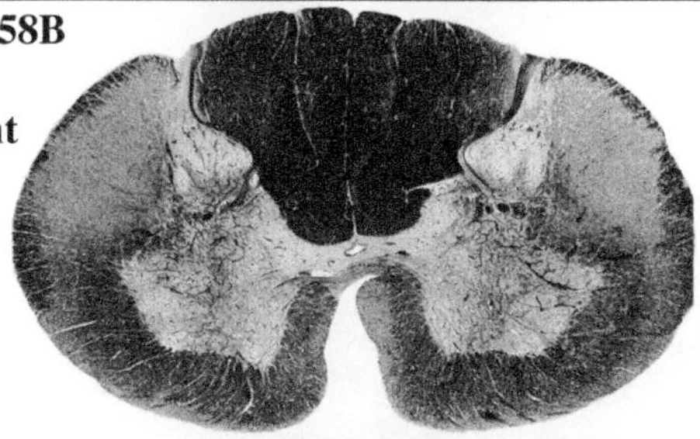

MATCHING SECTION NOT AVAILABLE

Plates 59A, 59B
Upper Thoracic
Total area:
10.017 mm^2

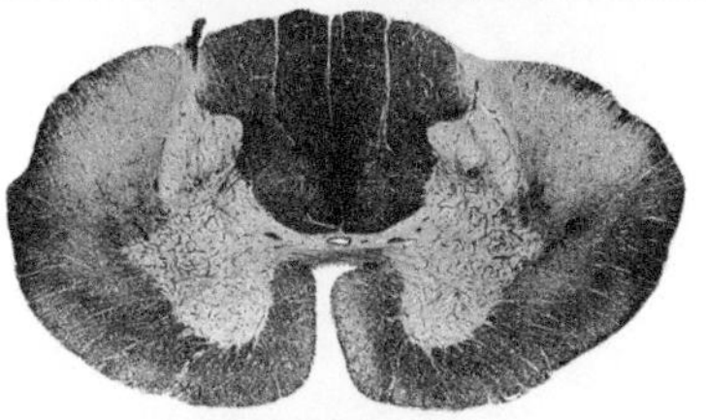

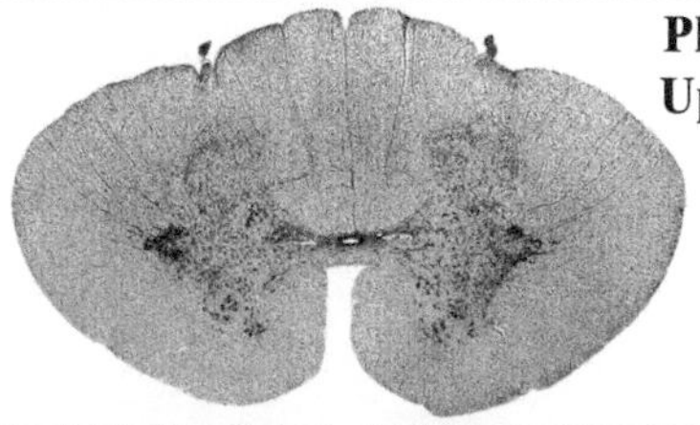

Plates 60A. 60B
Upper Thoracic
Total area:
9.8042 mm^2

Plates 61A, 61B
Lower Thoracic
Total area:
9.7045 mm^2

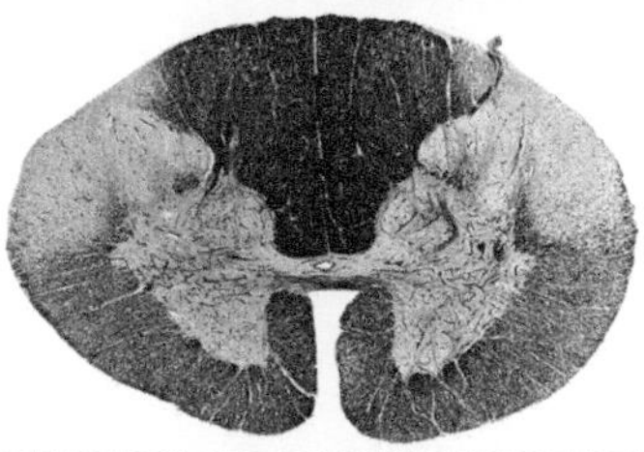

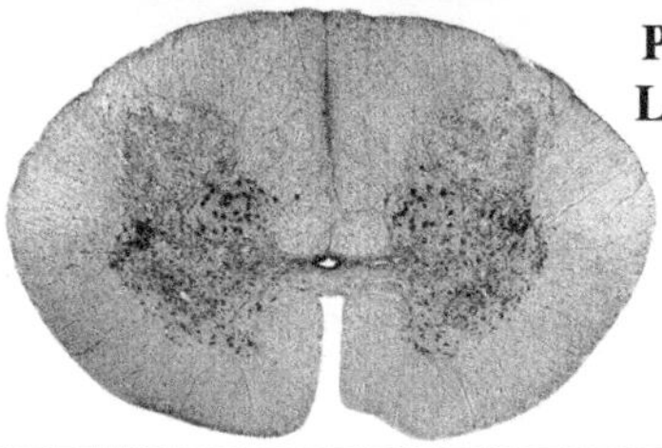

Plates 62A, 62B
Lower Thoracic
Total area:
10.856 mm^2

Plates 63A, 63B
Upper Lumbar
Total area:
11.213 mm^2

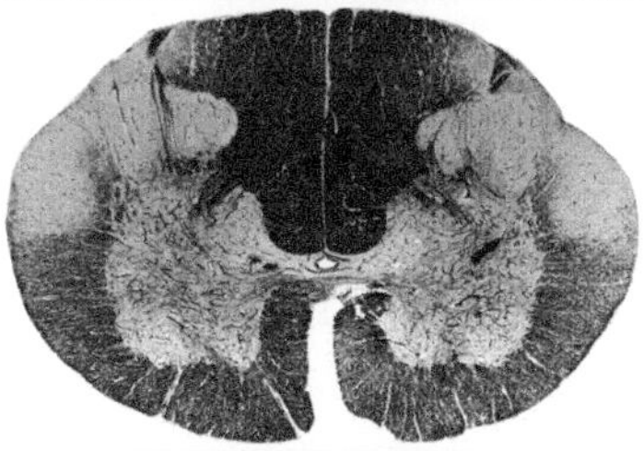

MATCHING SECTION NOT AVAILABLE

Plates 64A, 64B
Lumbar Enlargement
Total area:
14.299 mm^2

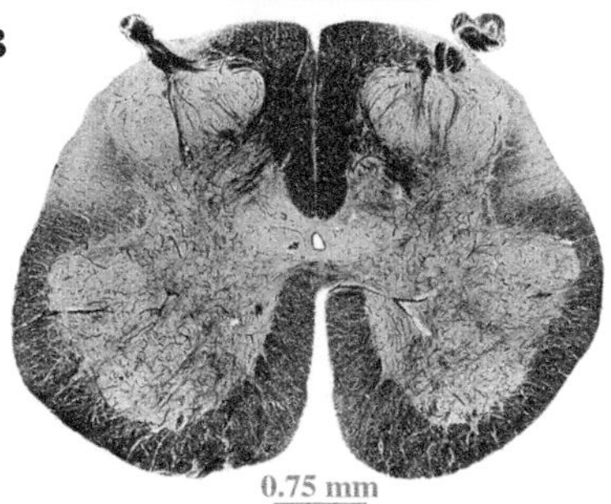

0.75 mm

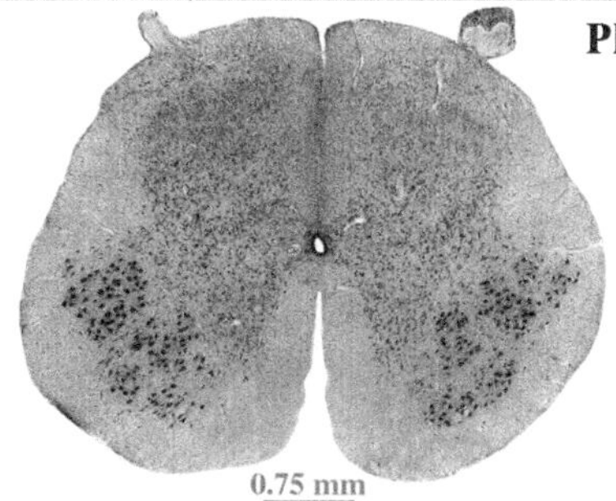

0.75 mm

Plates 65A, 65B
Lumbar Enlargement
Total area:
14.478 mm^2

PLATE 56A

CR 410 mm
Infant, 4 Weeks
Y23-60
Upper Cervical
Myelin stain

Areas (mm^2)

Central canal	.0054
Ependyma	.0115
Gray matter	3.5600
White matter	11.6090

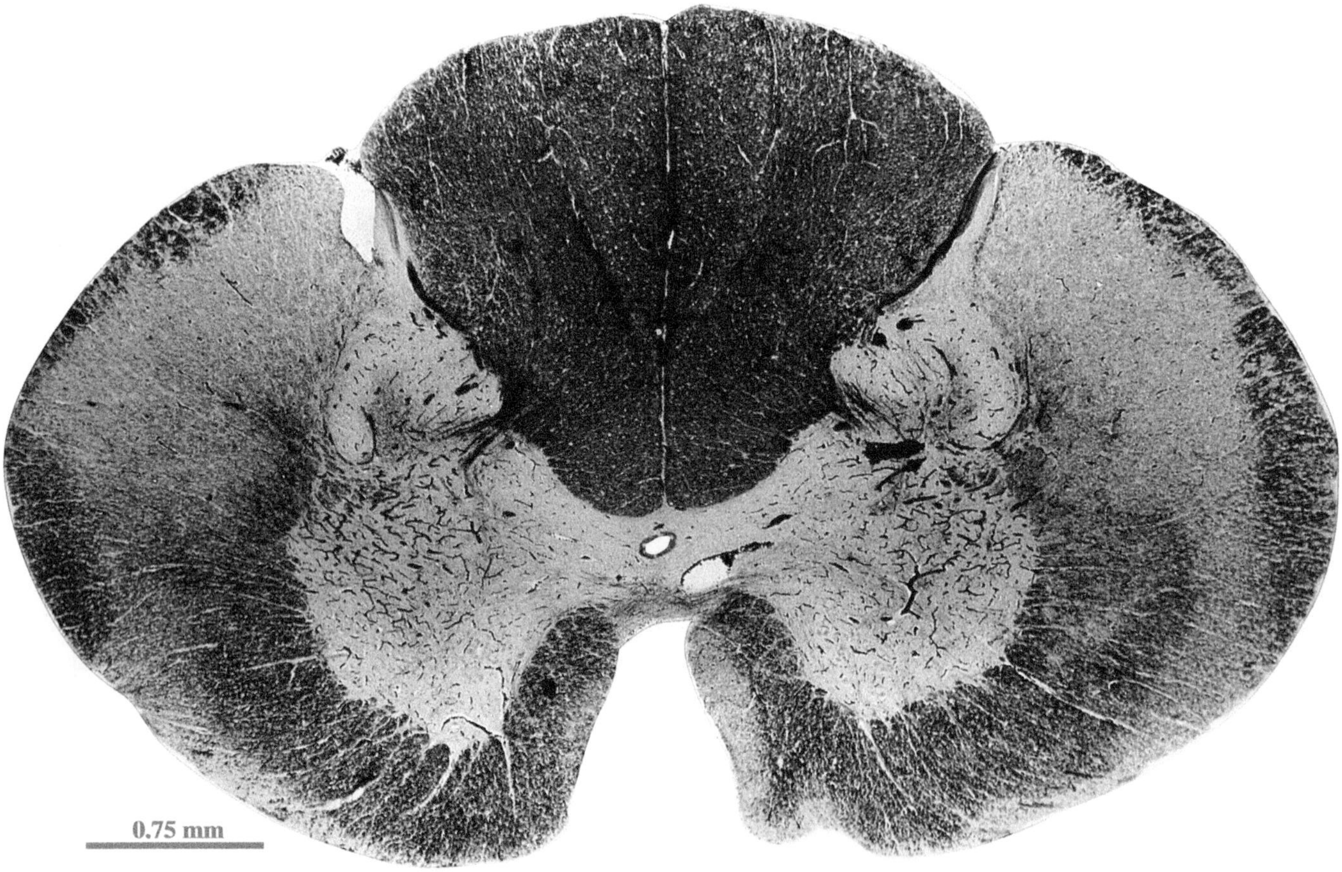

PLATE 56B

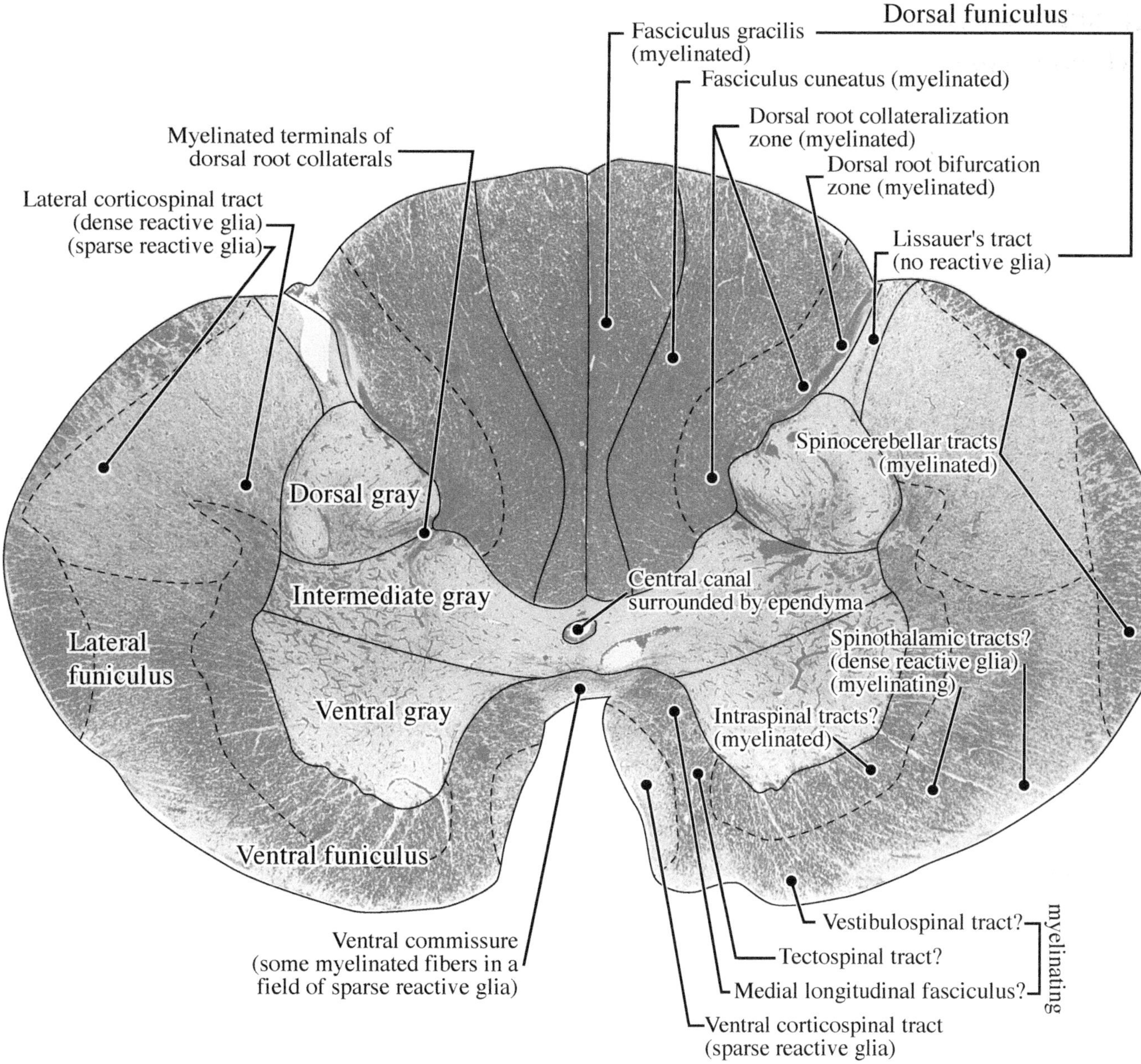
Dorsal funiculus
Fasciculus gracilis (myelinated)
Fasciculus cuneatus (myelinated)
Dorsal root collateralization zone (myelinated)
Dorsal root bifurcation zone (myelinated)
Lissauer's tract (no reactive glia)
Myelinated terminals of dorsal root collaterals
Lateral corticospinal tract
(dense reactive glia)
(sparse reactive glia)
Dorsal gray
Intermediate gray
Ventral gray
Lateral funiculus
Ventral funiculus
Spinocerebellar tracts (myelinated)
Central canal surrounded by ependyma
Spinothalamic tracts? (dense reactive glia) (myelinating)
Intraspinal tracts? (myelinated)
Vestibulospinal tract?
Tectospinal tract?
Medial longitudinal fasciculus?
myelinating
Ventral commissure (some myelinated fibers in a field of sparse reactive glia)
Ventral corticospinal tract (sparse reactive glia)

PLATE 57A

CR 410 mm
Infant, 4 Weeks
Y23-60
Upper Cervical
Cell body stain

Areas (mm^2)

Central canal	.0057
Ependyma	.0151
Gray matter	3.6685
White matter	12.2580

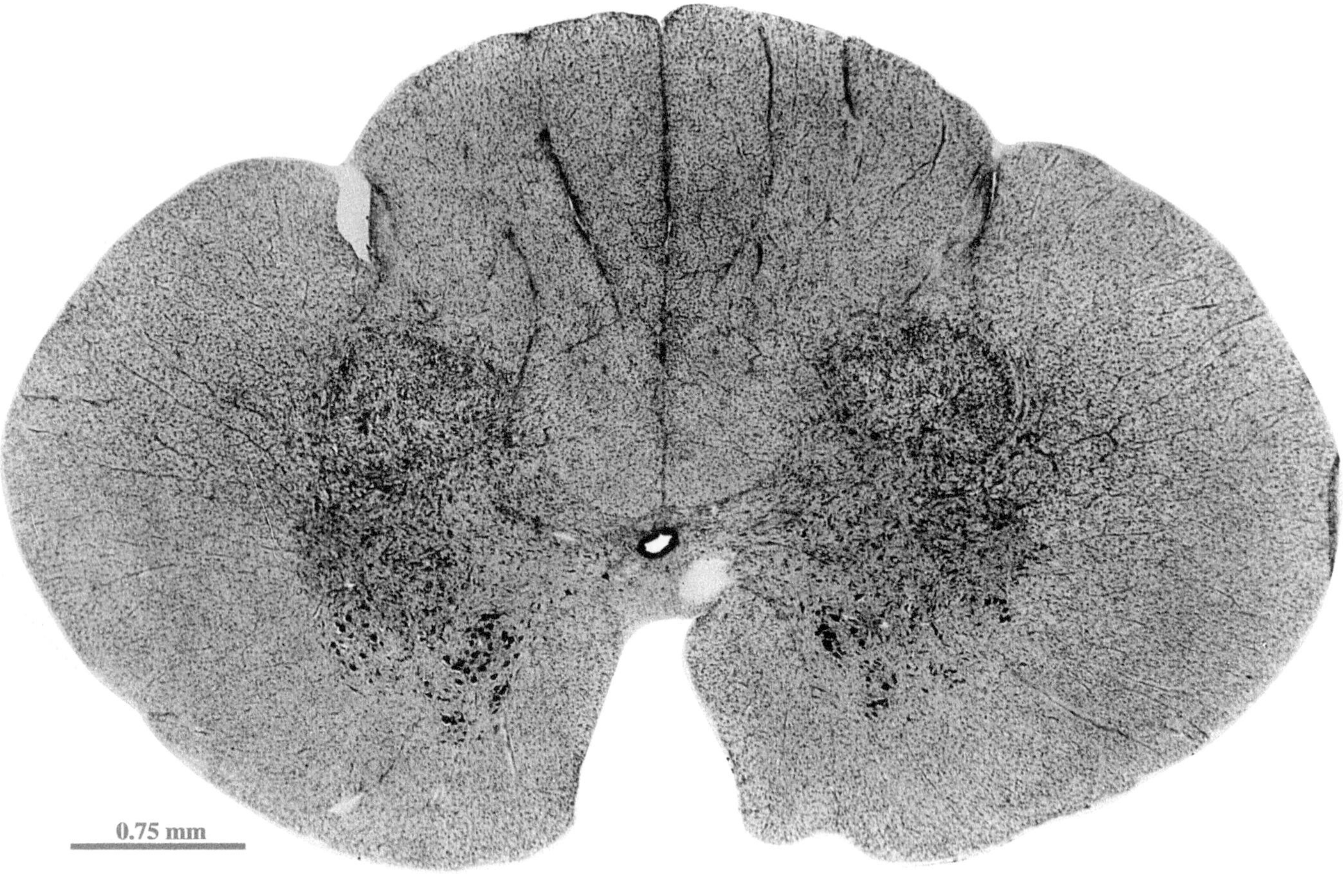

PLATE 57 B

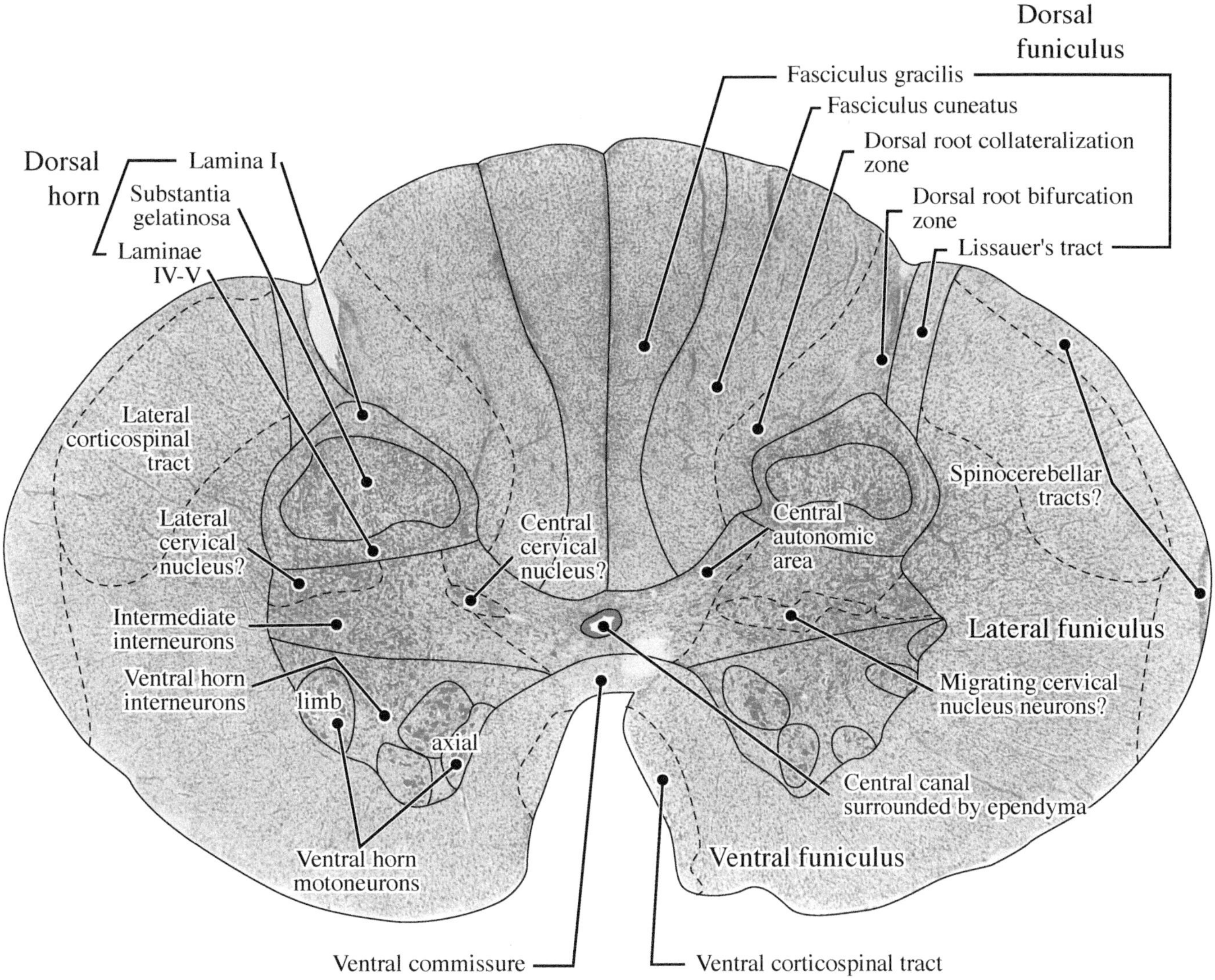
Dorsal funiculus
Fasciculus gracilis
Fasciculus cuneatus
Dorsal root collateralization zone
Dorsal root bifurcation zone
Lissauer's tract
Dorsal horn
Lamina I
Substantia gelatinosa
Laminae IV-V
Lateral corticospinal tract
Lateral cervical nucleus?
Intermediate interneurons
Ventral horn interneurons
limb
axial
Ventral horn motoneurons
Central cervical nucleus?
Central autonomic area
Spinocerebellar tracts?
Lateral funiculus
Migrating cervical nucleus neurons?
Central canal surrounded by ependyma
Ventral funiculus
Ventral commissure
Ventral corticospinal tract

PLATE 58A

CR 410 mm
Infant, 4 Weeks
Y23-60
Cervical Enlargement
Myelin stain

Areas (mm^2)

Central canal	.0062
Ependyma	.0143
Gray matter	4.7533
White matter	12.3620

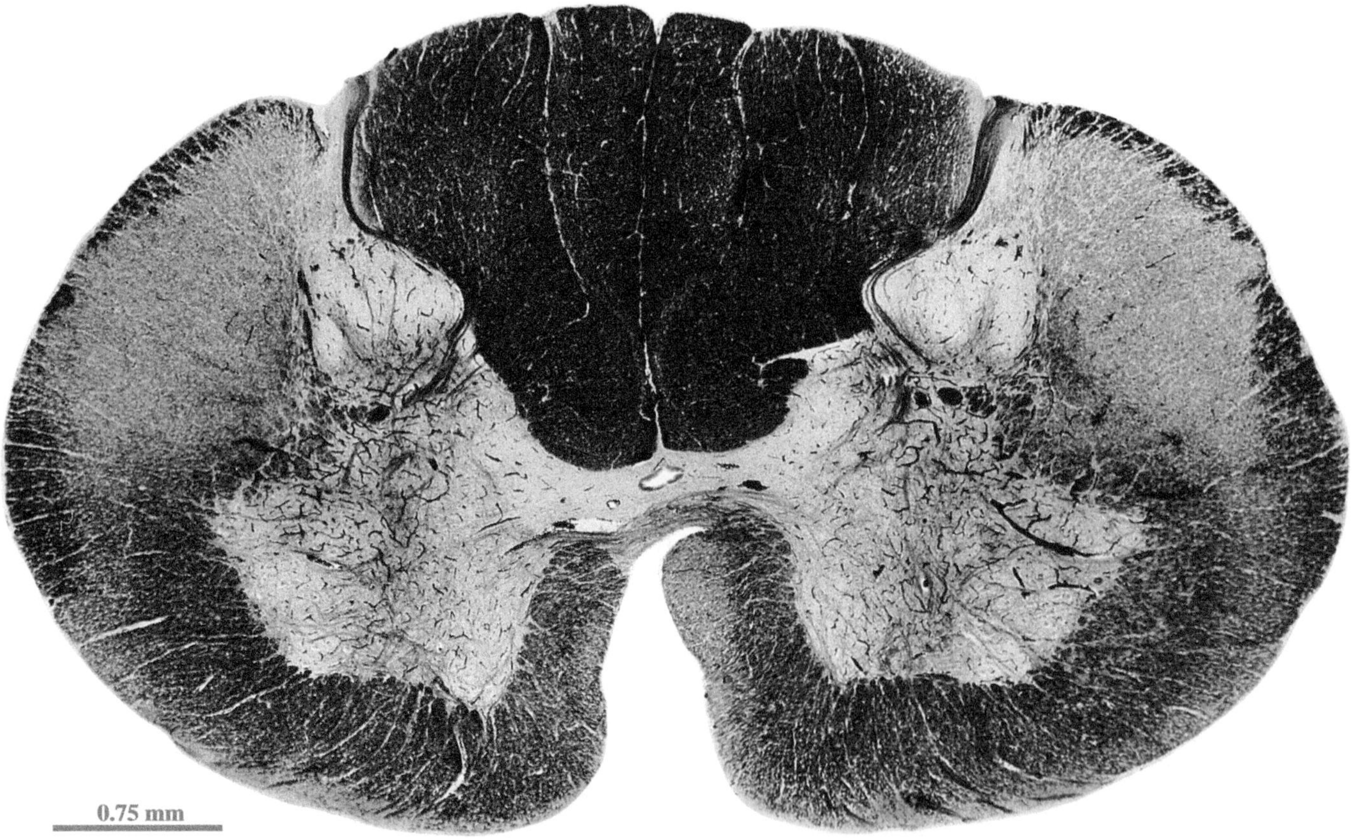

PLATE 58B

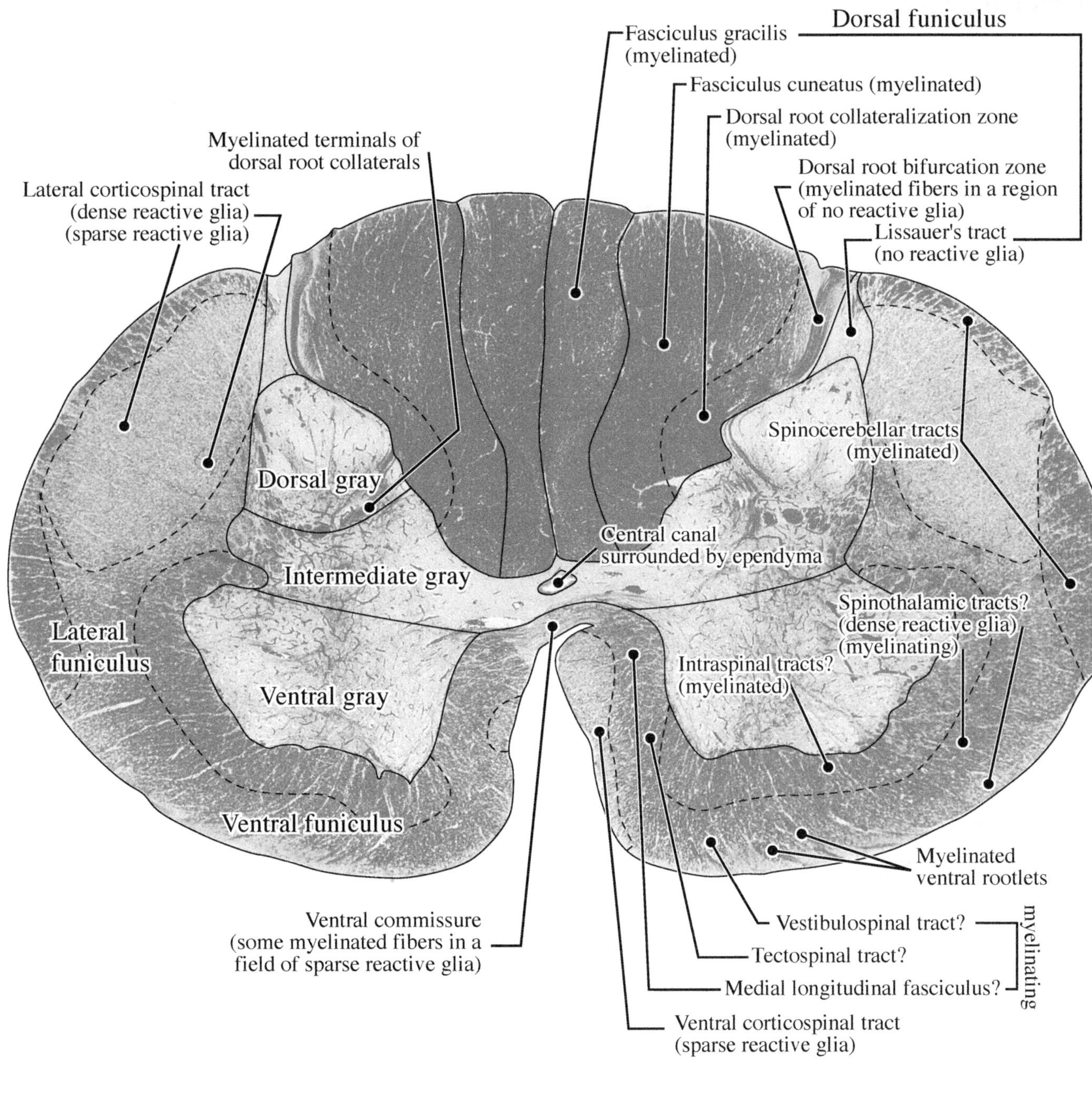
Dorsal funiculus
Fasciculus gracilis (myelinated)
Fasciculus cuneatus (myelinated)
Dorsal root collateralization zone (myelinated)
Myelinated terminals of dorsal root collaterals
Dorsal root bifurcation zone (myelinated fibers in a region of no reactive glia)
Lateral corticospinal tract (dense reactive glia) (sparse reactive glia)
Lissauer's tract (no reactive glia)
Spinocerebellar tracts (myelinated)
Dorsal gray
Central canal surrounded by ependyma
Intermediate gray
Spinothalamic tracts? (dense reactive glia) (myelinating)
Lateral funiculus
Intraspinal tracts? (myelinated)
Ventral gray
Ventral funiculus
Myelinated ventral rootlets
Ventral commissure (some myelinated fibers in a field of sparse reactive glia)
Vestibulospinal tract?
Tectospinal tract?
Medial longitudinal fasciculus?
myelinating
Ventral corticospinal tract (sparse reactive glia)

PLATE 59A

CR 410 mm
Infant, 4 Weeks
Y23-60
Upper Thoracic
Myelin stain

Areas (mm^2)

Central canal	.0032
Ependyma	.0094
Gray matter	2.3627
White matter	7.6419

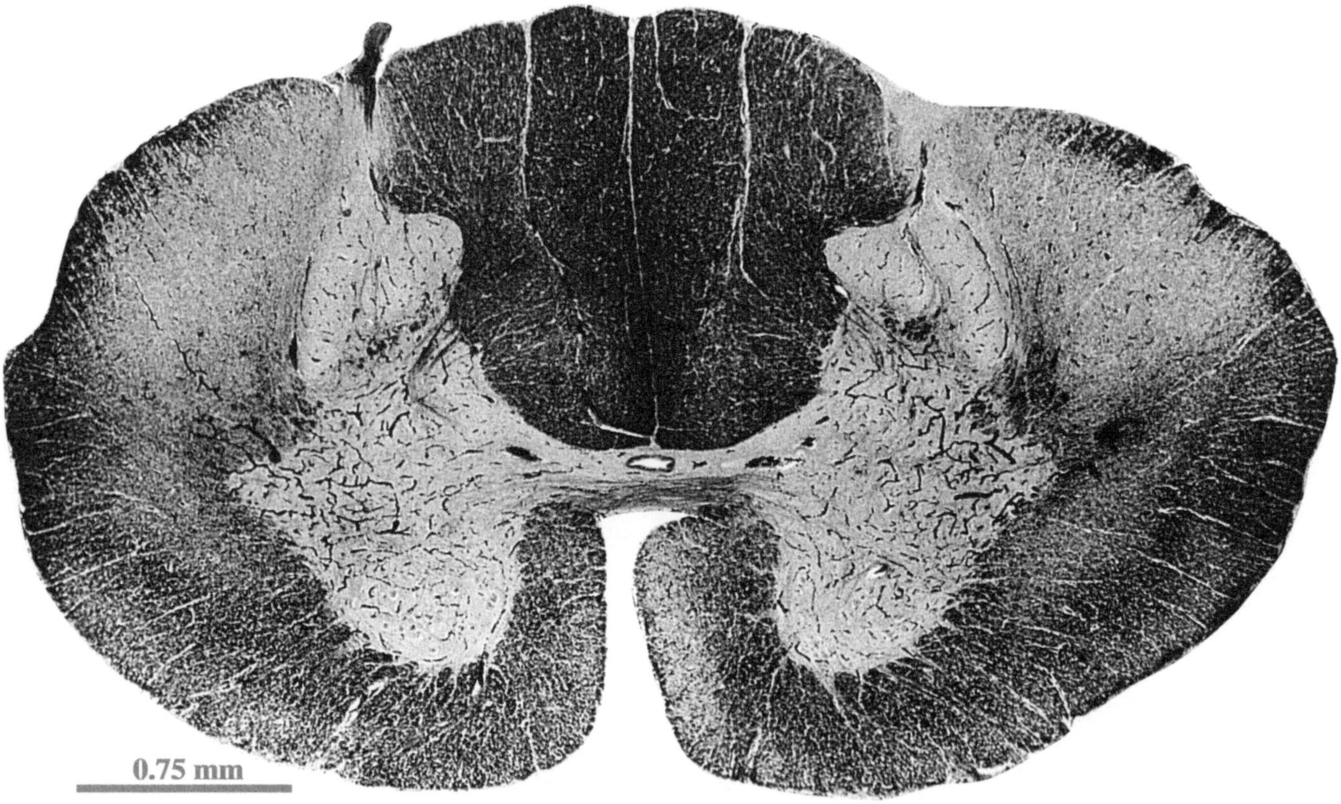

PLATE 59B

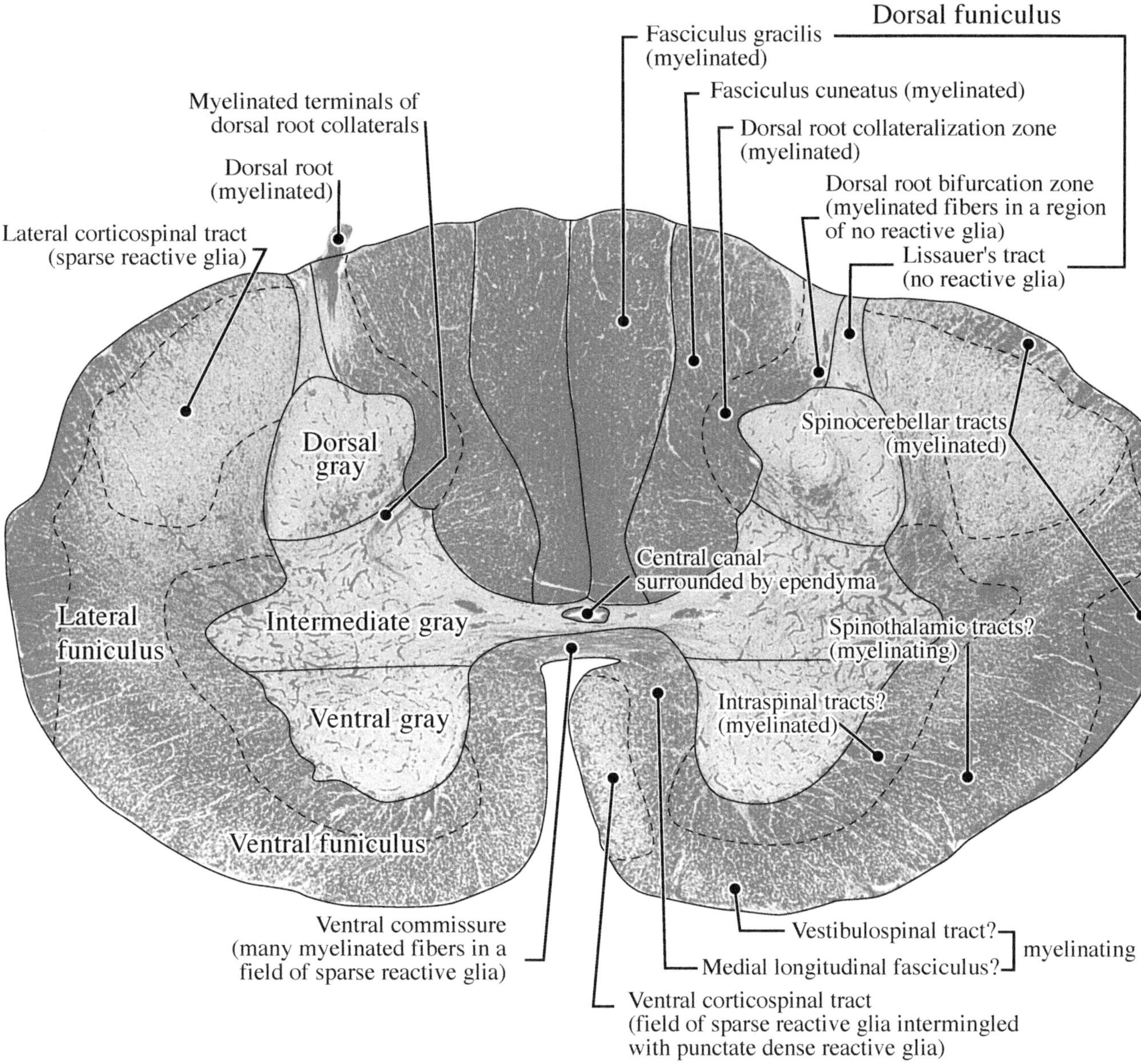
Dorsal funiculus
Fasciculus gracilis (myelinated)
Fasciculus cuneatus (myelinated)
Dorsal root collateralization zone (myelinated)
Dorsal root bifurcation zone (myelinated fibers in a region of no reactive glia)
Lissauer's tract (no reactive glia)
Myelinated terminals of dorsal root collaterals
Dorsal root (myelinated)
Lateral corticospinal tract (sparse reactive glia)
Dorsal gray
Spinocerebellar tracts (myelinated)
Central canal surrounded by ependyma
Lateral funiculus
Intermediate gray
Spinothalamic tracts? (myelinating)
Intraspinal tracts? (myelinated)
Ventral gray
Ventral funiculus
Ventral commissure (many myelinated fibers in a field of sparse reactive glia)
Vestibulospinal tract?
Medial longitudinal fasciculus?
myelinating
Ventral corticospinal tract (field of sparse reactive glia intermingled with punctate dense reactive glia)

PLATE 60A

CR 410 mm
Infant, 4 Weeks
Y23-60
Upper Thoracic
Cell body stain

Areas (mm^2)

Central canal	.0026
Ependyma	.0113
Gray matter	2.2909
White matter	7.4994

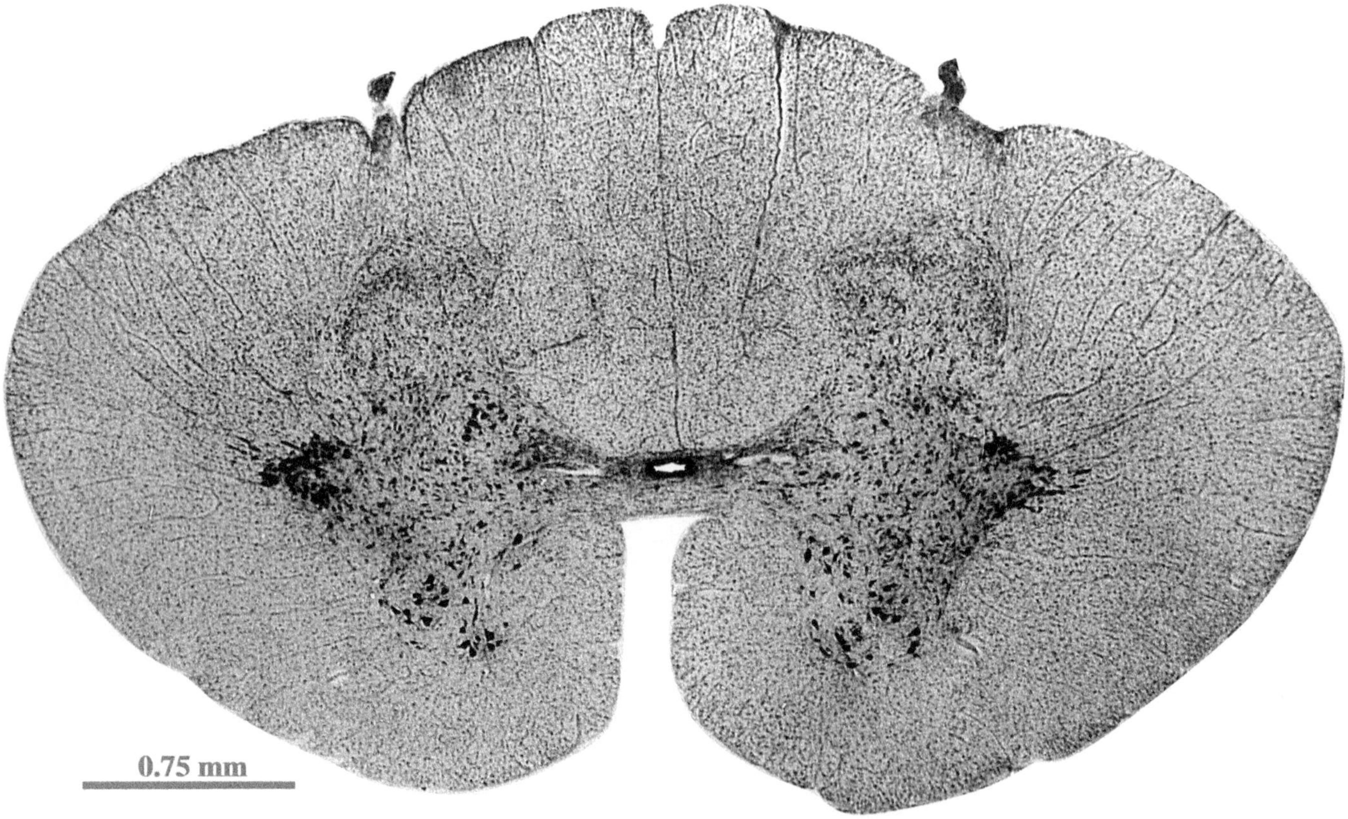

PLATE 60B

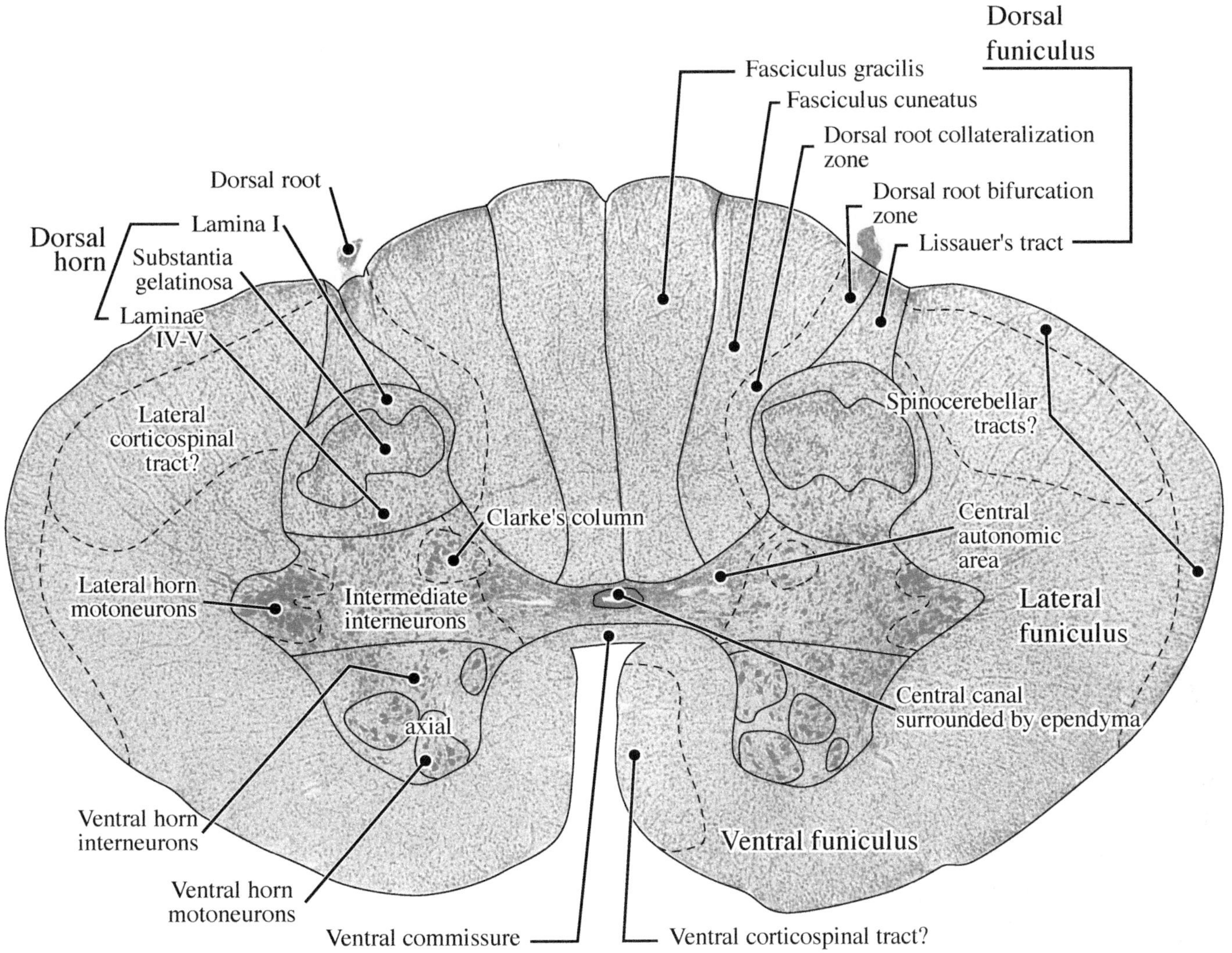
Dorsal funiculus
Fasciculus gracilis
Fasciculus cuneatus
Dorsal root collateralization zone
Dorsal root bifurcation zone
Lissauer's tract
Dorsal root
Dorsal horn
Lamina I
Substantia gelatinosa
Laminae IV-V
Lateral corticospinal tract?
Spinocerebellar tracts?
Clarke's column
Central autonomic area
Lateral horn motoneurons
Intermediate interneurons
Lateral funiculus
Central canal surrounded by ependyma
axial
Ventral horn interneurons
Ventral horn motoneurons
Ventral funiculus
Ventral commissure
Ventral corticospinal tract?

PLATE 61A

CR 410 mm
Infant, 4 Weeks
Y23-60
Lower Thoracic
Myelin stain

Areas (mm^2)

Central canal	.0029
Ependyma	.0083
Gray matter	2.9576
White matter	6.7358

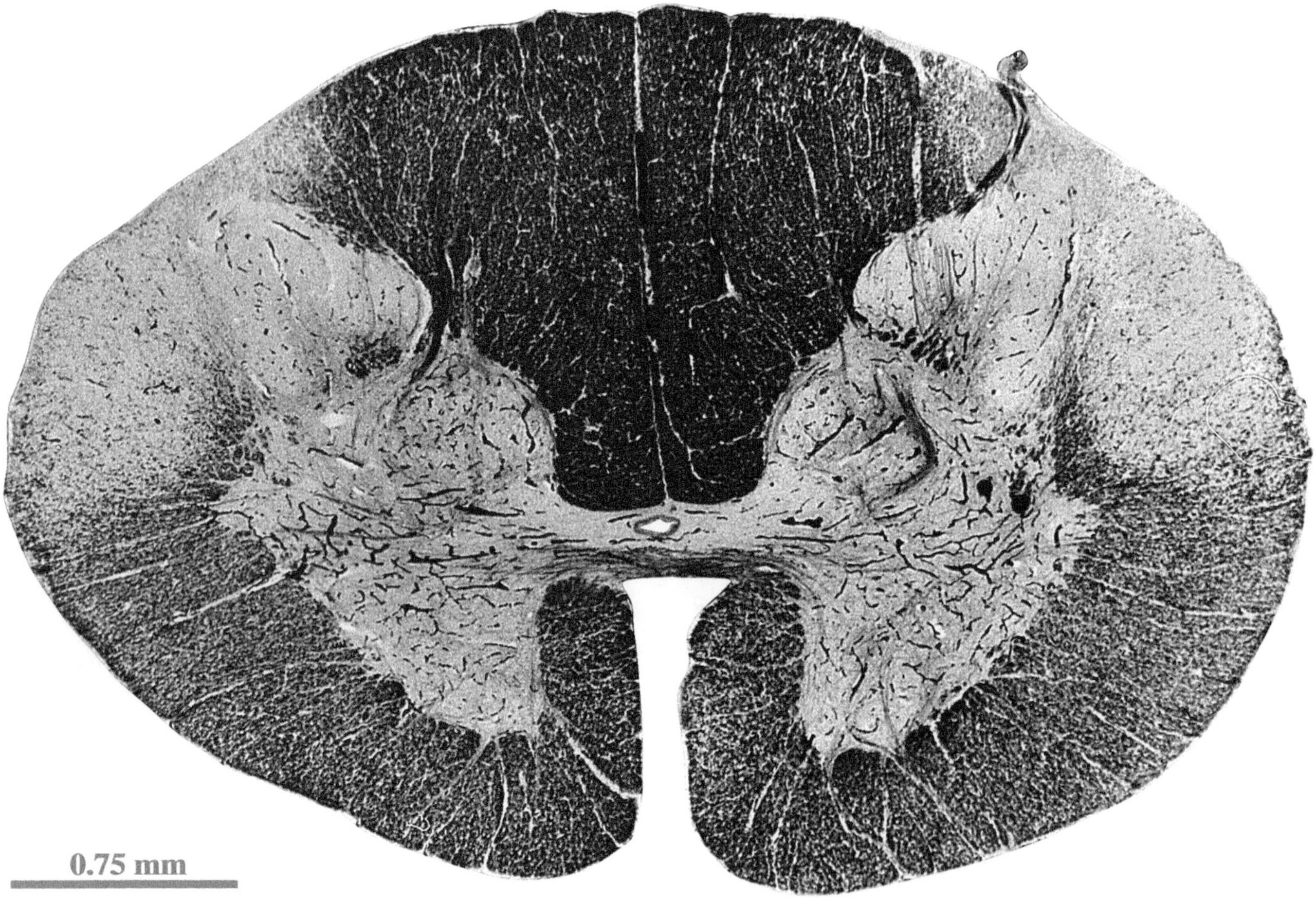

PLATE 61B

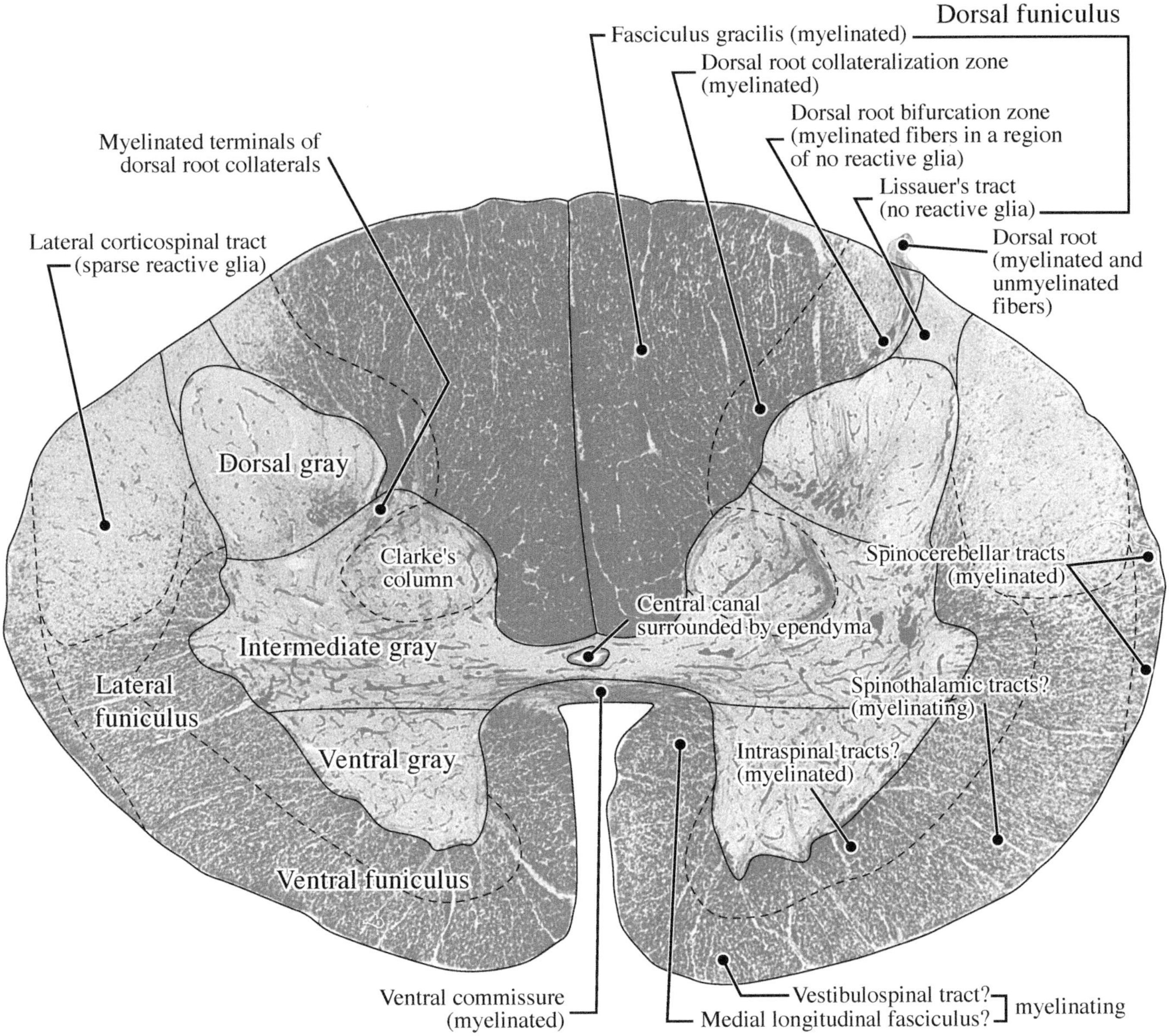
Dorsal funiculus
Fasciculus gracilis (myelinated)
Dorsal root collateralization zone (myelinated)
Dorsal root bifurcation zone (myelinated fibers in a region of no reactive glia)
Lissauer's tract (no reactive glia)
Dorsal root (myelinated and unmyelinated fibers)
Myelinated terminals of dorsal root collaterals
Lateral corticospinal tract (sparse reactive glia)
Dorsal gray
Clarke's column
Intermediate gray
Central canal surrounded by ependyma
Spinocerebellar tracts (myelinated)
Spinothalamic tracts? (myelinating)
Lateral funiculus
Ventral gray
Intraspinal tracts? (myelinated)
Ventral funiculus
Ventral commissure (myelinated)
Vestibulospinal tract?
Medial longitudinal fasciculus?
myelinating

PLATE 62A

CR 410 mm
Infant, 4 Weeks
Y23-60
Lower Thoracic
Cell body stain

Areas (mm^2)

Central canal	.0045
Ependyma	.0132
Gray matter	3.5220
White matter	7.3162

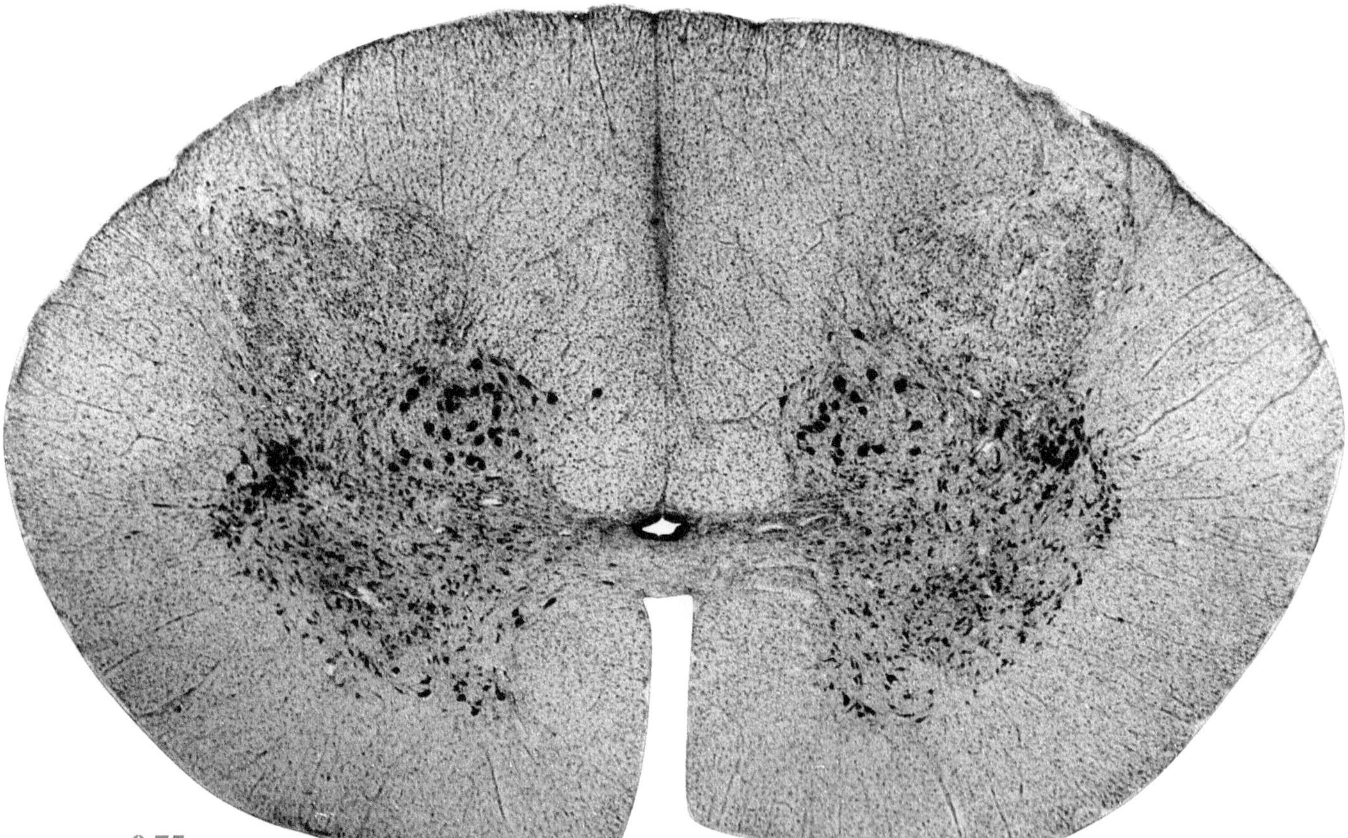

0.75 mm

PLATE 62B

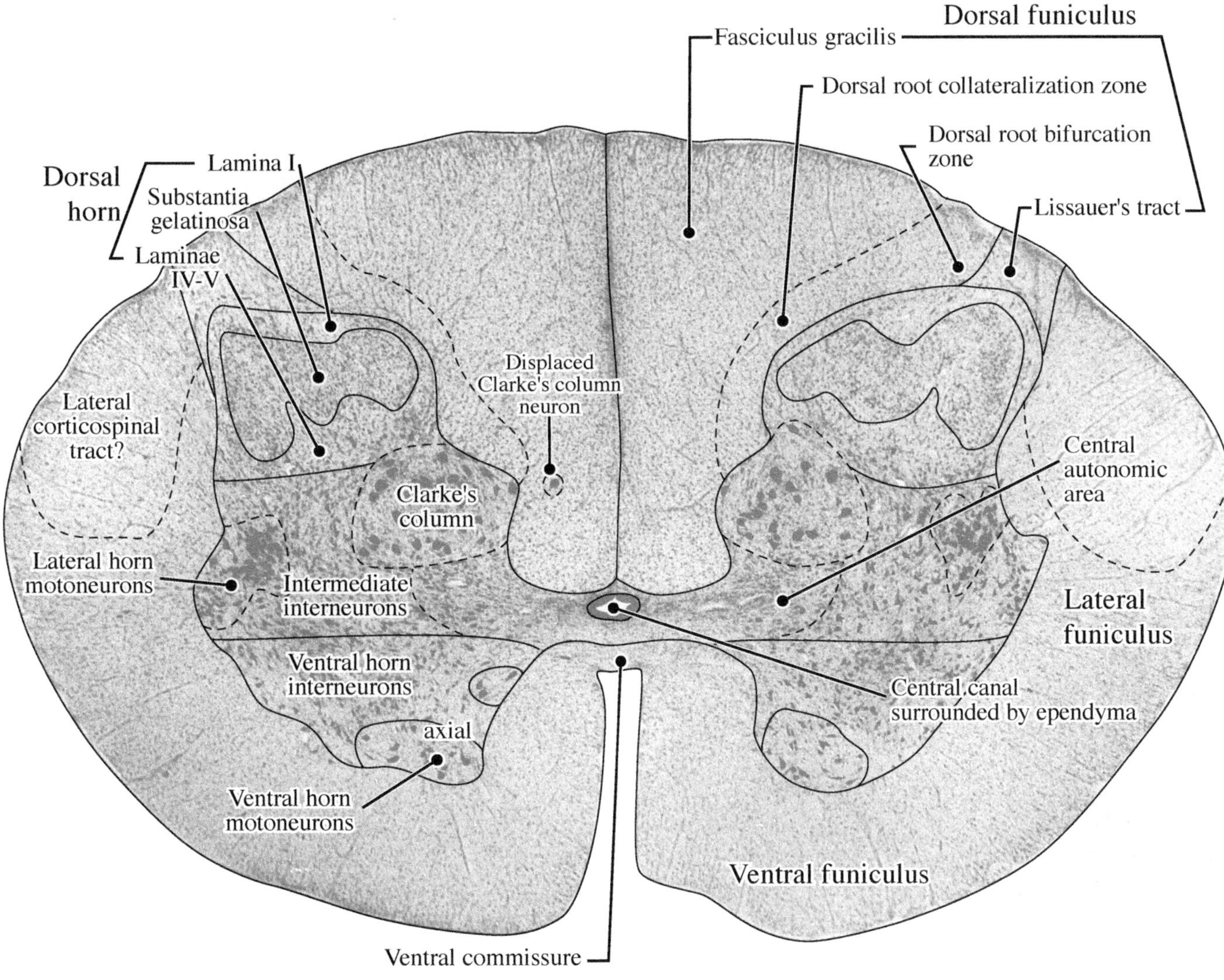
Dorsal funiculus
Fasciculus gracilis
Dorsal root collateralization zone
Dorsal root bifurcation zone
Lissauer's tract
Dorsal horn
Lamina I
Substantia gelatinosa
Laminae IV-V
Displaced Clarke's column neuron
Lateral corticospinal tract?
Clarke's column
Central autonomic area
Lateral horn motoneurons
Intermediate interneurons
Lateral funiculus
Ventral horn interneurons
Central canal surrounded by ependyma
axial
Ventral horn motoneurons
Ventral funiculus
Ventral commissure

PLATE 63A

CR 410 mm
Infant, 4 Weeks
Y23-60
Upper Lumbar
Myelin stain

Areas (mm²)

Central canal	.0070
Ependyma	.0070
Gray matter	4.1280
White matter	7.0713

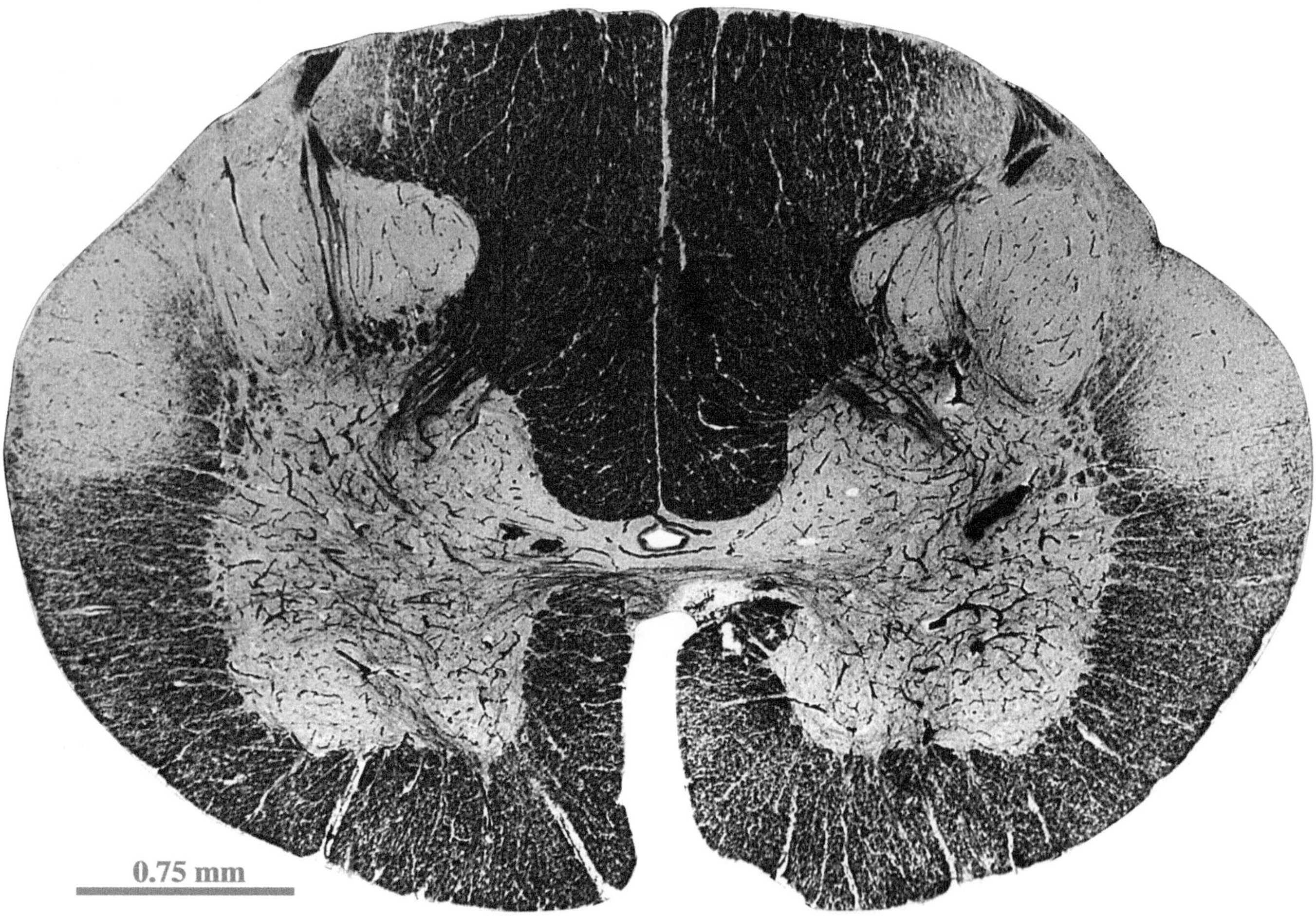

PLATE 63B

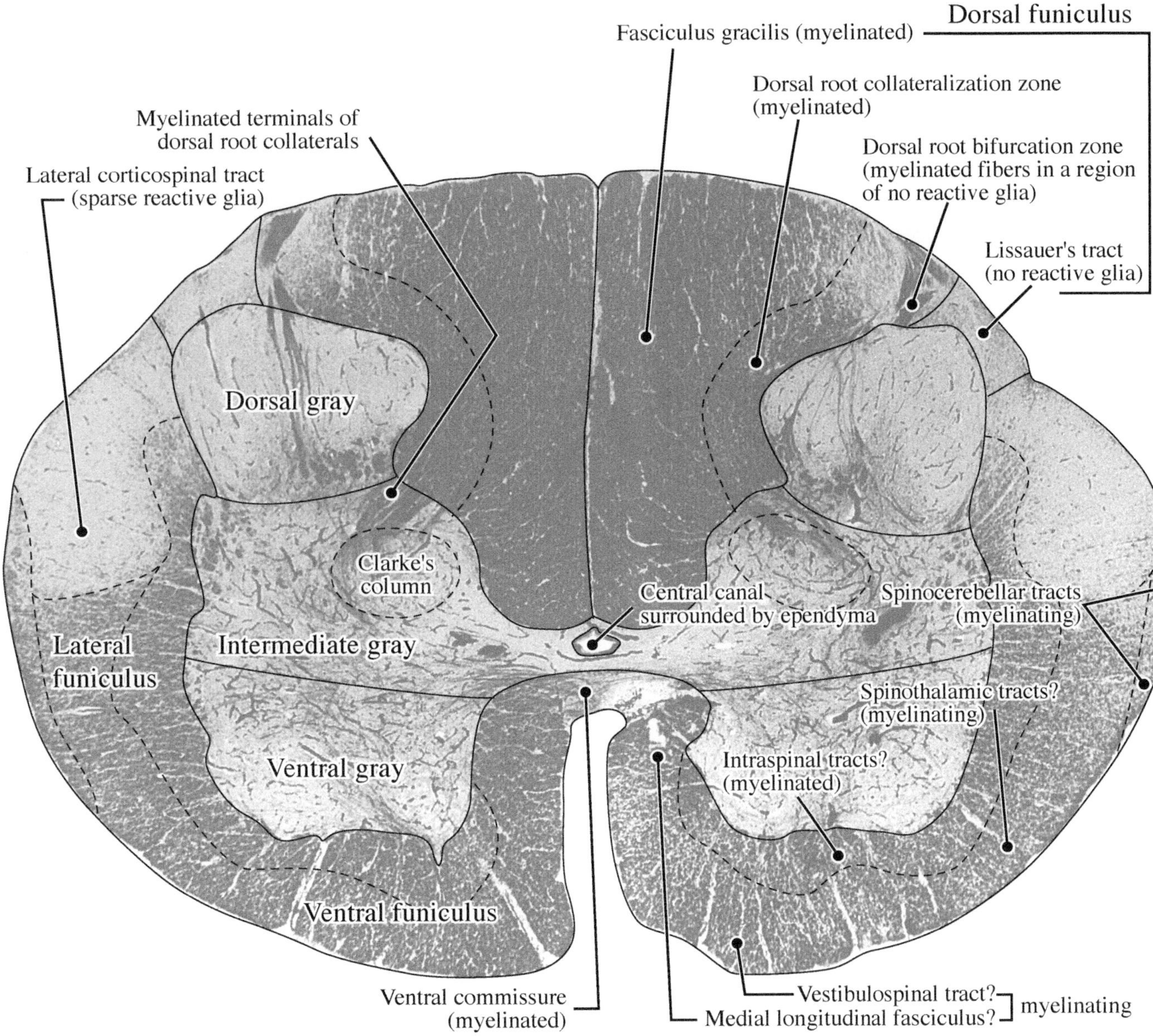
Dorsal funiculus
Fasciculus gracilis (myelinated)
Dorsal root collateralization zone (myelinated)
Myelinated terminals of dorsal root collaterals
Dorsal root bifurcation zone (myelinated fibers in a region of no reactive glia)
Lateral corticospinal tract (sparse reactive glia)
Lissauer's tract (no reactive glia)
Dorsal gray
Clarke's column
Central canal surrounded by ependyma
Spinocerebellar tracts (myelinating)
Lateral funiculus
Intermediate gray
Spinothalamic tracts? (myelinating)
Intraspinal tracts? (myelinated)
Ventral gray
Ventral funiculus
Ventral commissure (myelinated)
Vestibulospinal tract?
Medial longitudinal fasciculus?
myelinating

PLATE 64A

CR 410 mm
Infant, 4 Weeks
Y23-60
Lumbar Enlargement
Myelin stain

Areas (mm²)

Central canal	.0090
Ependyma	.0086
Gray matter	7.8739
White matter	6.4072

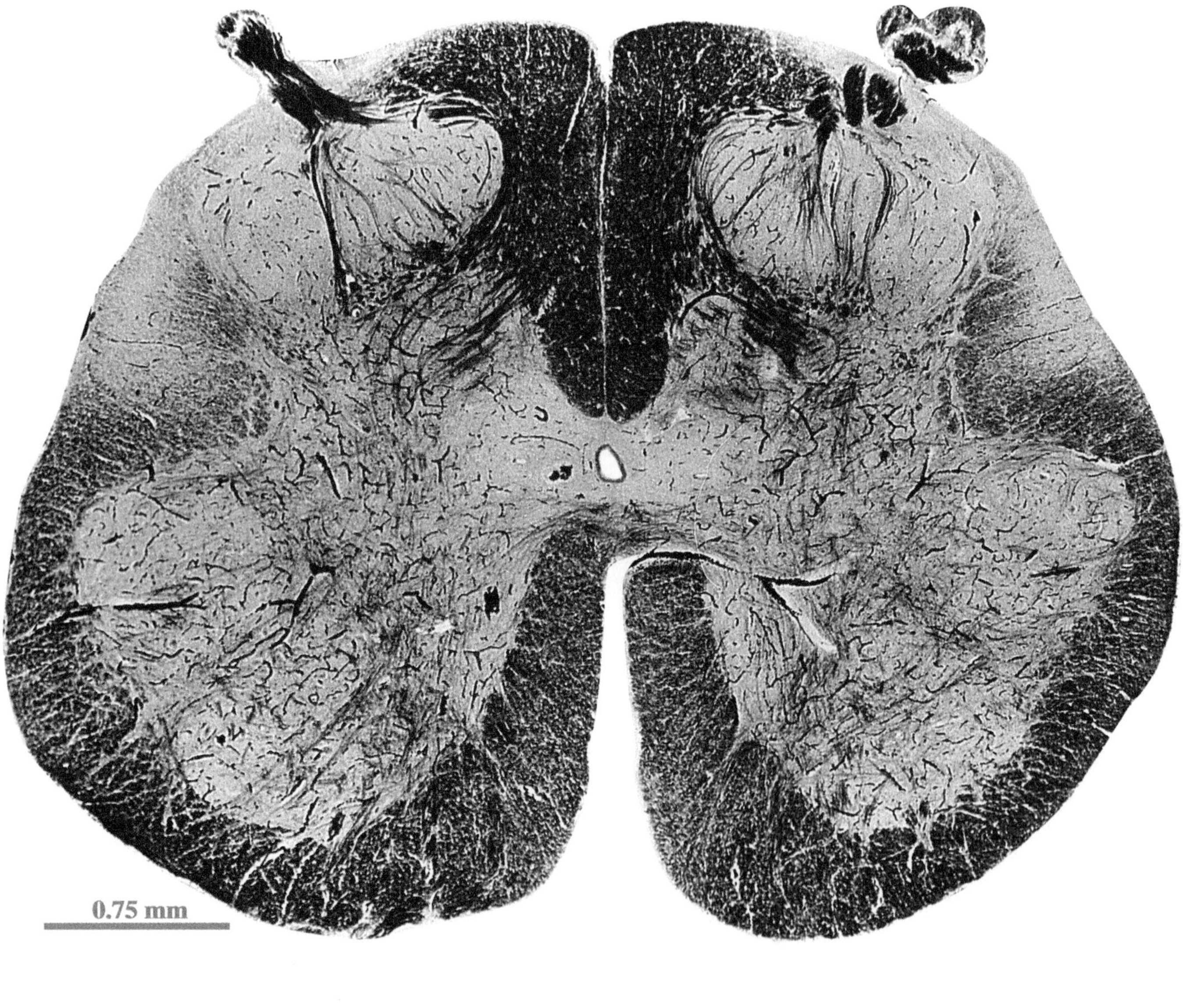

PLATE 64B

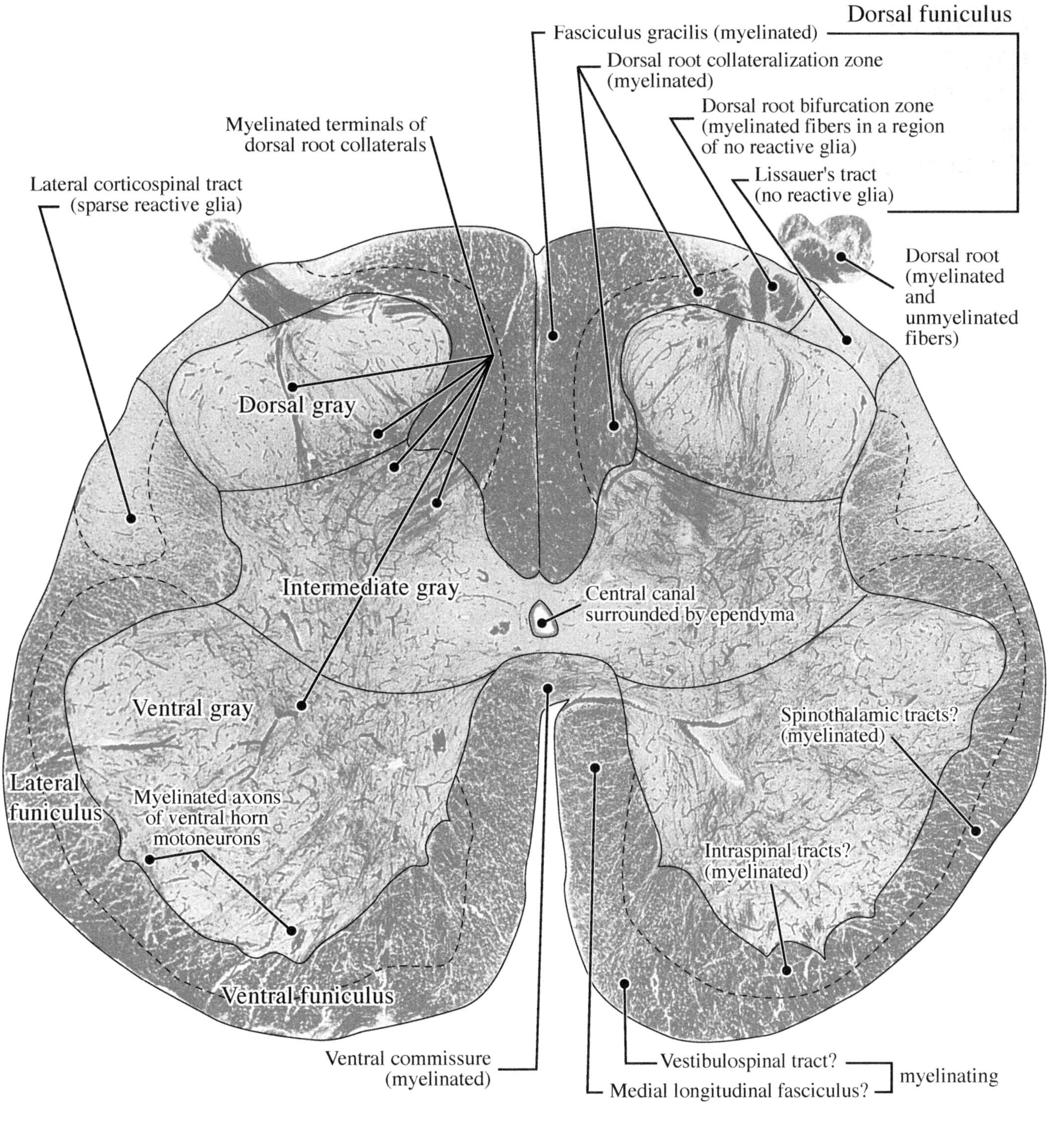
Dorsal funiculus
Fasciculus gracilis (myelinated)
Dorsal root collateralization zone (myelinated)
Dorsal root bifurcation zone (myelinated fibers in a region of no reactive glia)
Lissauer's tract (no reactive glia)
Myelinated terminals of dorsal root collaterals
Lateral corticospinal tract (sparse reactive glia)
Dorsal root (myelinated and unmyelinated fibers)
Dorsal gray
Intermediate gray
Central canal surrounded by ependyma
Ventral gray
Spinothalamic tracts? (myelinated)
Lateral funiculus
Myelinated axons of ventral horn motoneurons
Intraspinal tracts? (myelinated)
Ventral funiculus
Ventral commissure (myelinated)
Vestibulospinal tract?
Medial longitudinal fasciculus?
myelinating

PLATE 65A

CR 410 mm
Infant, 4 Weeks
Y23-60
Lumbar Enlargement
Cell body stain

Areas (mm²)

Central canal	.0075
Ependyma	.0148
Gray matter	7.9194
White matter	6.5366

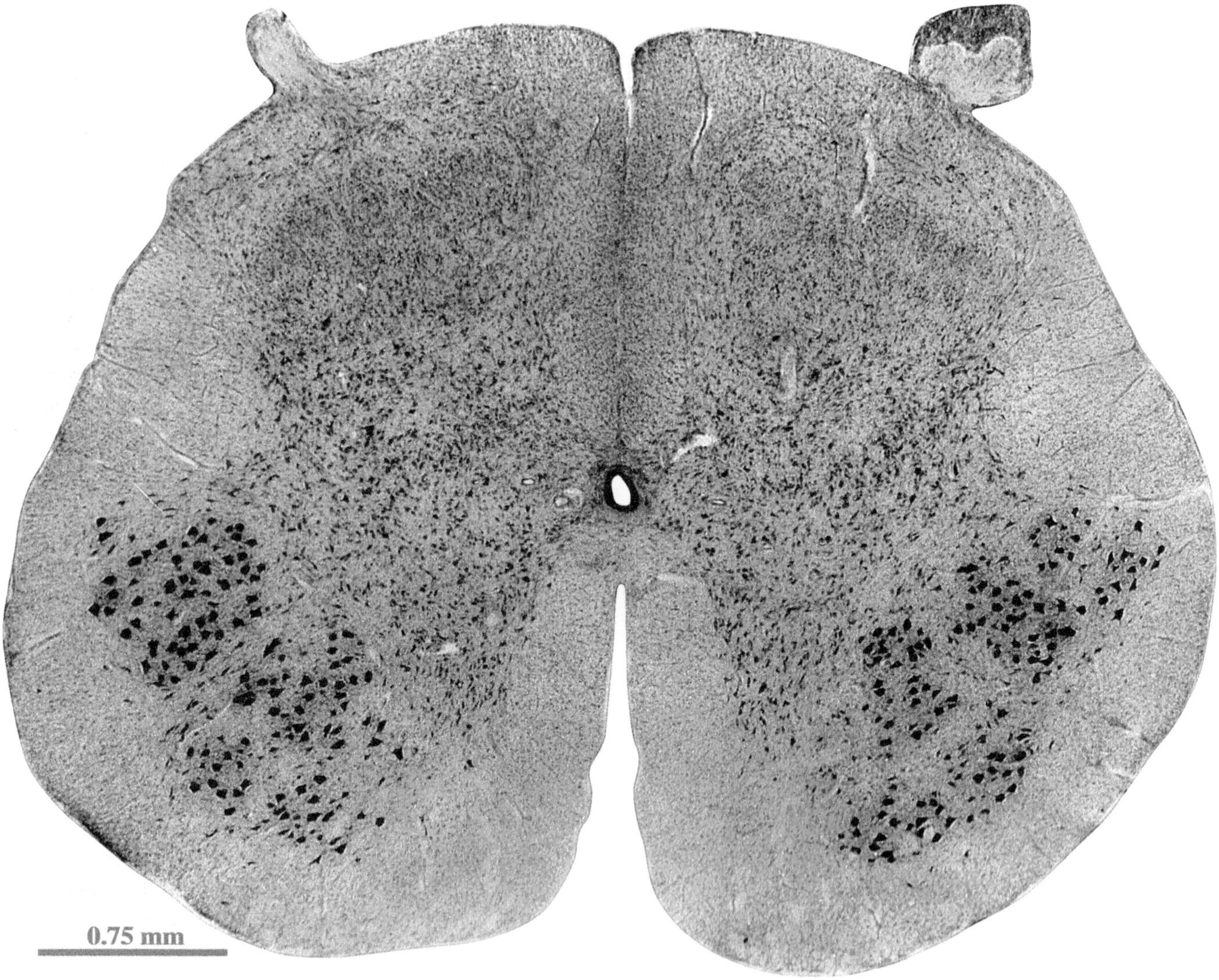

PLATE 65B

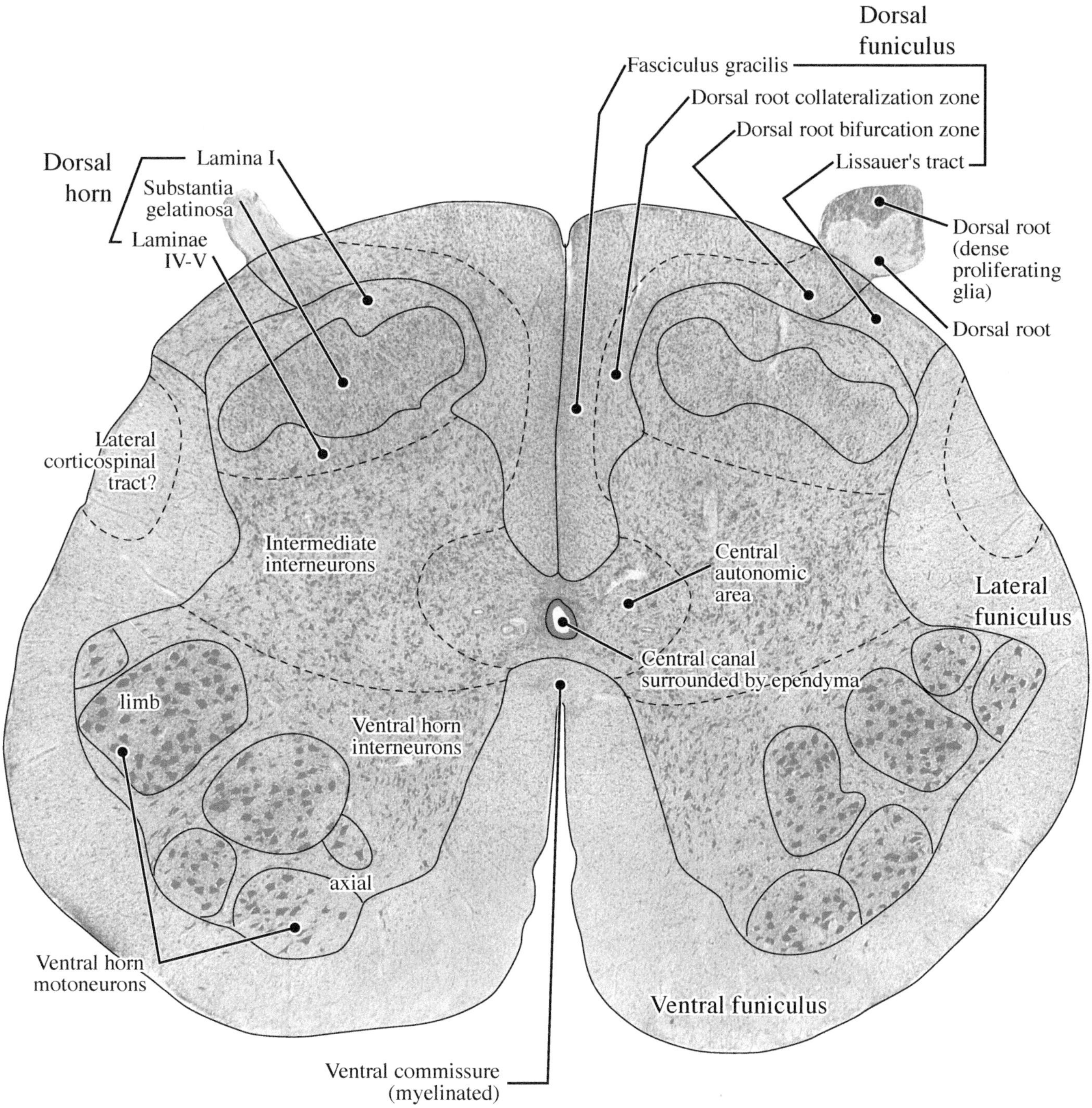
Dorsal funiculus
Fasciculus gracilis
Dorsal root collateralization zone
Dorsal root bifurcation zone
Lissauer's tract
Dorsal horn
Lamina I
Substantia gelatinosa
Laminae IV-V
Dorsal root (dense proliferating glia)
Dorsal root
Lateral corticospinal tract?
Intermediate interneurons
Central autonomic area
Lateral funiculus
Central canal surrounded by ependyma
limb
Ventral horn interneurons
axial
Ventral horn motoneurons
Ventral funiculus
Ventral commissure (myelinated)

PART VIII: Y286-62 CR 440 mm (4 Months)

Plate 66 is a survey of matched myelin-stained and cell-body-stained sections from Y286-62, a specimen in the Yakovlev Collection. All sections are shown at the same scale. The boxes enclosing each section list the approximate level and the total area of the section in square millimeters (mm^2). Full-page normal-contrast photographs of each section are in **Plates 67A–80A.** Low contrast photographs with superimposed labels and outlines of structural details are in **Plates 67B–80B.** In this specimen, the myelin-stained and cell-body-stained sections were preserved on separate large glass plates without any section numbers. Forty one myelin-stained sections and 38 cell-body-stained sections were photographed ranging from upper cervical to lumbar enlargement levels. Sections beyond the lumbar enlargement were damaged. The 79 photographic prints were intuitively arranged from upper cervical to lumbar enlargement levels, using internal features such as the size of the corticospinal tracts, and the width of the ventral horn. Then, myelin- and cell-body-stained sections were matched and 7 different levels for analysis.

As in the previous specimens, the cross-sectional area of a myelin-stained section is smaller than the matching cell-body-stained section with the exception of the cervidal enlargement. At that level, the cell-body and myelin-stained sections are damaged with a cut across the dorsal funiculus; the cell-body-stained section may not have expanded normally during processing. Using the total areas of the myelin-stained sections, the overall size differences between levels indicate the following comparisons, The cervical enlargement has the largest cross-sectional area, but it is only 5% larger than the lumbar enlargement. The middle thoracic level has the smallest cross sectional area, 54% smaller than the cervical enlargement and 52% smaller than the lumbar enlargement.

Myelination in this specimen is approaching completion, and is approaching the adult pattern (**Table 7**). All parts of the dorsal funiculus are myelinated except Lissauer's tract, which remains unmyelinated in maturity. The lateral corticospinal tract has a gradient of myelination. At cervical levels, medial parts are myelinated and a thin lateral crescent adjacent to the spinocerebellar tract is still myelinating. At thoracic levels, the gradient exists, except the medial parts are myelinating, but lateral parts contain dense reactive glia. At lumbar levels, there is no myelination, but it has dense to very sparse reactive glia. The ventral corticospinal tract may or may not exist in this specimen. If it is present, it is completely myelinated throughout its length (*see* Chapter 9, Figures 9-25 through 9-32, Altman and Bayer, 2001). There is some indication of a myelination gradient in the spinothalamic tracts at the upper cervical level, because the outer fibers appear to myelinate later than the inner fibers (those closer to the intraspinal tracts). Caudally, the spinocephalic tract is myelinated. It is important to note that the myelination gradients in the lateral corticospinal and spinocephalic tracts give clear evidence that axons myelinate first proximally (near the cell body) then distally (near the axon terminals). In addition, the mediolateral myelination gradient in the lateral corticospinal tract is most probably related to the retarded entry of axons from later-generated neurons in medial parts of the primary motor cortex (destined to terminate in the ventral horn at lumbar and sccral levels).

The gray matter in the myelin-stained sections shows heavy fascicles of myelinated fibers penetrating the dorsal horn that were pointed out in the previous specimen. Myelination in the subgelatinosal plexus and the reticulated area continues to progress.

Table 7: Glia types and concentration in the white matter in a 4–month infant

Name	Myelination	Reactive glia	Proliferating glia
DORSAL ROOT	myelinated	---	sparse
VENTRAL ROOT	myelinated	---	---
DORSAL FUNICULUS:			
dorsal root bif. zone	*many fibers	---	sparse
dorsal root col. zone	myelinated	---	sparse
fasciculus gracilis	myelinated	---	sparse
fasciculuc cuneatus	myelinated	---	sparse
Lissauer's tract	---	none	sparse
LATERAL and VENTRAL FUNICULI:			
lat. corticospinal tract	---	†gradient	sparse
ven. corticospinal tract	**myelinated	---	sparse
spinocerebellar tracts	myelinated	---	sparse
ventral commissure	myelinated	---	sparse
intraspinal tract	myelinated	---	sparse
spinocephalic tract	††gradient	††gradient	sparse
med. long. fasciculus	some fibers	---	sparse
vestibulospinal tract	some fibers	---	sparse

* intermingled with unmyelinated fibers
† myelinated to myelinating at cervical and thoracic levels, dense reactive glia to sparse reactive glia at upper lumbar level, sparse reactive glia at lumbar enlargement
** the presence of this tract is assumed. If present, it is myelinated.
†† at upper cervical level: myelinating internally, dense reactive glia externally; at other levels: myelinated or myelinating

PLATE 66

CR 440 mm, INFANT, 4 MONTHS, Y286-62

MYELIN STAIN		CELL BODY STAIN	
Plates 67A, 67B **Upper Cervical** Total area: 27.489 mm^2	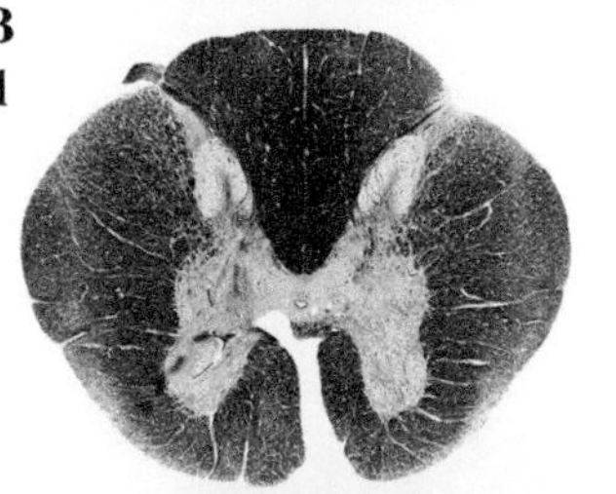	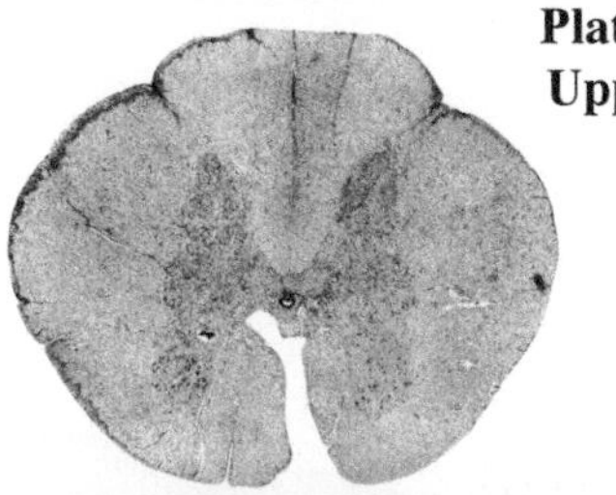	**Plates 68A, 68B** **Upper Cervical** Total area: 28.327 mm^2
Plates 69A, 69B **Cervical Enlargement** Total area: 33.314 mm^2	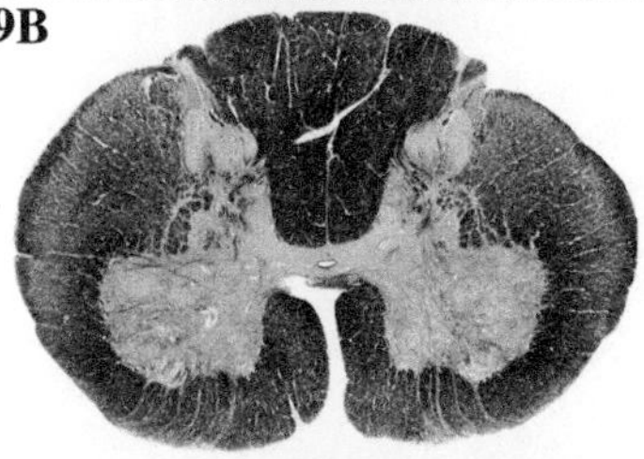	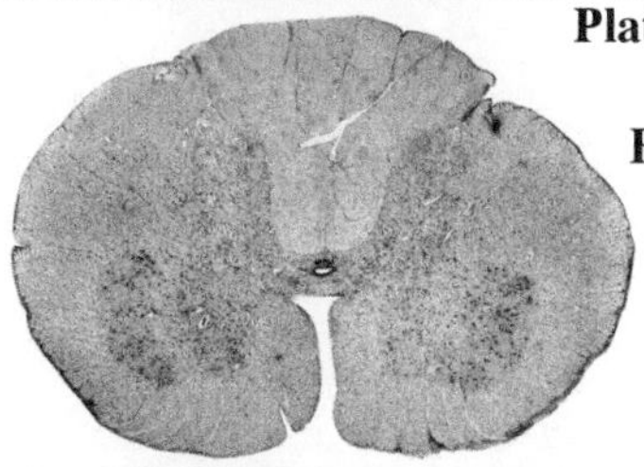	**Plates 70A, 70B** **Cervical Enlargement** Total area: 33.211 mm^2
Plates 71A, 71B **Upper Thoracic** Total area: 22.398 mm^2	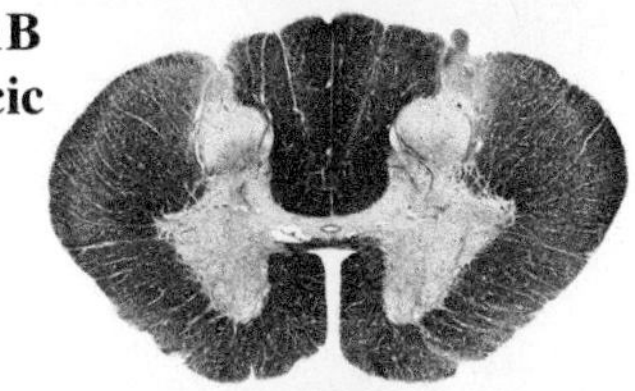	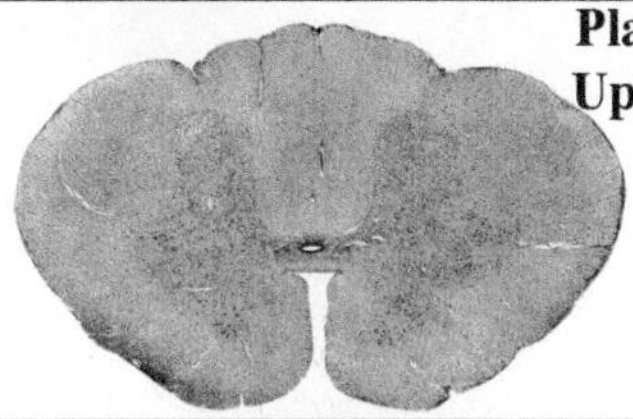	**Plates 72A, 72B** **Upper Thoracic** Total area: 24.760 mm^2
Plates 73A, 73B **Middle Thoracic** Total area: 15.291 mm^2	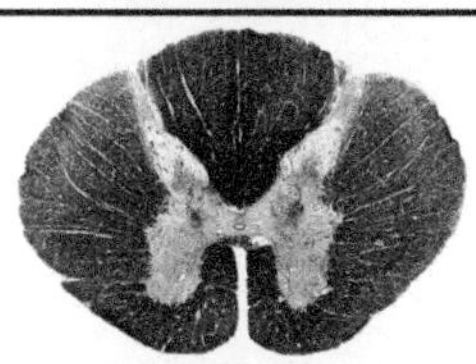	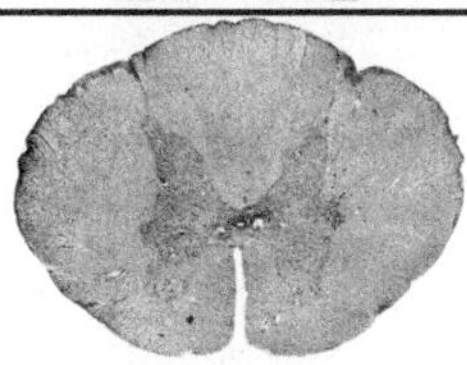	**Plates 74A, 74B** **Middle Thoracic** Total area: 16.112 mm^2
Plates 75A, 75B **Lower Thoracic** Total area: 17.098 mm^2	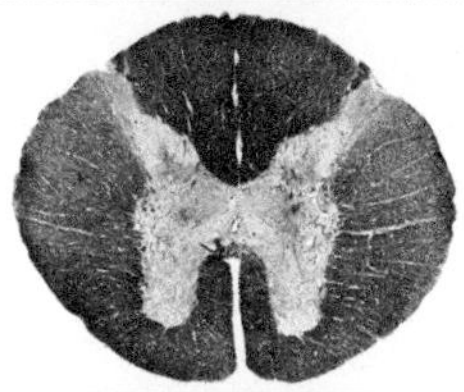	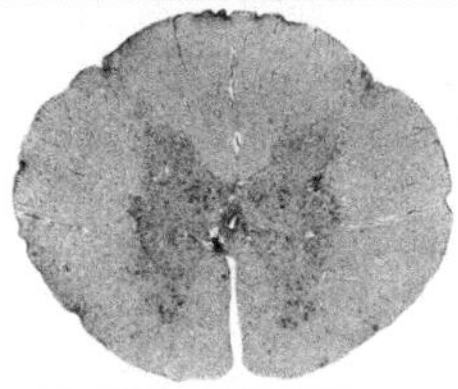	**Plates 76A, 76B** **Lower Thoracic** Total area: 17.300 mm^2
Plates 77A, 77B **Upper Lumbar** Total area: 24.774 mm^2	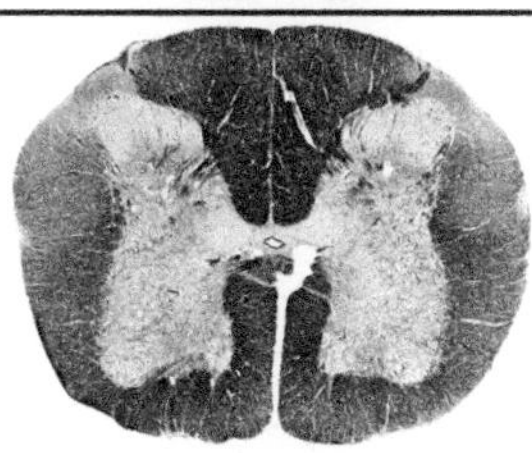	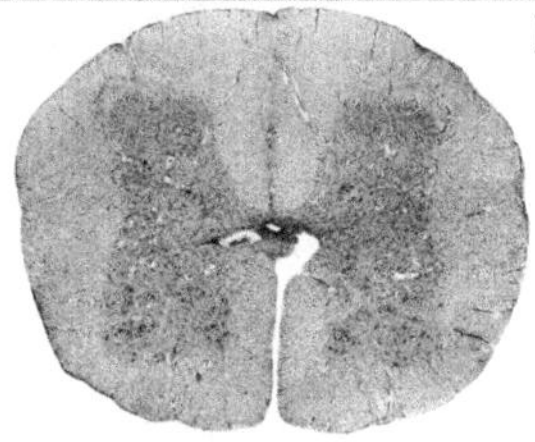	**Plates 78A, 78B** **Upper Lumbar** Total area: 24.914 mm^2
Plates 79A, 79B **Lumbar Enlargement** Total area: 31.708 mm^2	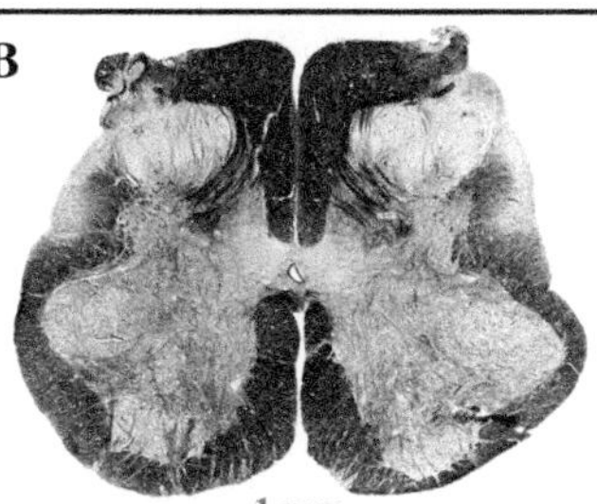 1 mm	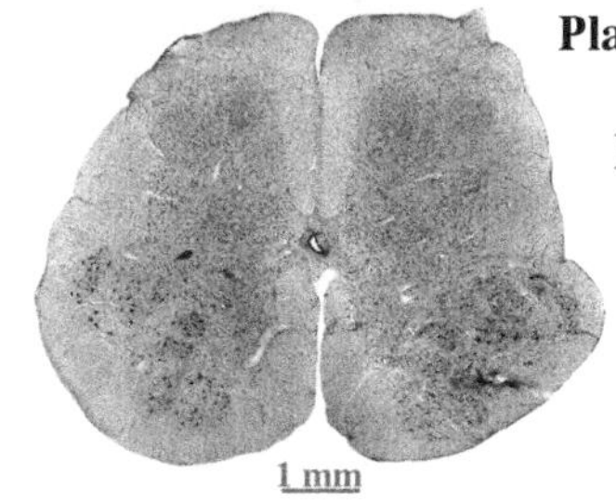 1 mm	**Plates 80A, 80B** **Lumbar Enlargement** Total area: 32.275 mm^2

PLATE 67A

CR 440 mm
Infant, 4 Months
Y286-62
Upper Cervical
Myelin stain

Areas (mm²)

Central canal	.0066
Ependyma	.0082
Gray matter	6.0358
White matter	21.4380

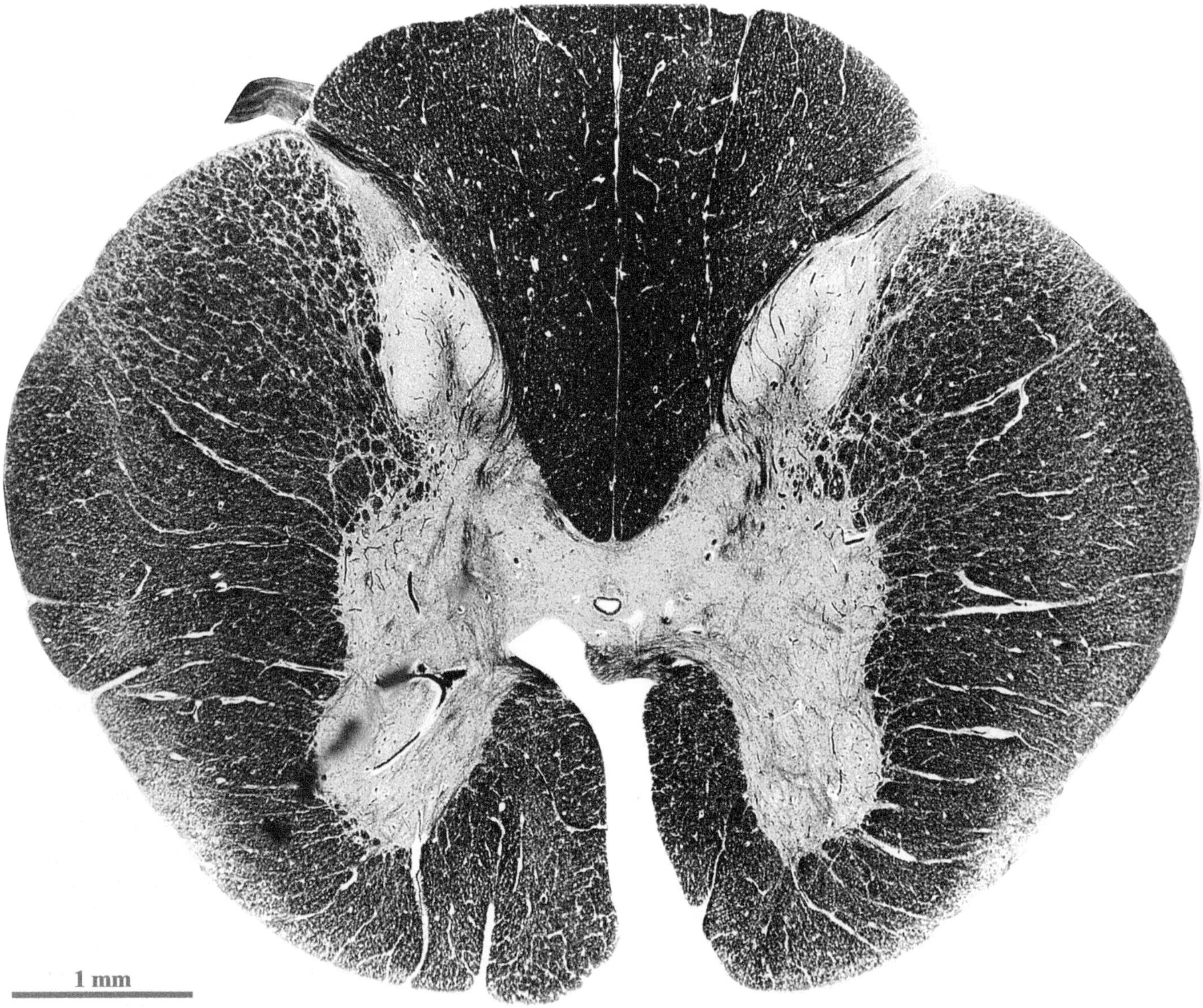

PLATE 67B

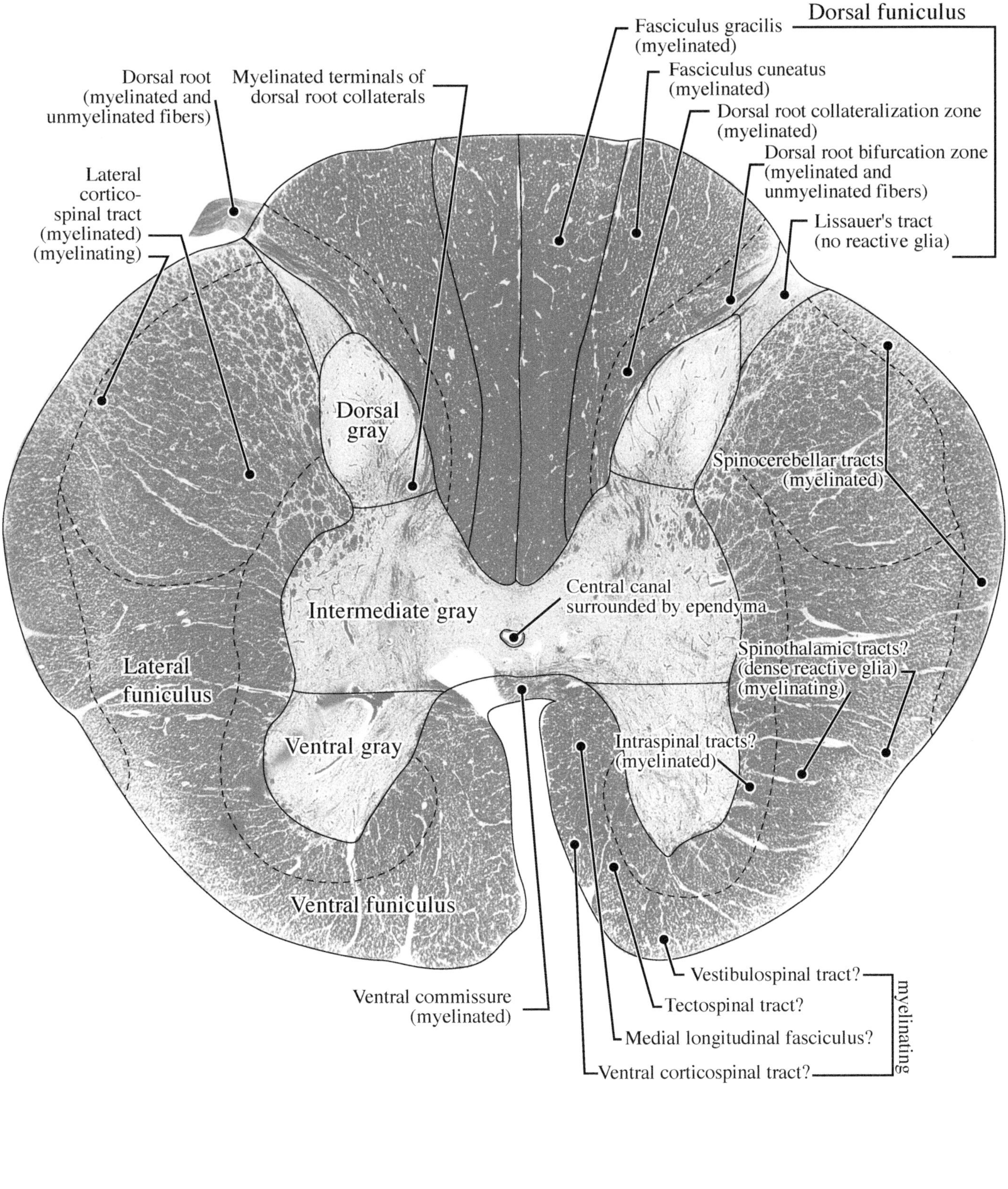
Dorsal funiculus
Fasciculus gracilis (myelinated)
Fasciculus cuneatus (myelinated)
Dorsal root collateralization zone (myelinated)
Dorsal root bifurcation zone (myelinated and unmyelinated fibers)
Lissauer's tract (no reactive glia)
Dorsal root (myelinated and unmyelinated fibers)
Myelinated terminals of dorsal root collaterals
Lateral cortico-spinal tract (myelinated) (myelinating)
Dorsal gray
Spinocerebellar tracts (myelinated)
Central canal surrounded by ependyma
Intermediate gray
Lateral funiculus
Spinothalamic tracts? (dense reactive glia) (myelinating)
Intraspinal tracts? (myelinated)
Ventral gray
Ventral funiculus
Ventral commissure (myelinated)
Vestibulospinal tract?
Tectospinal tract?
Medial longitudinal fasciculus?
Ventral corticospinal tract?
myelinating

PLATE 68A

CR 440 mm
Infant, 4 Months
Y286-62
Upper Cervical
Cell body stain

Areas (mm^2)

Central canal	.0045
Ependyma	.0145
Gray matter	6.0685
White matter	22.2390

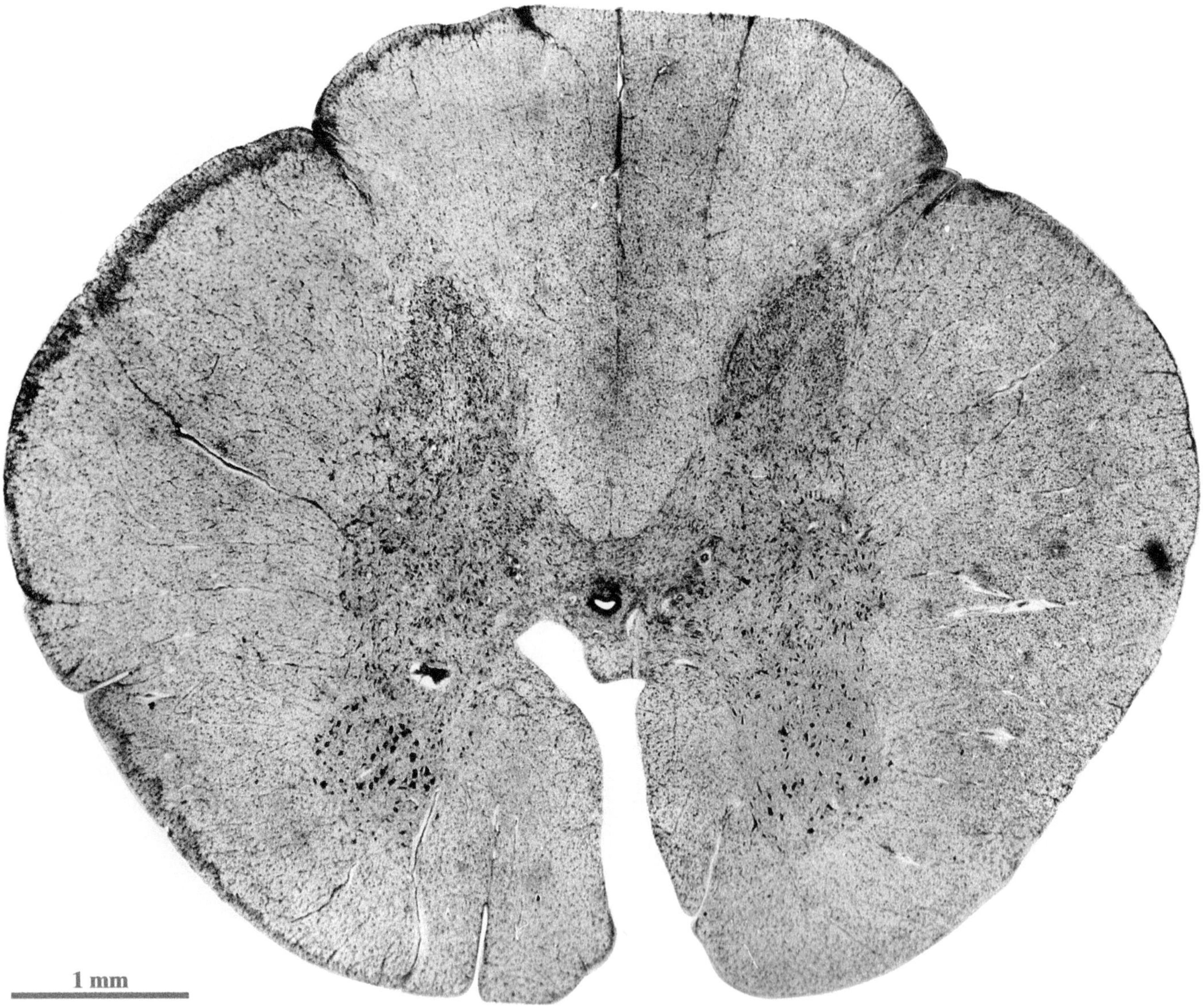

PLATE 68B

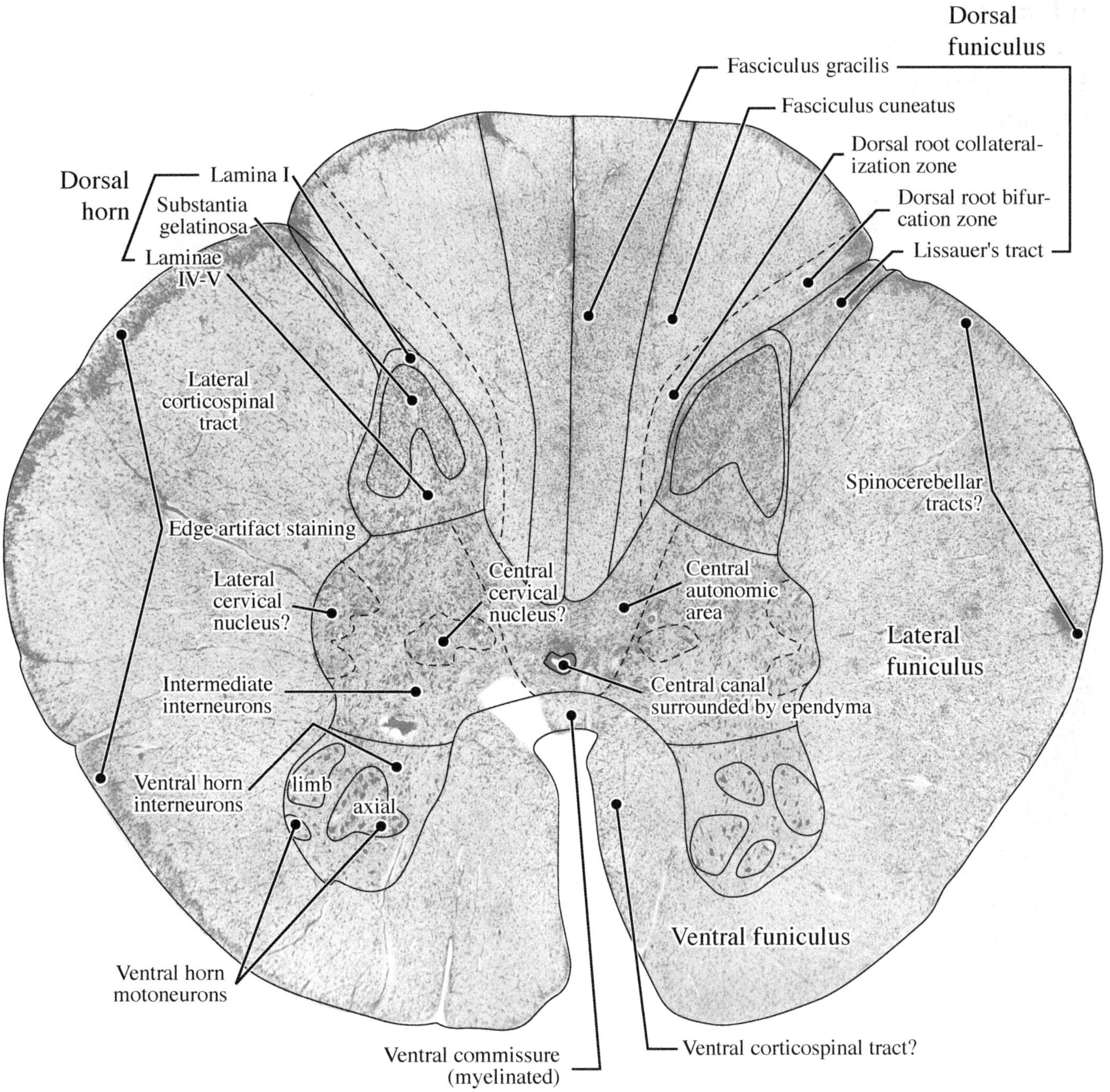
Dorsal funiculus
Fasciculus gracilis
Fasciculus cuneatus
Dorsal root collateral-ization zone
Dorsal root bifur-cation zone
Lissauer's tract
Dorsal horn
Lamina I
Substantia gelatinosa
Laminae IV-V
Lateral corticospinal tract
Edge artifact staining
Lateral cervical nucleus?
Central cervical nucleus?
Central autonomic area
Spinocerebellar tracts?
Lateral funiculus
Intermediate interneurons
Central canal surrounded by ependyma
Ventral horn interneurons
limb
axial
Ventral funiculus
Ventral horn motoneurons
Ventral commissure (myelinated)
Ventral corticospinal tract?

PLATE 69A

CR 440 mm
Infant, 4 Months
Y286-62
Cervical Enlargement
Myelin stain

Areas (mm^2)

Central canal	.0124
Ependyma	.0160
Gray matter	10.2870
White matter	22.9990

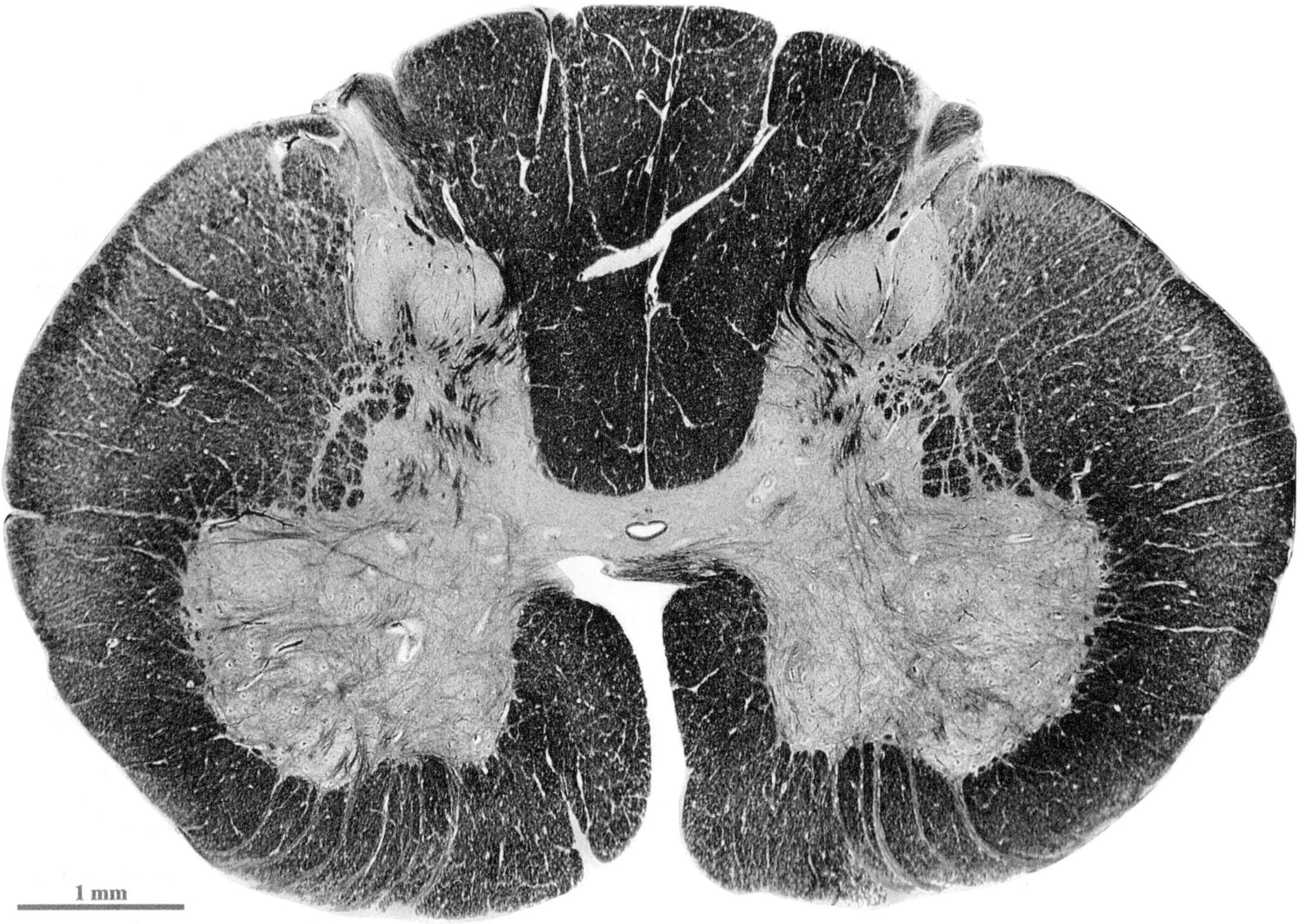

PLATE 69B

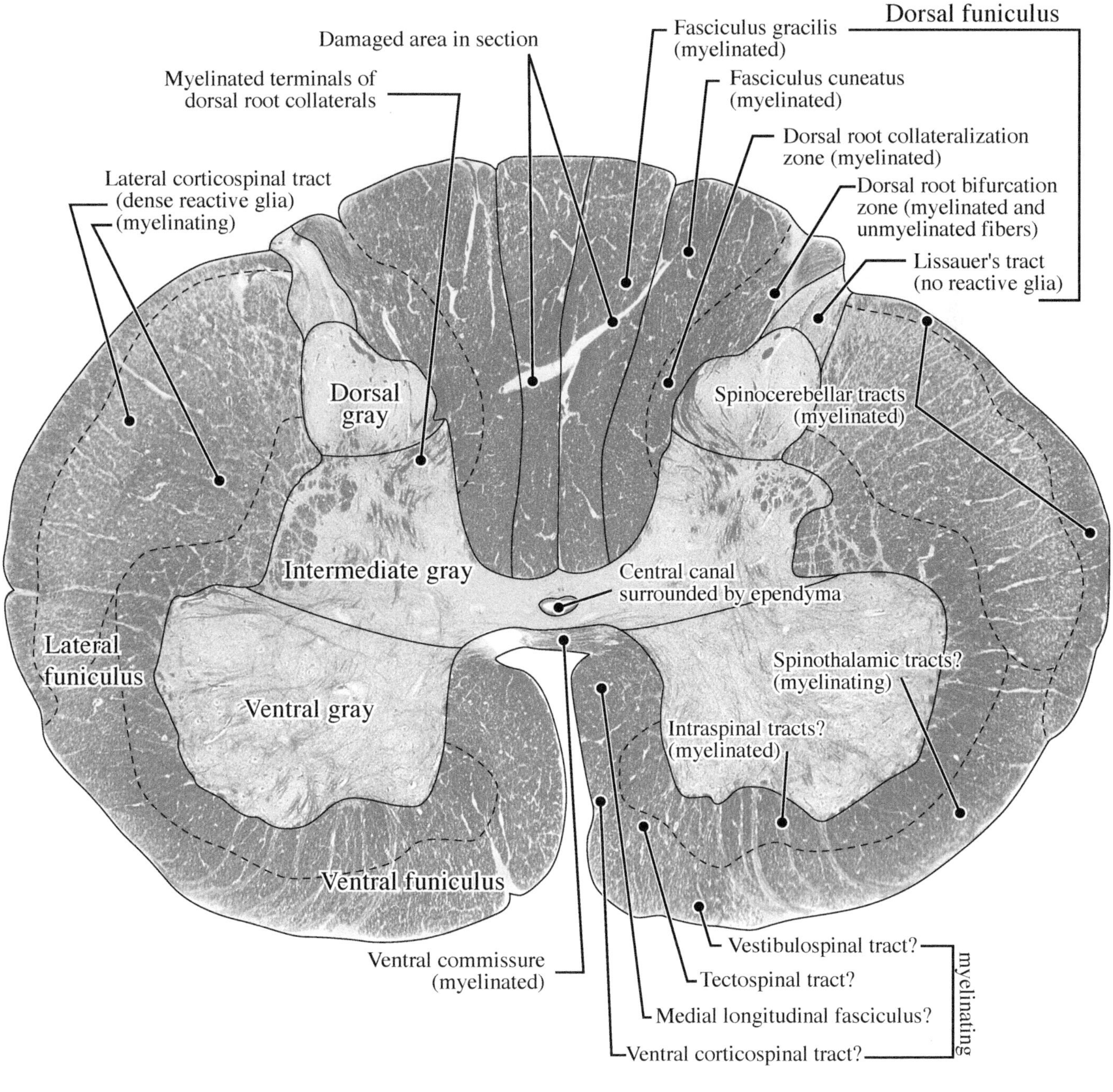
Dorsal funiculus
Damaged area in section
Fasciculus gracilis (myelinated)
Myelinated terminals of dorsal root collaterals
Fasciculus cuneatus (myelinated)
Dorsal root collateralization zone (myelinated)
Lateral corticospinal tract (dense reactive glia) (myelinating)
Dorsal root bifurcation zone (myelinated and unmyelinated fibers)
Lissauer's tract (no reactive glia)
Dorsal gray
Spinocerebellar tracts (myelinated)
Intermediate gray
Central canal surrounded by ependyma
Lateral funiculus
Spinothalamic tracts? (myelinating)
Ventral gray
Intraspinal tracts? (myelinated)
Ventral funiculus
Vestibulospinal tract?
Ventral commissure (myelinated)
Tectospinal tract?
Medial longitudinal fasciculus?
Ventral corticospinal tract?
myelinating

PLATE 70A

CR 440 mm
Infant, 4 Months
Y286-62
Cervical Enlargement
Cell body stain

Areas (mm^2)

Central canal	.0010
Ependyma	.0256
Gray matter	10.1850
White matter	22.9900

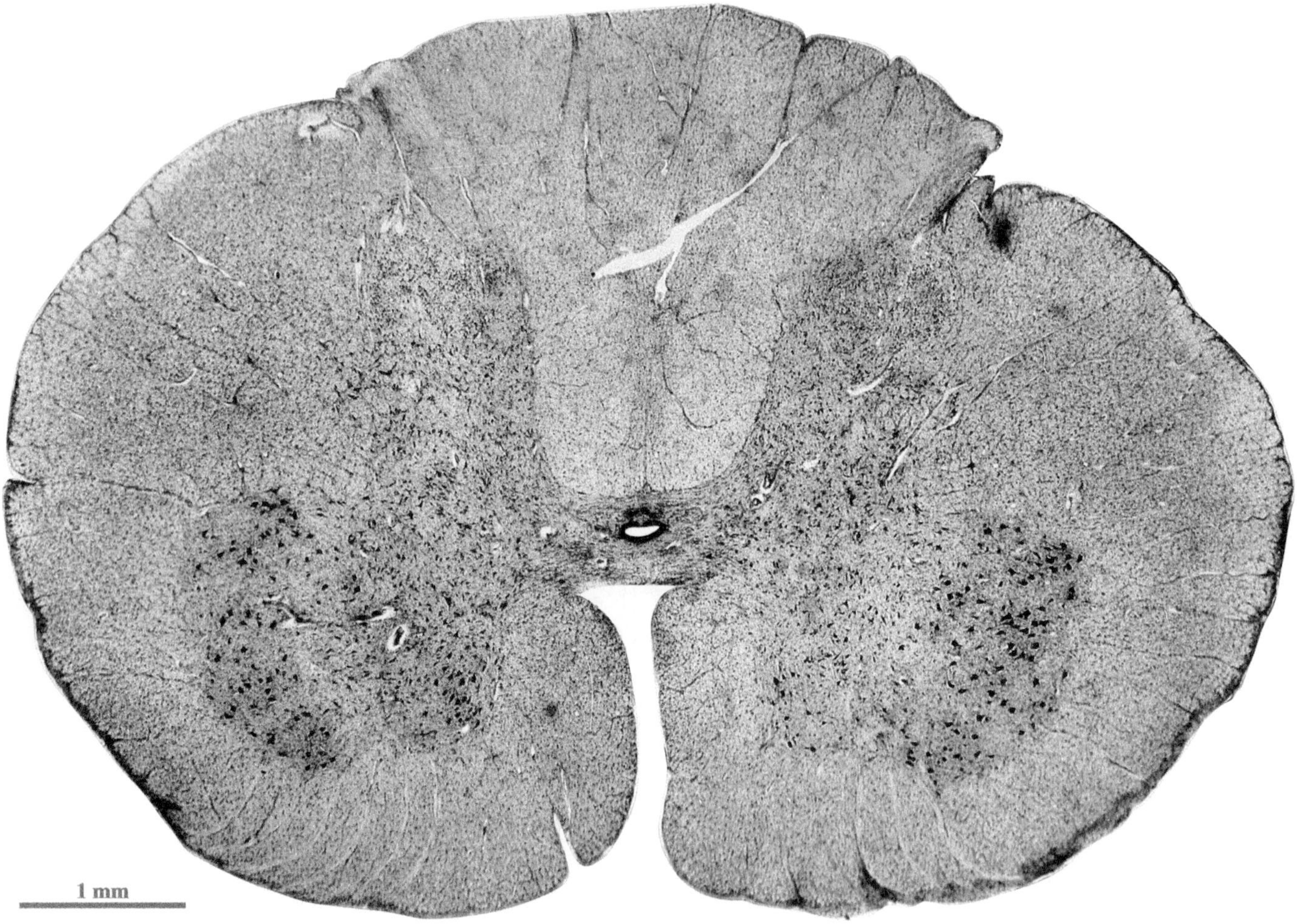

PLATE 70B

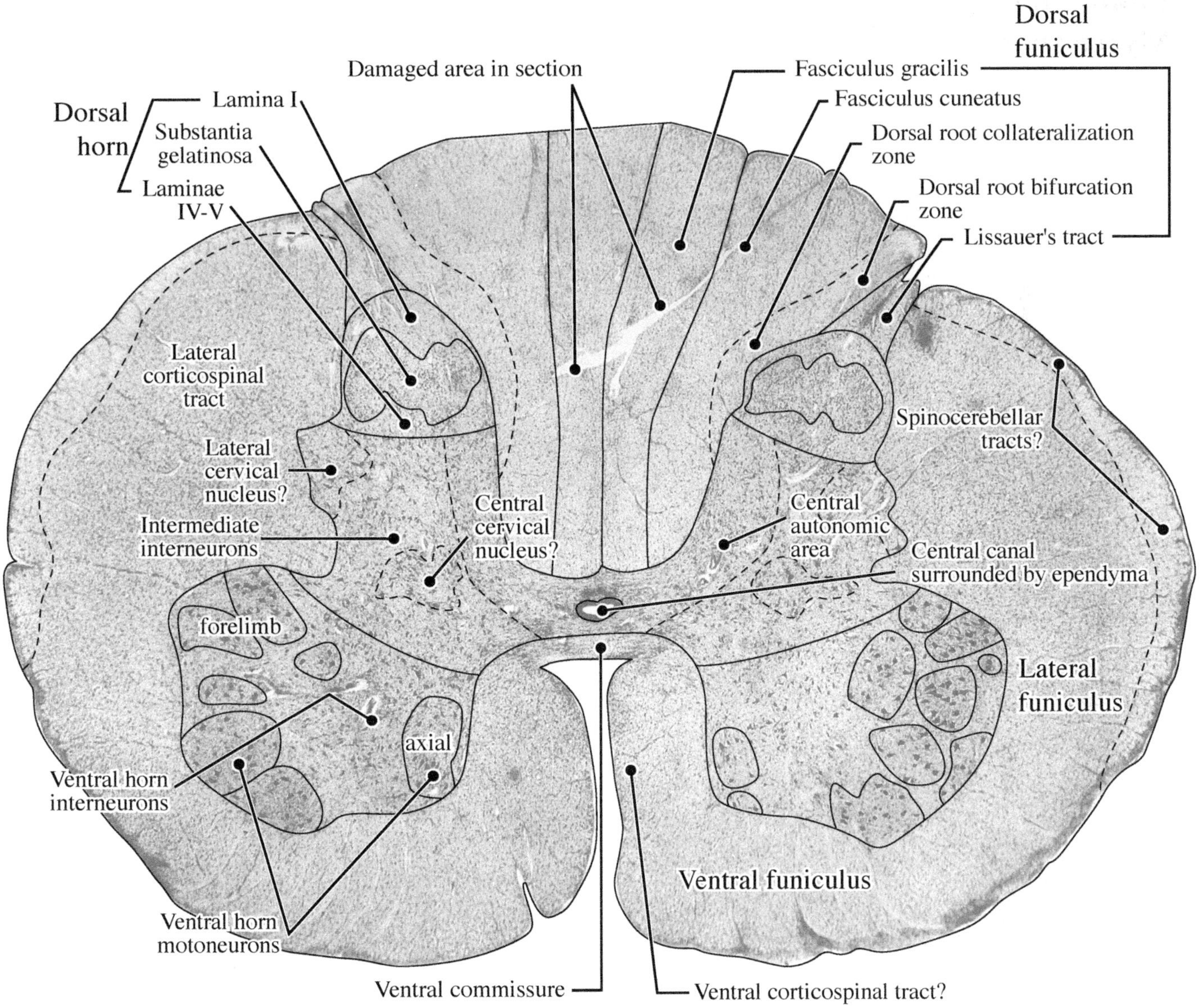
Dorsal funiculus
Damaged area in section
Fasciculus gracilis
Fasciculus cuneatus
Dorsal root collateralization zone
Dorsal root bifurcation zone
Lissauer's tract
Dorsal horn
Lamina I
Substantia gelatinosa
Laminae IV-V
Lateral corticospinal tract
Lateral cervical nucleus?
Intermediate interneurons
Central cervical nucleus?
Central autonomic area
Spinocerebellar tracts?
Central canal surrounded by ependyma
forelimb
axial
Lateral funiculus
Ventral horn interneurons
Ventral funiculus
Ventral horn motoneurons
Ventral commissure
Ventral corticospinal tract?

PLATE 71A

CR 440 mm
Infant, 4 Months
Y286-62
Upper Thoracic
Myelin stain

Areas (mm^2)

Central canal	.0088
Ependyma	.0136
Gray matter	5.6424
White matter	16.7340

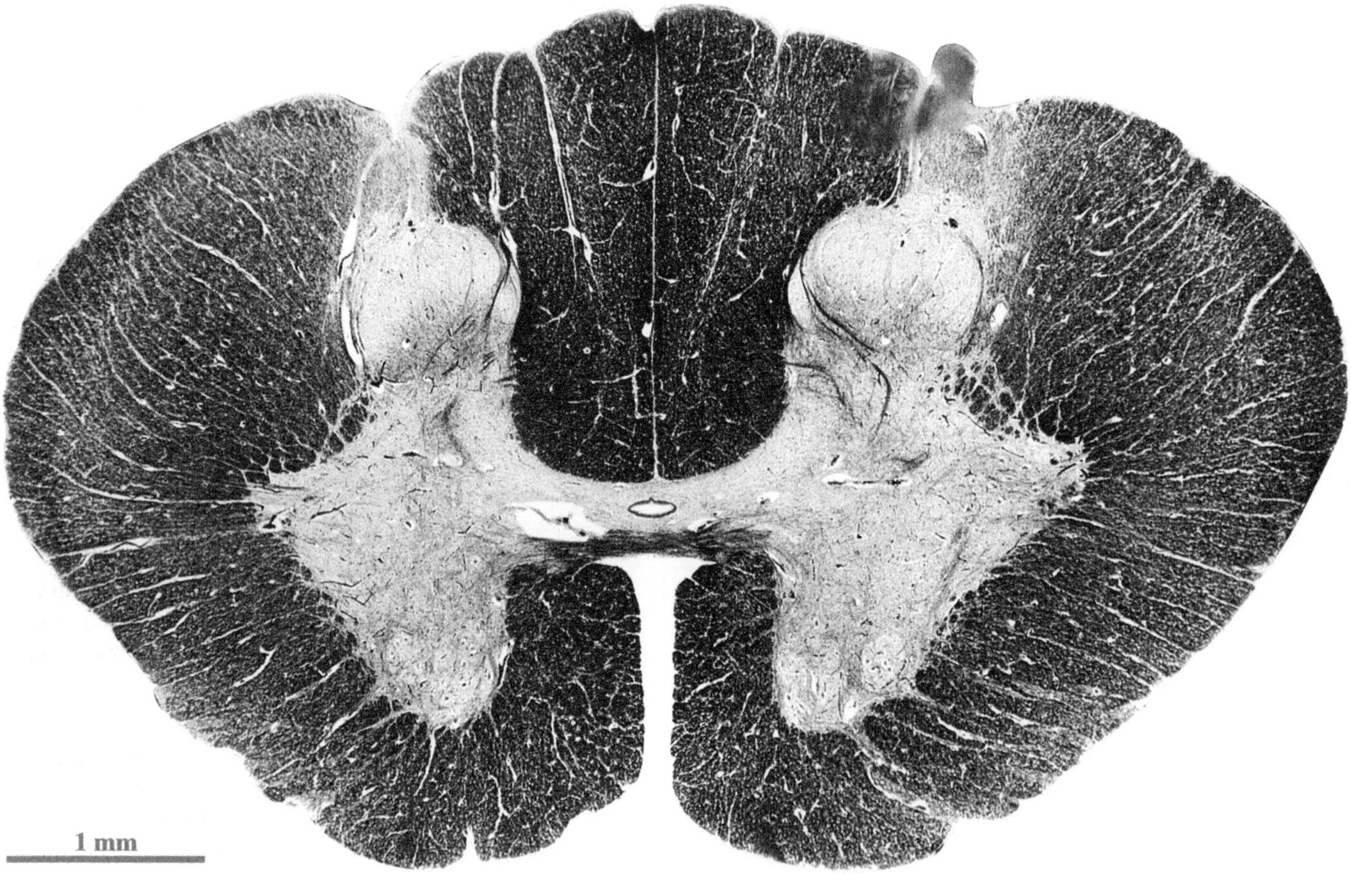

1 mm

PLATE 71B

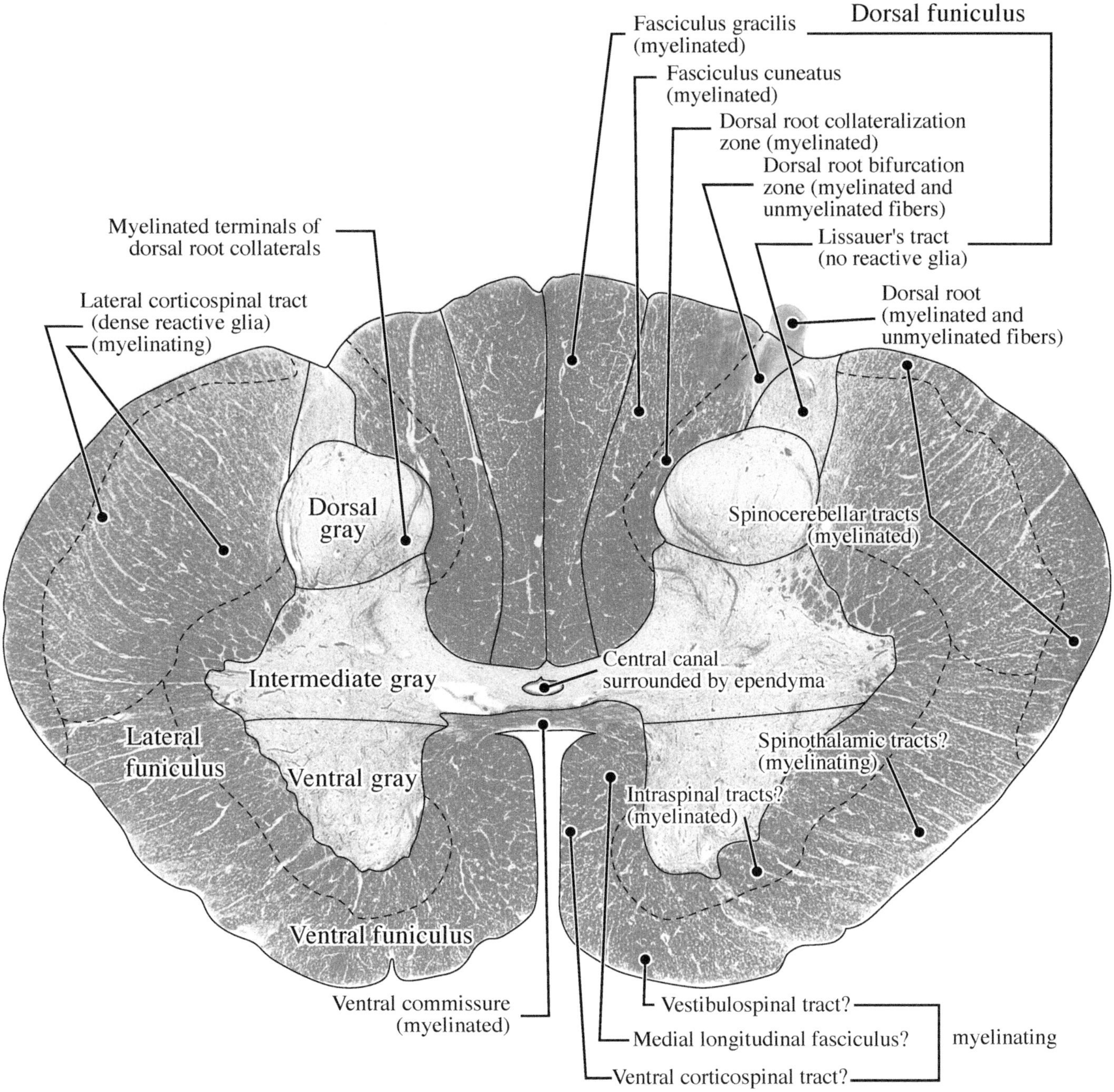
Fasciculus gracilis
(myelinated)
Dorsal funiculus
Fasciculus cuneatus
(myelinated)
Dorsal root collateralization
zone (myelinated)
Dorsal root bifurcation
zone (myelinated and
unmyelinated fibers)
Myelinated terminals of
dorsal root collaterals
Lissauer's tract
(no reactive glia)
Lateral corticospinal tract
(dense reactive glia)
(myelinating)
Dorsal root
(myelinated and
unmyelinated fibers)
Dorsal
gray
Spinocerebellar tracts
(myelinated)
Central canal
surrounded by ependyma
Intermediate gray
Lateral
funiculus
Spinothalamic tracts?
(myelinating)
Ventral gray
Intraspinal tracts?
(myelinated)
Ventral funiculus
Ventral commissure
(myelinated)
Vestibulospinal tract?
Medial longitudinal fasciculus?
myelinating
Ventral corticospinal tract?

PLATE 72A

CR 440 mm
Infant, 4 Months
Y286-62
Upper Thoracic
Cell body stain

Areas (mm²)

Central canal	.0099
Ependyma	.0241
Gray matter	6.5545
White matter	18.1720

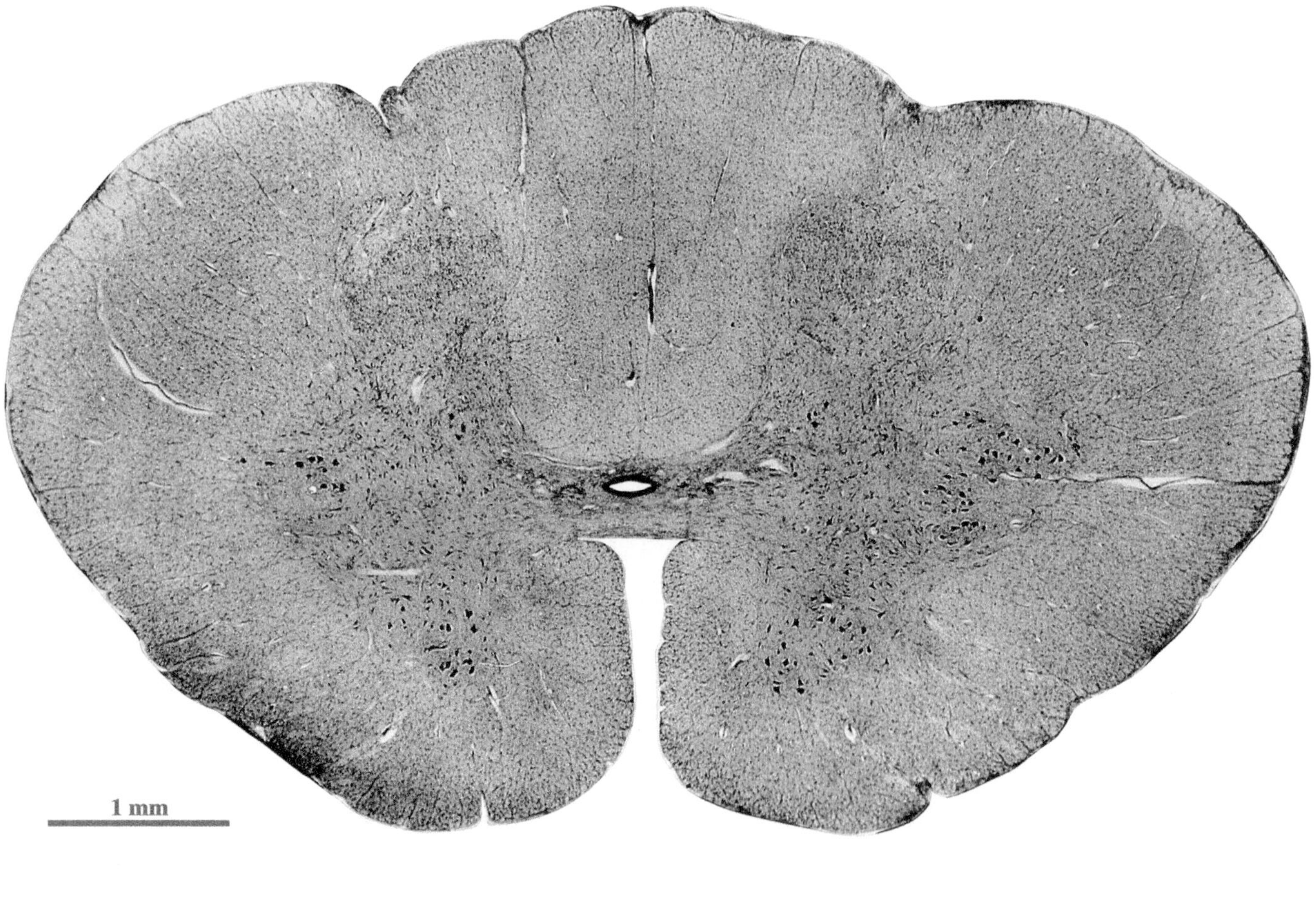

PLATE 72B

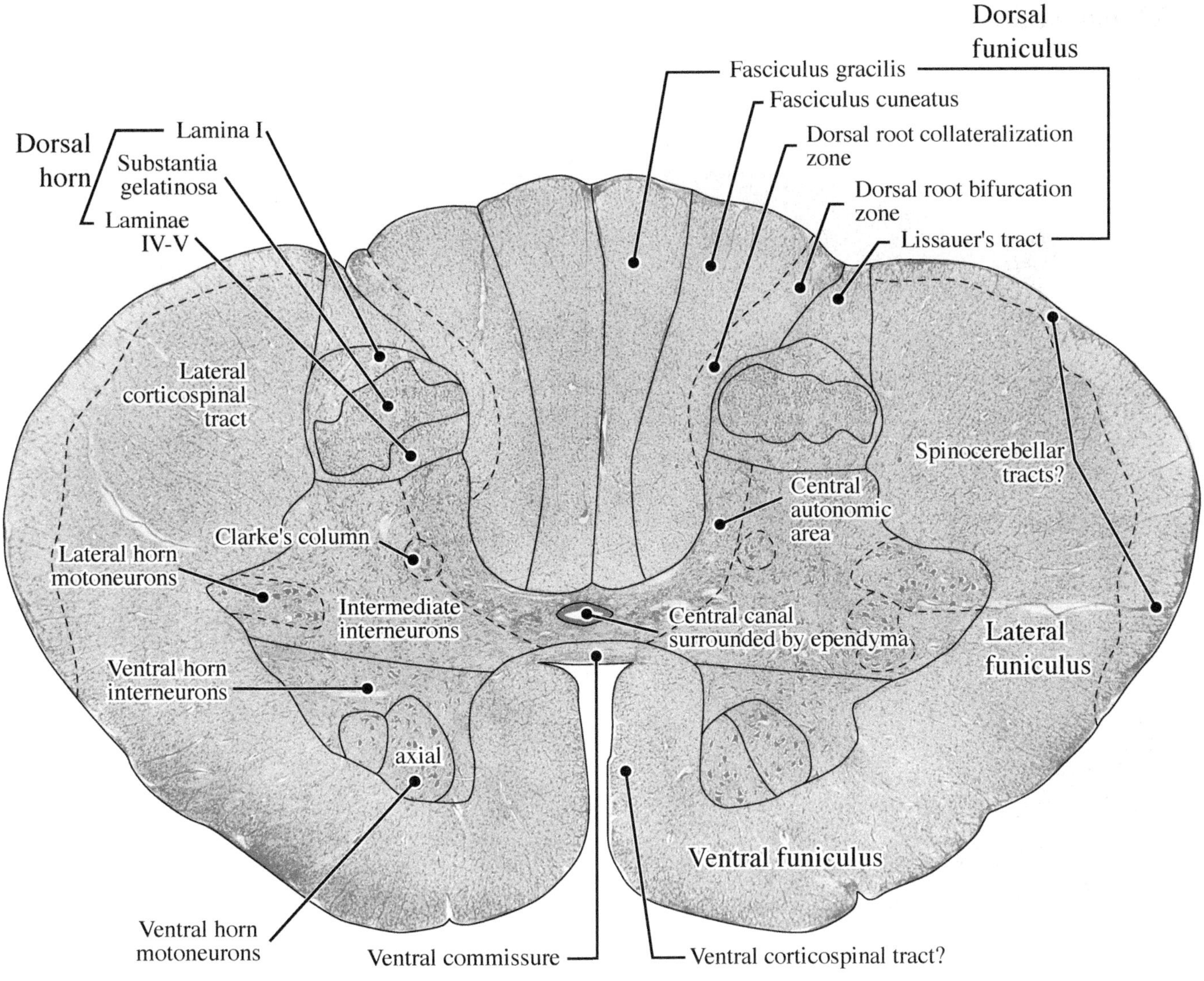
Dorsal funiculus
Fasciculus gracilis
Fasciculus cuneatus
Dorsal root collateralization zone
Dorsal root bifurcation zone
Lissauer's tract
Dorsal horn
Lamina I
Substantia gelatinosa
Laminae IV-V
Lateral corticospinal tract
Spinocerebellar tracts?
Central autonomic area
Clarke's column
Lateral horn motoneurons
Intermediate interneurons
Central canal surrounded by ependyma
Lateral funiculus
Ventral horn interneurons
axial
Ventral funiculus
Ventral horn motoneurons
Ventral commissure
Ventral corticospinal tract?

PLATE 73A

CR 440 mm
Infant, 4 Months
Y286-62
Middle Thoracic
Myelin stain

Areas (mm^2)

Central canal	.0040
Ependyma	.0062
Gray matter	2.8387
White matter	12.4420

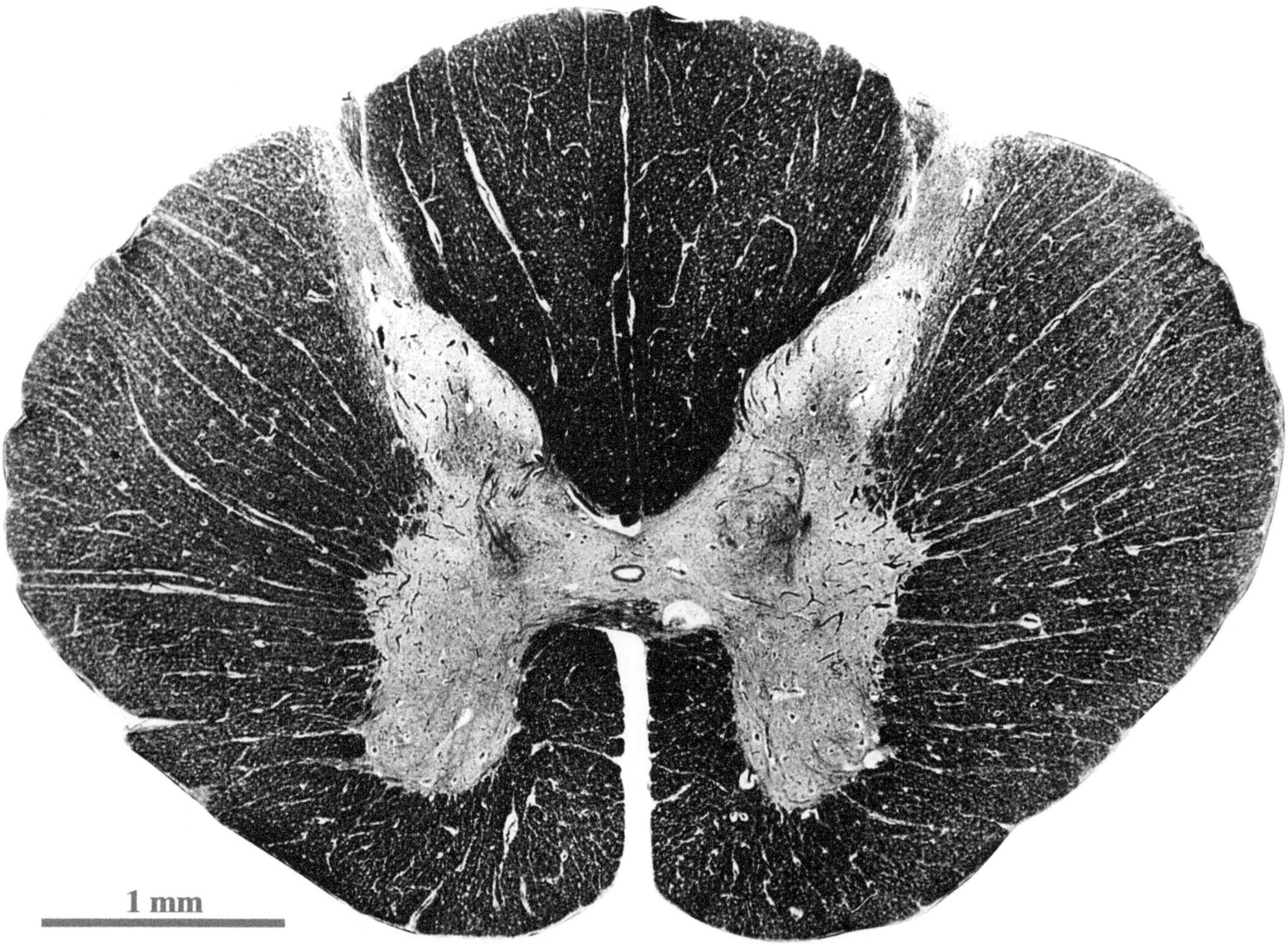

PLATE 73B

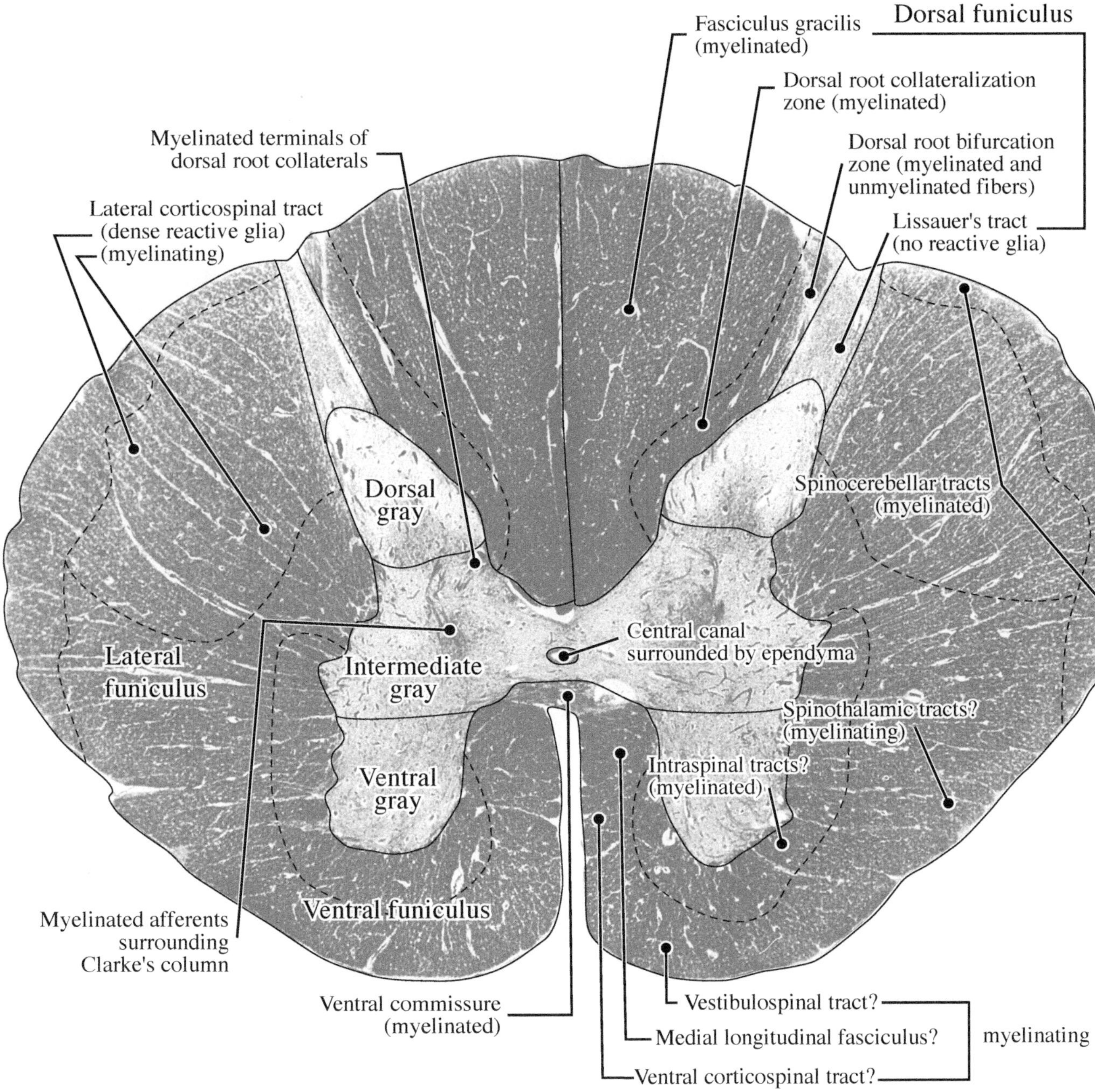
Fasciculus gracilis
(myelinated)
Dorsal funiculus
Dorsal root collateralization
zone (myelinated)
Myelinated terminals of
dorsal root collaterals
Dorsal root bifurcation
zone (myelinated and
unmyelinated fibers)
Lateral corticospinal tract
(dense reactive glia)
(myelinating)
Lissauer's tract
(no reactive glia)
Dorsal
gray
Spinocerebellar tracts
(myelinated)
Central canal
surrounded by ependyma
Lateral
funiculus
Intermediate
gray
Spinothalamic tracts?
(myelinating)
Ventral
gray
Intraspinal tracts?
(myelinated)
Ventral funiculus
Myelinated afferents
surrounding
Clarke's column
Ventral commissure
(myelinated)
Vestibulospinal tract?
Medial longitudinal fasciculus?
Ventral corticospinal tract?
myelinating

PLATE 74A

CR 440 mm
Infant, 4 Months
Y286-62
Middle Thoracic
Cell body stain

Areas (mm^2)

Central canal	.0031
Ependyma	.0117
Gray matter	2.8799
White matter	13.2170

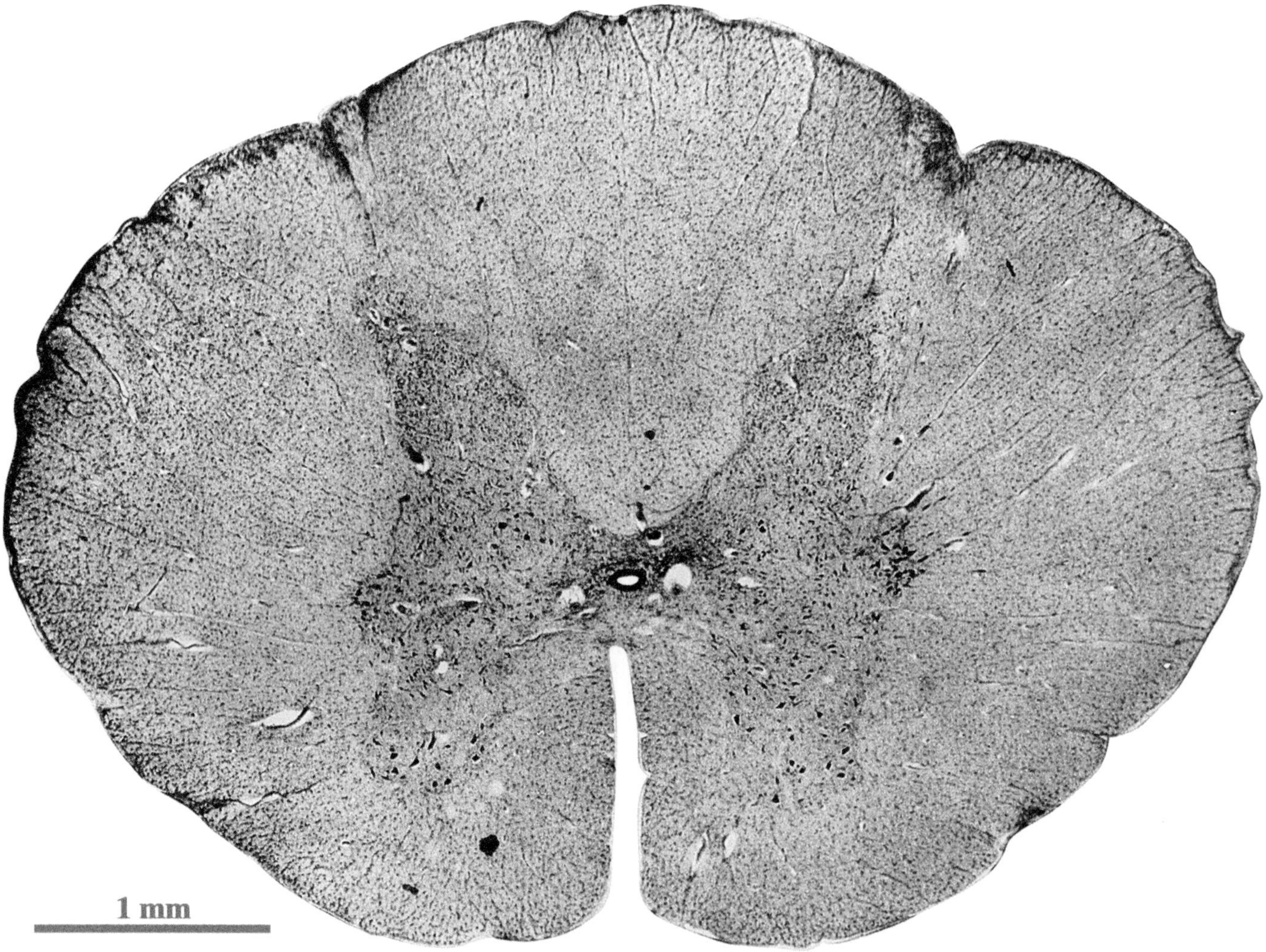

PLATE 74B

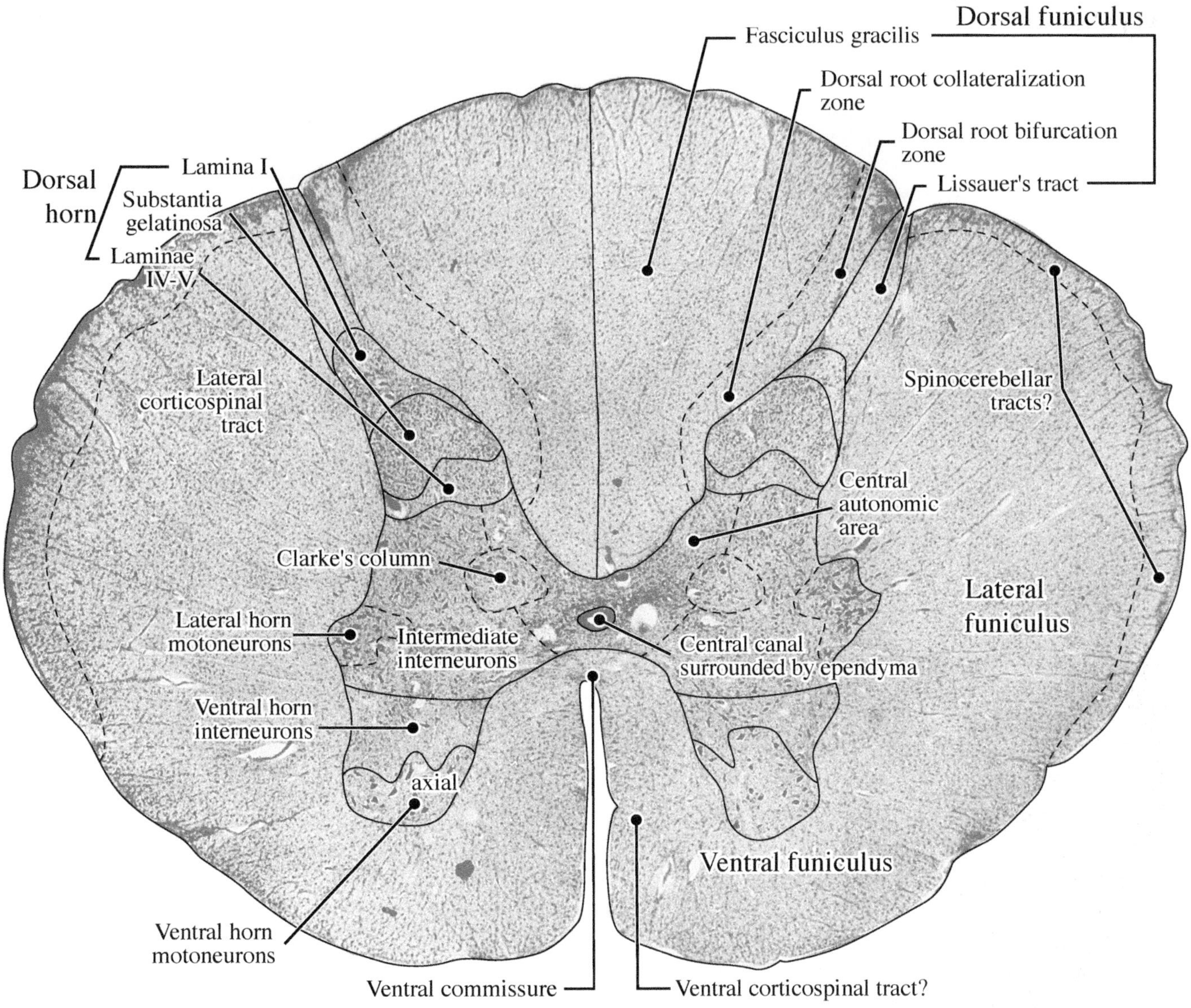
Dorsal funiculus
Fasciculus gracilis
Dorsal root collateralization zone
Dorsal root bifurcation zone
Lissauer's tract
Dorsal horn
Lamina I
Substantia gelatinosa
Laminae IV-V
Lateral corticospinal tract
Spinocerebellar tracts?
Central autonomic area
Clarke's column
Lateral funiculus
Lateral horn motoneurons
Intermediate interneurons
Central canal surrounded by ependyma
Ventral horn interneurons
axial
Ventral funiculus
Ventral horn motoneurons
Ventral commissure
Ventral corticospinal tract?

PLATE 75A

CR 440 mm
Infant, 4 Months
Y286-62
Lower Thoracic
Myelin stain

Areas (mm²)

Central canal	.0090
Ependyma	.0091
Gray matter	4.0248
White matter	13.0550

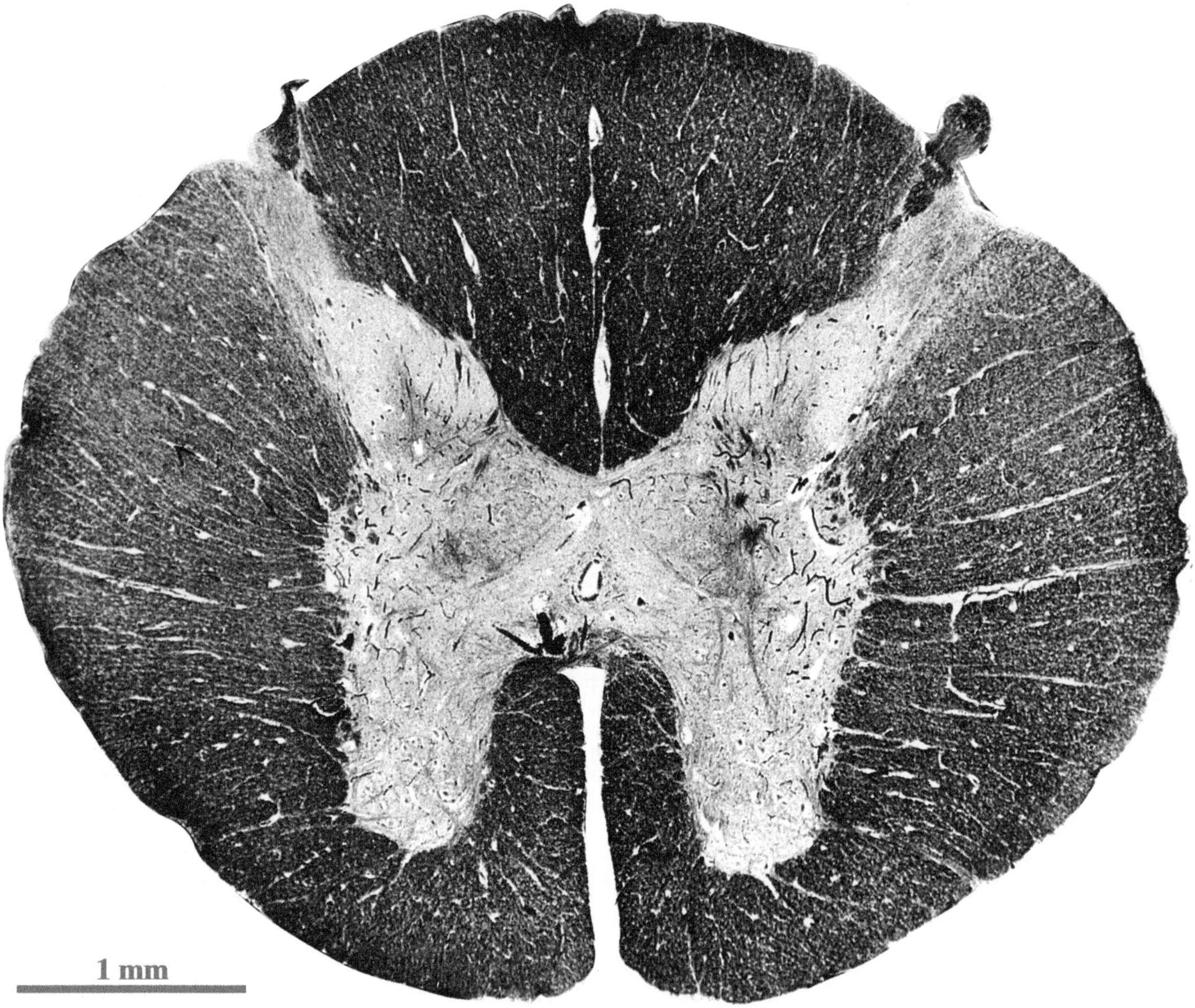

PLATE 75B

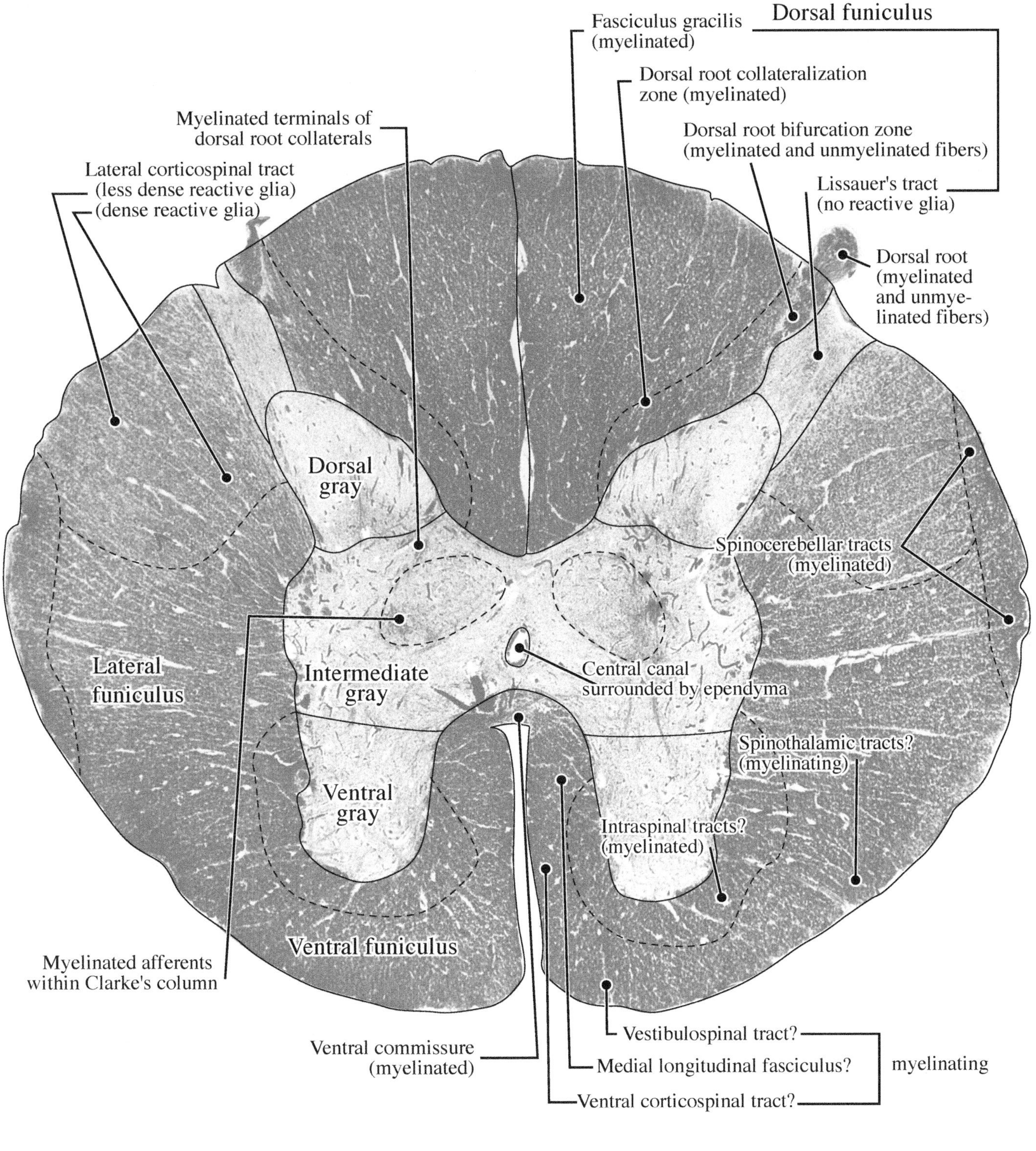
Dorsal funiculus
Fasciculus gracilis
(myelinated)
Dorsal root collateralization
zone (myelinated)
Dorsal root bifurcation zone
(myelinated and unmyelinated fibers)
Lissauer's tract
(no reactive glia)
Dorsal root
(myelinated
and unmye-
linated fibers)
Myelinated terminals of
dorsal root collaterals
Lateral corticospinal tract
(less dense reactive glia)
(dense reactive glia)
Dorsal
gray
Spinocerebellar tracts
(myelinated)
Lateral
funiculus
Intermediate
gray
Central canal
surrounded by ependyma
Spinothalamic tracts?
(myelinating)
Ventral
gray
Intraspinal tracts?
(myelinated)
Ventral funiculus
Myelinated afferents
within Clarke's column
Vestibulospinal tract?
Medial longitudinal fasciculus?
Ventral corticospinal tract?
myelinating
Ventral commissure
(myelinated)

PLATE 76A

CR 440 mm
Infant, 4 Months
Y286-62
Lower Thoracic
Cell body stain

Areas (mm^2)

Central canal	.0073
Ependyma	.0160
Gray matter	4.2476
White matter	13.0290

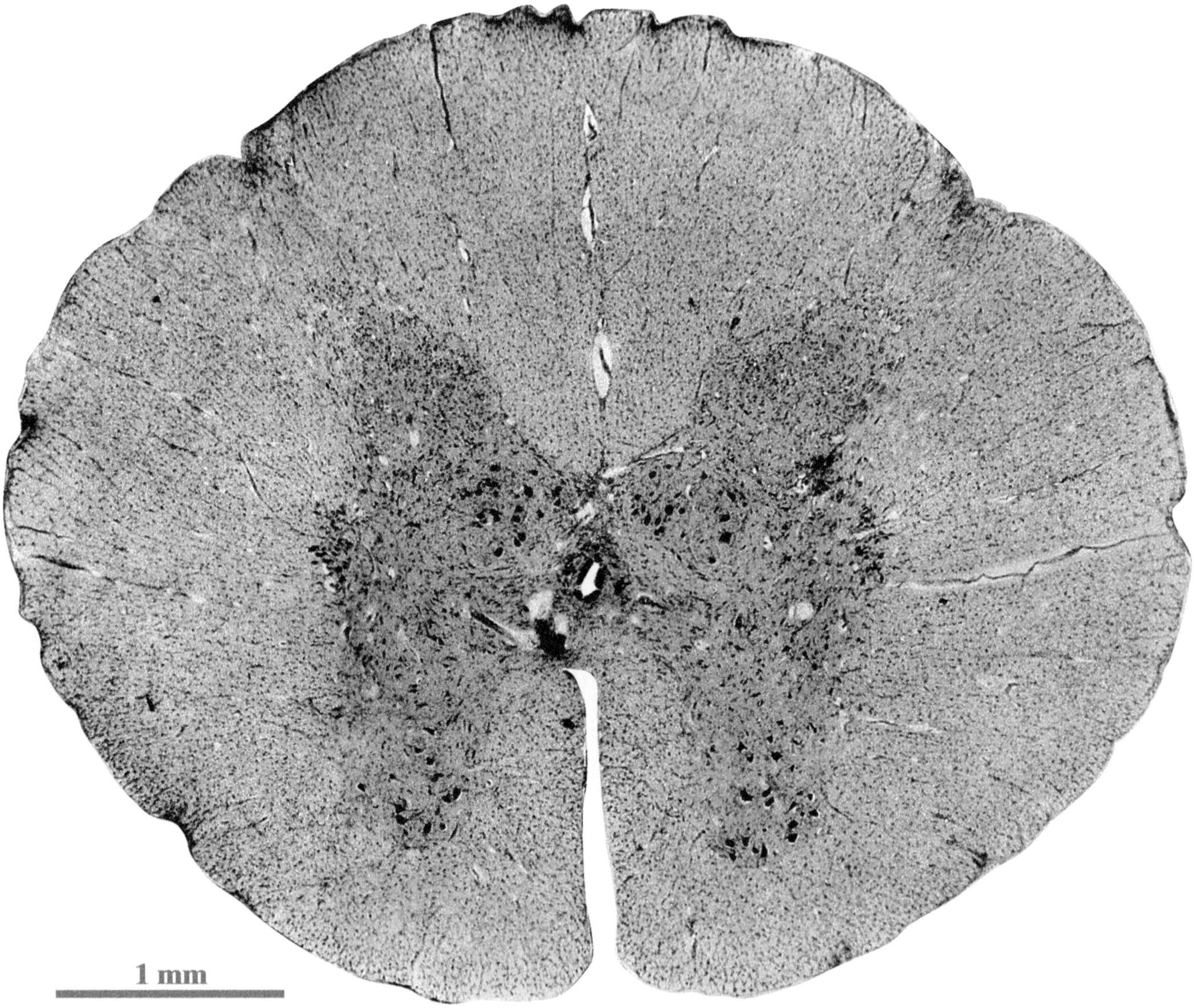

PLATE 76B

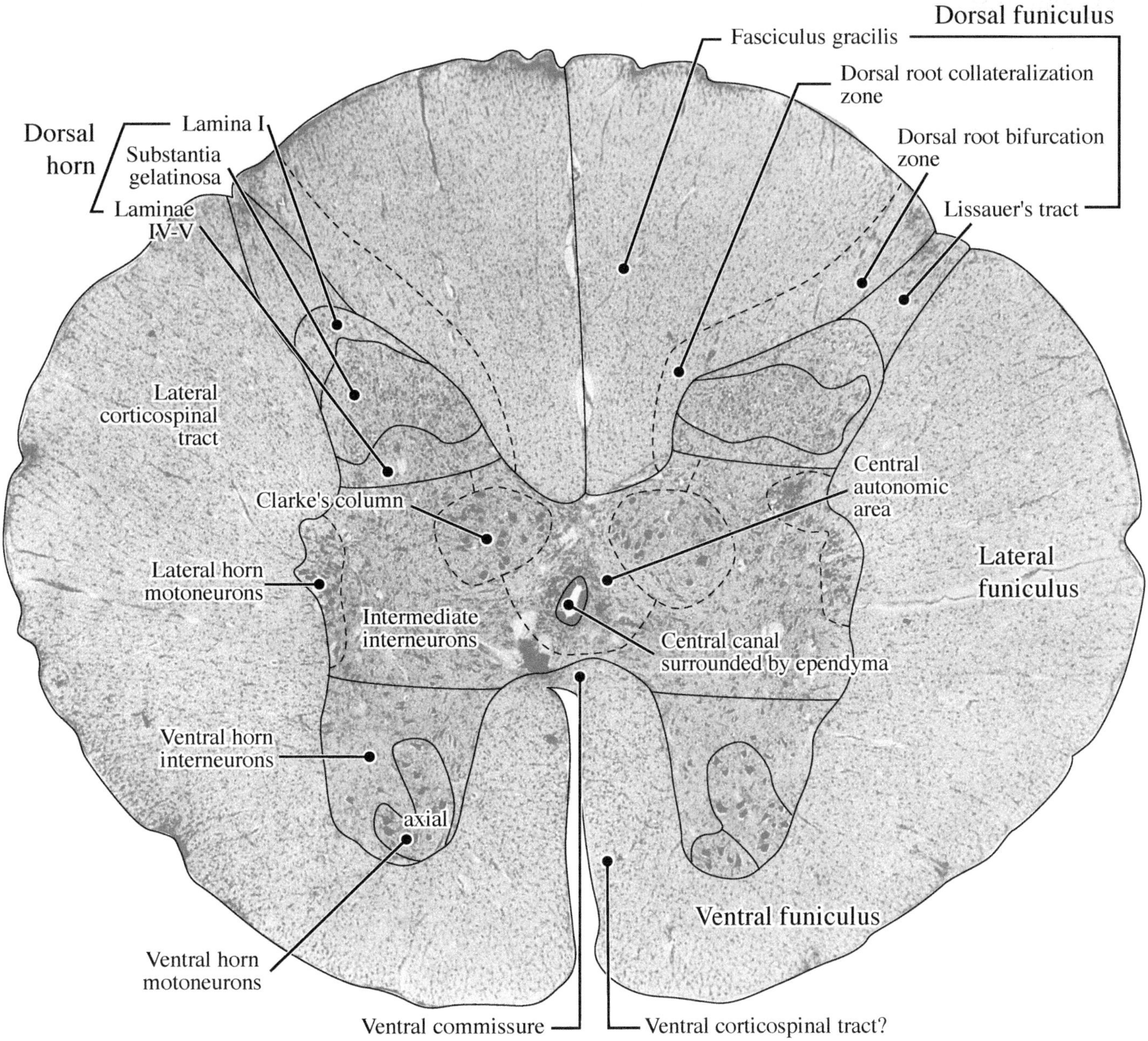
Dorsal funiculus
Fasciculus gracilis
Dorsal root collateralization zone
Dorsal root bifurcation zone
Lissauer's tract
Dorsal horn
Lamina I
Substantia gelatinosa
Laminae IV-V
Lateral corticospinal tract
Clarke's column
Central autonomic area
Lateral horn motoneurons
Intermediate interneurons
Central canal surrounded by ependyma
Lateral funiculus
Ventral horn interneurons
axial
Ventral horn motoneurons
Ventral funiculus
Ventral commissure
Ventral corticospinal tract?

PLATE 77A

CR 440 mm
Infant, 4 Months
Y286-62
Upper Lumbar
Myelin stain

Areas (mm²)

Central canal	.0140
Ependyma	.0155
Gray matter	9.5065
White matter	15.2380

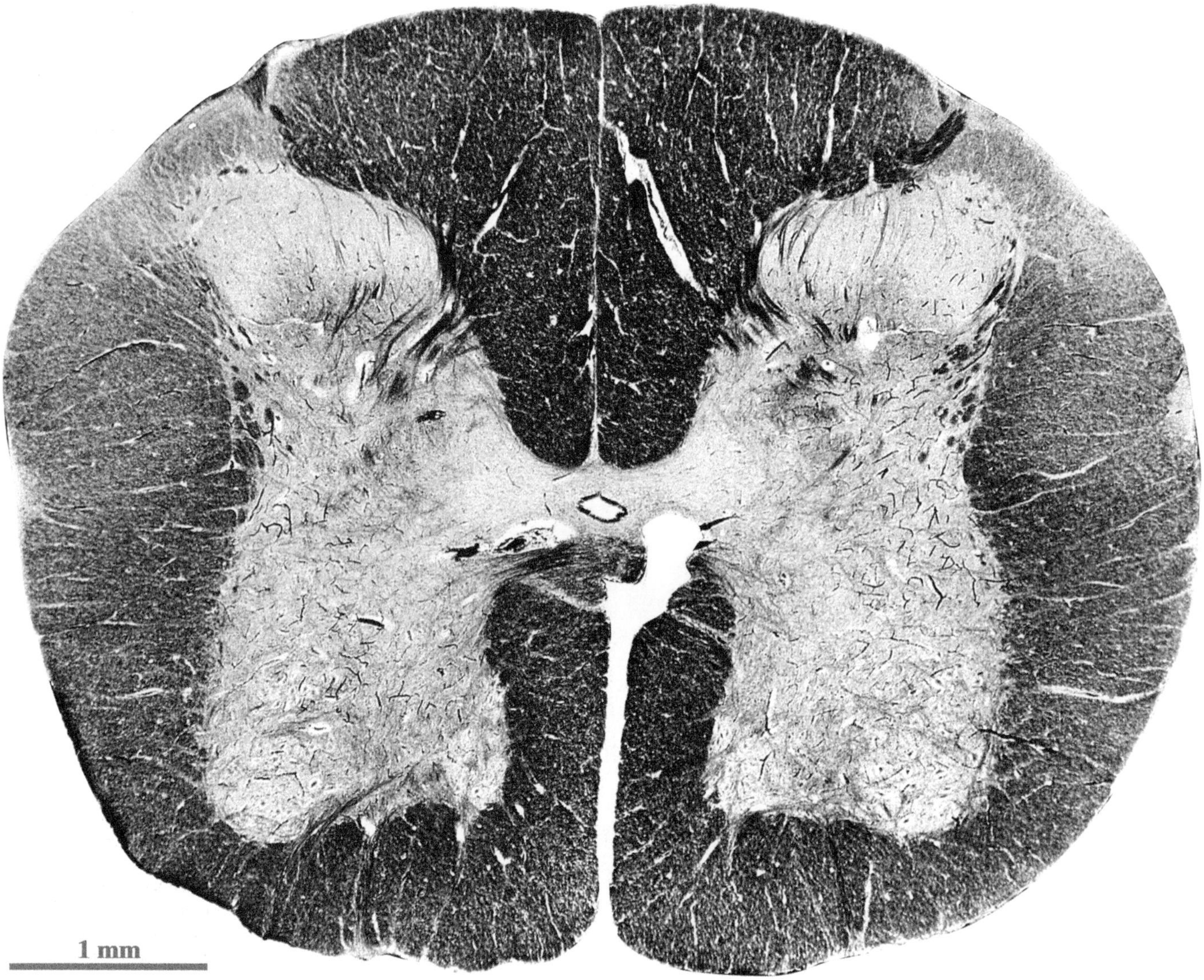

PLATE 77B

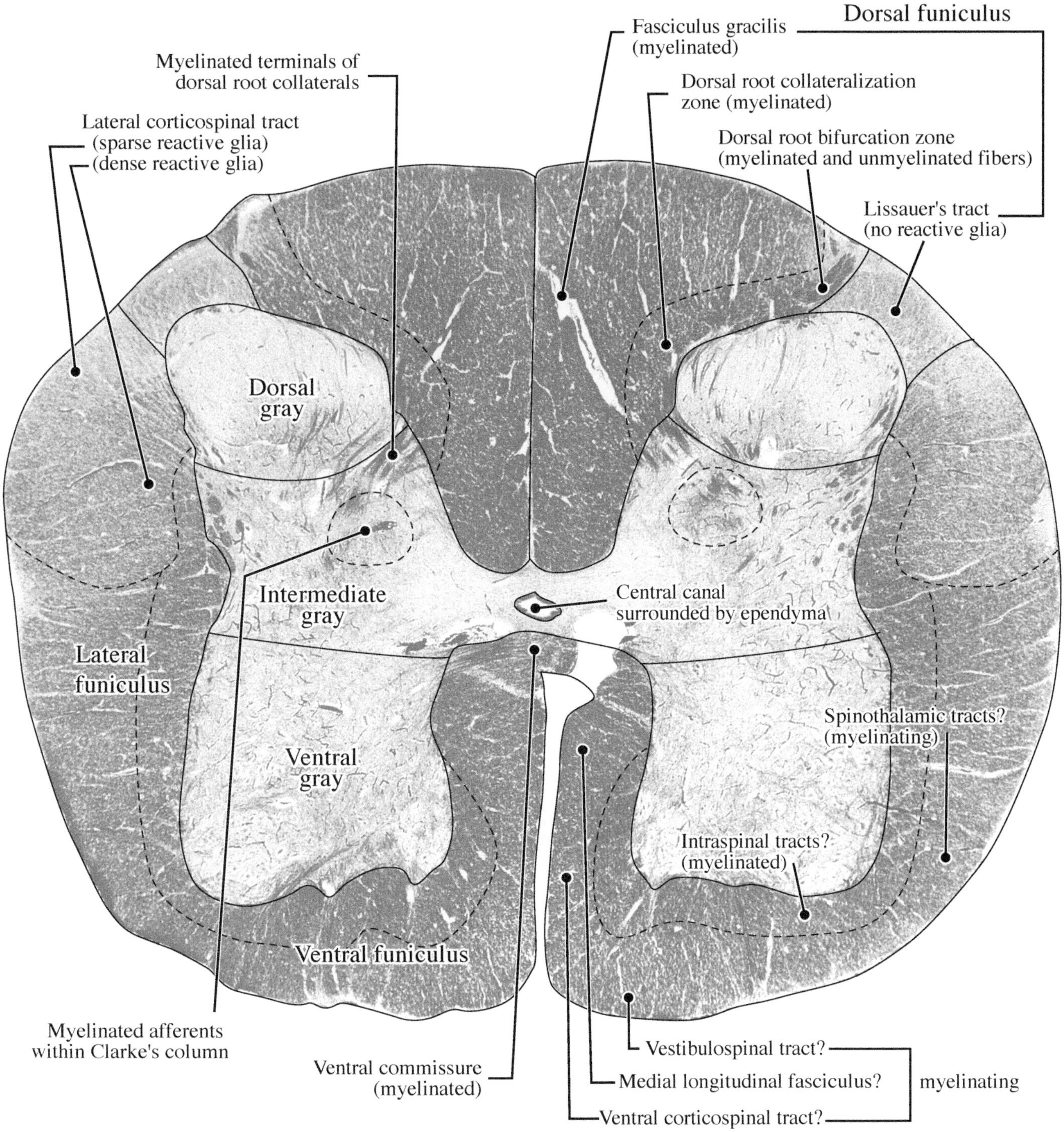
Dorsal funiculus
Fasciculus gracilis
(myelinated)
Myelinated terminals of
dorsal root collaterals
Dorsal root collateralization
zone (myelinated)
Lateral corticospinal tract
(sparse reactive glia)
(dense reactive glia)
Dorsal root bifurcation zone
(myelinated and unmyelinated fibers)
Lissauer's tract
(no reactive glia)
Dorsal
gray
Intermediate
gray
Central canal
surrounded by ependyma
Lateral
funiculus
Ventral
gray
Spinothalamic tracts?
(myelinating)
Intraspinal tracts?
(myelinated)
Ventral funiculus
Myelinated afferents
within Clarke's column
Ventral commissure
(myelinated)
Vestibulospinal tract?
Medial longitudinal fasciculus?
Ventral corticospinal tract?
myelinating

PLATE 78A

CR 440 mm
Infant, 4 Months
Y286-62
Upper Lumbar
Cell body stain

Areas (mm²)

Central canal	.0086
Ependyma	.0270
Gray matter	9.8606
White matter	15.0180

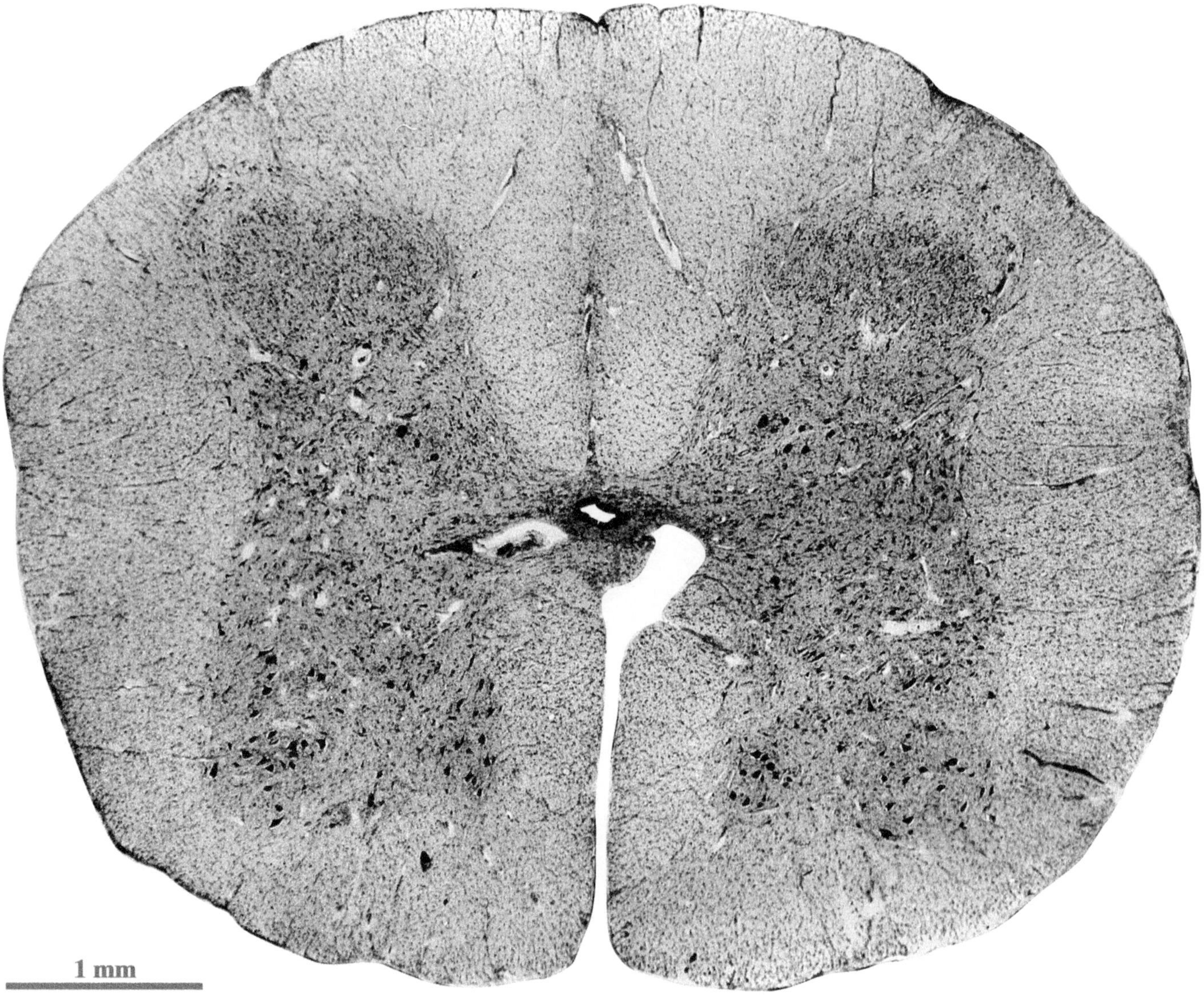

PLATE 78B

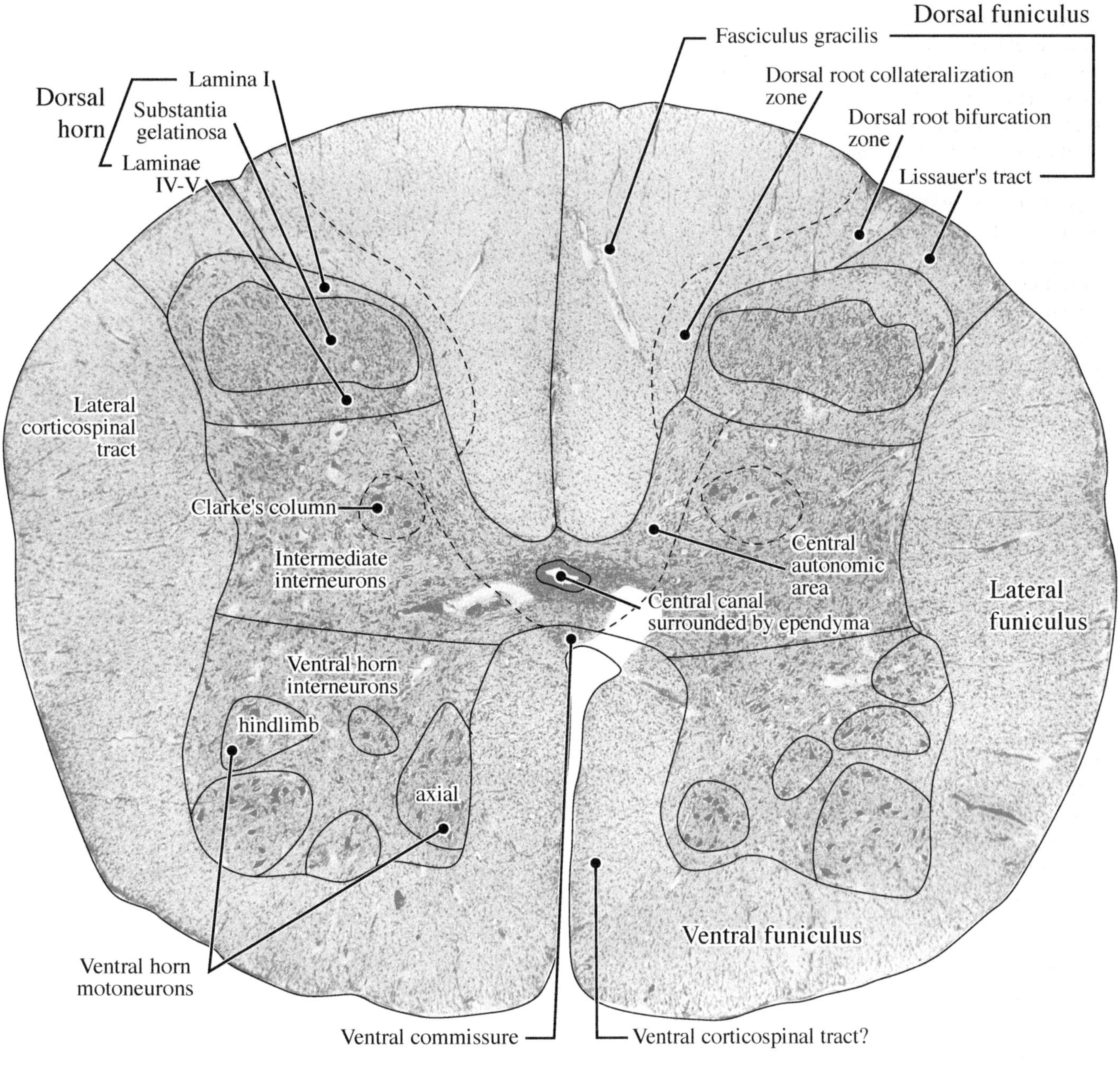
Dorsal funiculus
Fasciculus gracilis
Dorsal root collateralization zone
Dorsal root bifurcation zone
Lissauer's tract
Dorsal horn
Lamina I
Substantia gelatinosa
Laminae IV-V
Lateral corticospinal tract
Clarke's column
Intermediate interneurons
Central autonomic area
Central canal surrounded by ependyma
Lateral funiculus
Ventral horn interneurons
hindlimb
axial
Ventral funiculus
Ventral horn motoneurons
Ventral commissure
Ventral corticospinal tract?

PLATE 79A

CR 440 mm
Infant, 4 Months
Y286-62
Lumbar Enlargement
Myelin stain

Areas (mm^2)

Central canal	.0180
Ependyma	.0187
Gray matter	17.7370
White matter	13.9340

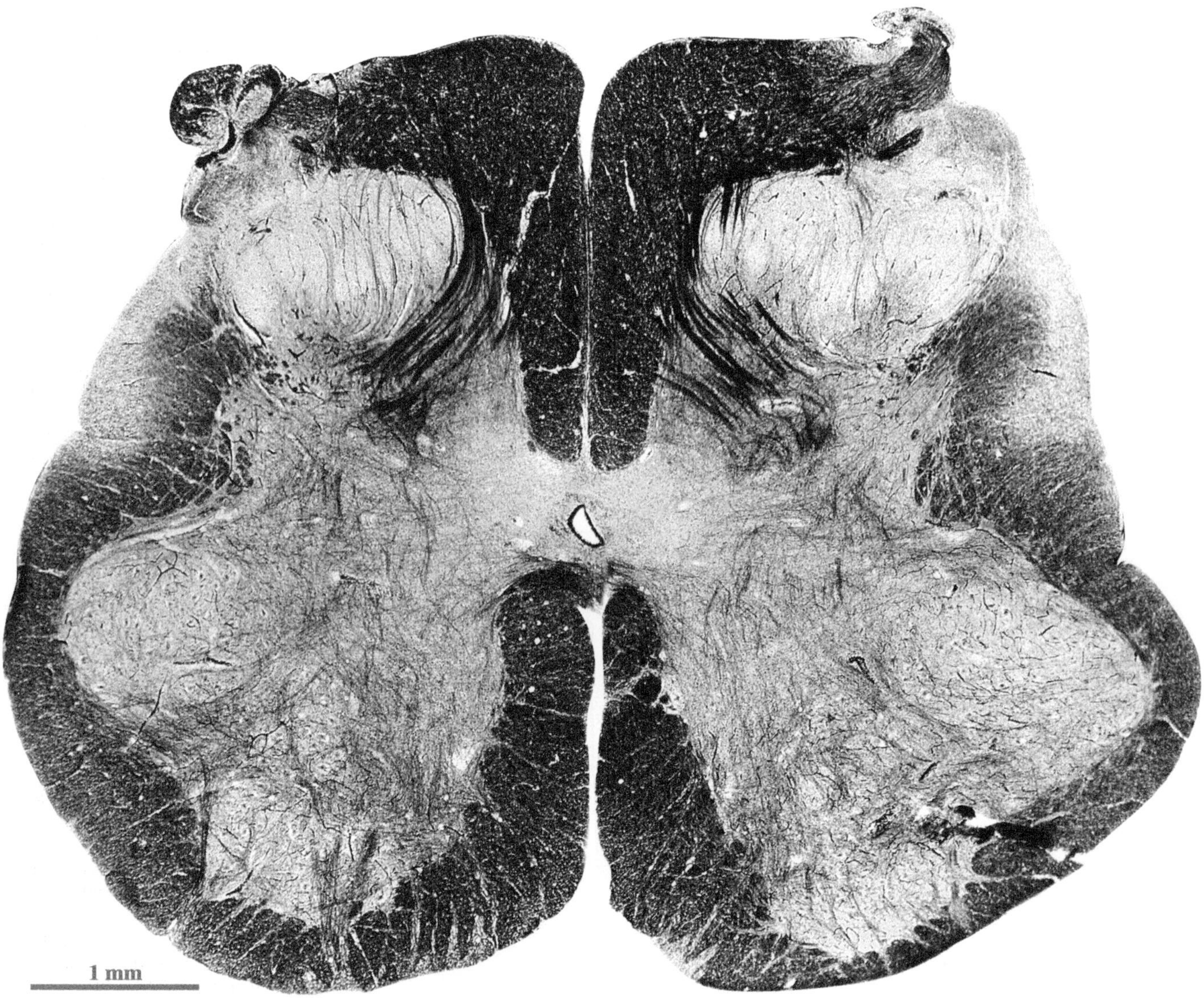

PLATE 79B

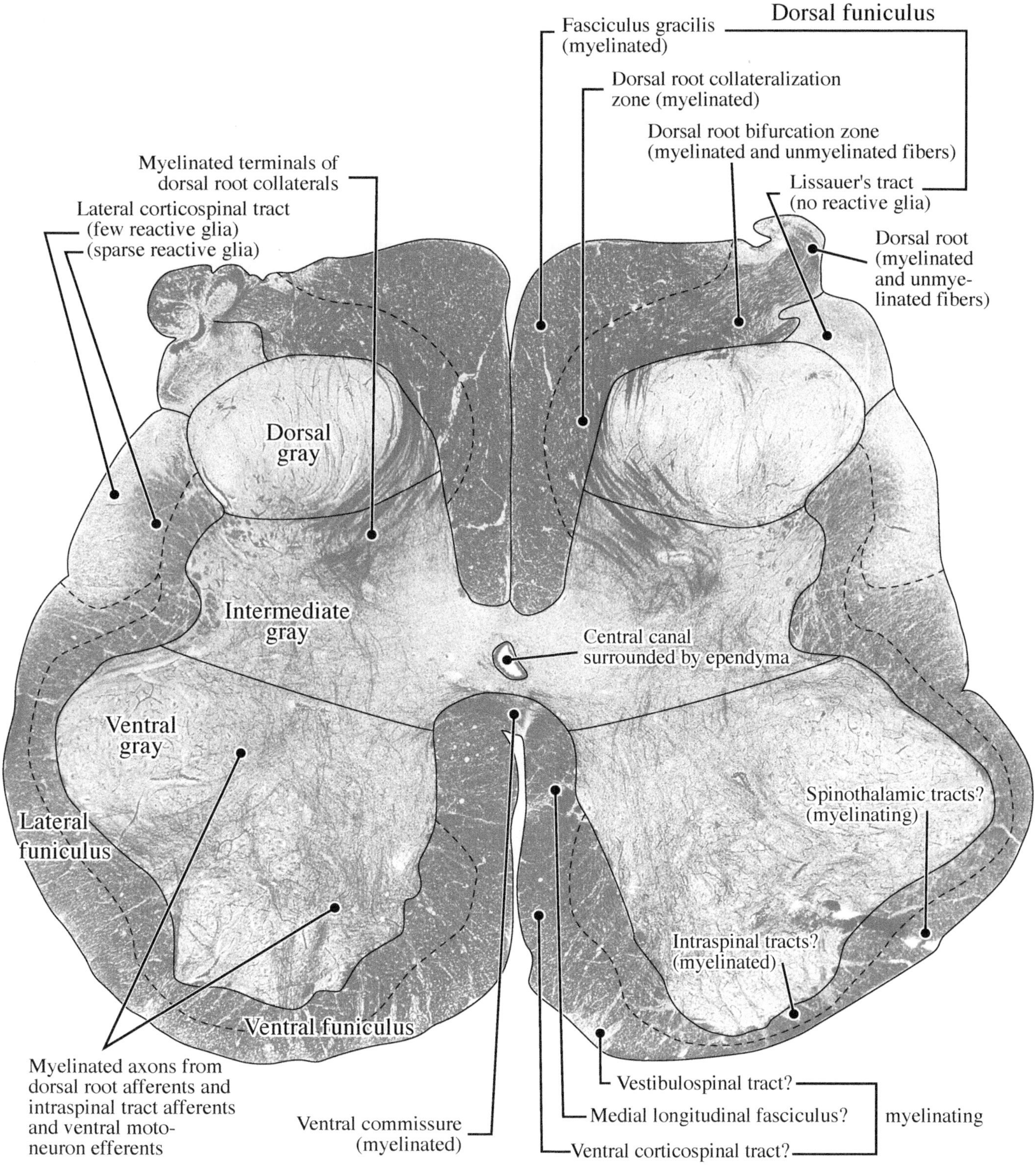
Dorsal funiculus
Fasciculus gracilis
(myelinated)
Dorsal root collateralization
zone (myelinated)
Dorsal root bifurcation zone
(myelinated and unmyelinated fibers)
Lissauer's tract
(no reactive glia)
Dorsal root
(myelinated
and unmye-
linated fibers)
Myelinated terminals of
dorsal root collaterals
Lateral corticospinal tract
(few reactive glia)
(sparse reactive glia)
Dorsal
gray
Intermediate
gray
Central canal
surrounded by ependyma
Ventral
gray
Lateral
funiculus
Spinothalamic tracts?
(myelinating)
Intraspinal tracts?
(myelinated)
Ventral funiculus
Myelinated axons from
dorsal root afferents and
intraspinal tract afferents
and ventral moto-
neuron efferents
Ventral commissure
(myelinated)
Vestibulospinal tract?
Medial longitudinal fasciculus?
Ventral corticospinal tract?
myelinating

PLATE 80A

CR 440 mm
Infant, 4 Months
Y286-62
Lumbar Enlargement
Cell body stain

Areas (mm^2)

Central canal	.0115
Ependyma	.0231
Gray matter	18.3550
White matter	13.8850

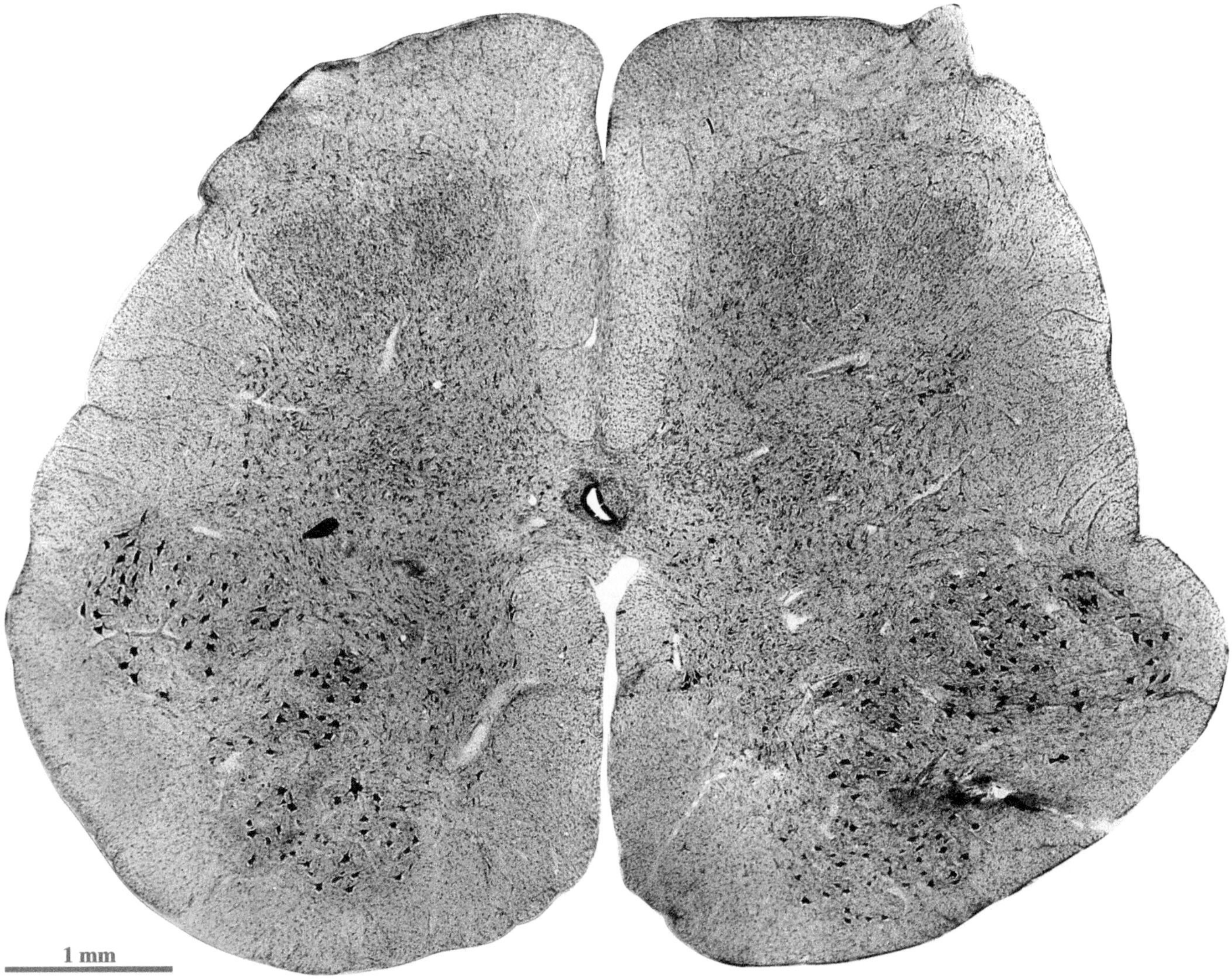

PLATE 80B

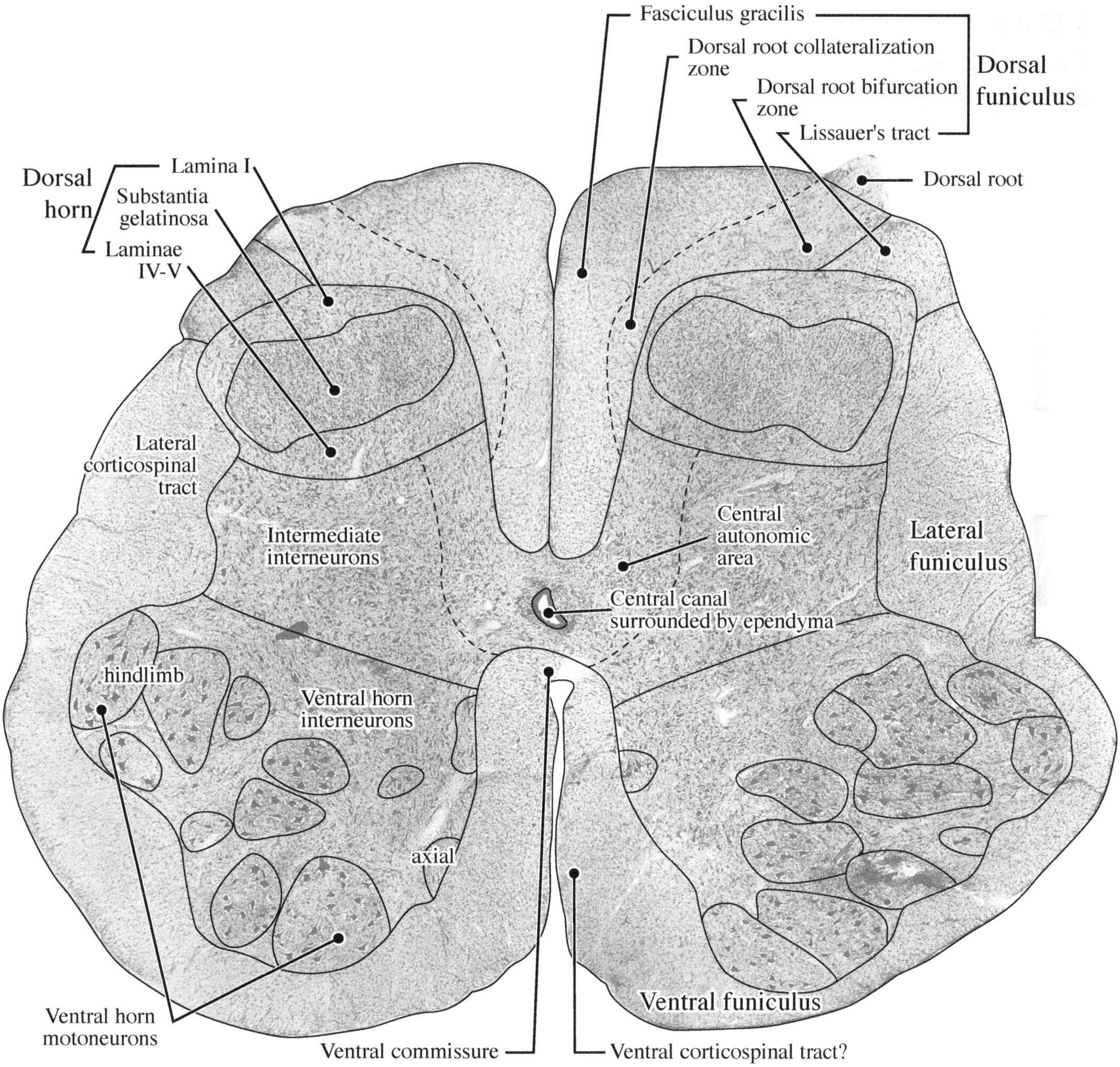
Fasciculus gracilis
Dorsal root collateralization zone
Dorsal root bifurcation zone
Lissauer's tract
Dorsal funiculus
Dorsal root
Dorsal horn
Lamina I
Substantia gelatinosa
Laminae IV-V
Lateral corticospinal tract
Intermediate interneurons
Central autonomic area
Lateral funiculus
Central canal surrounded by ependyma
hindlimb
Ventral horn interneurons
axial
Ventral funiculus
Ventral horn motoneurons
Ventral commissure
Ventral corticospinal tract?

For Product Safety Concerns and Information please contact our EU representative GPSR@taylorandfrancis.com Taylor & Francis Verlag GmbH, Kaufingerstraße 24, 80331 München, Germany

Batch number: 10731517

Printed by Printforce, the Netherlands